COSMIC RAIN

The Controversial Discovery of Small Comets

By Dr. Louis A. Frank

ANOMALIST BOOKS
Charlottesville, Virginia

Cosmic Rain: The Controversial Discovery of Small Comets
Copyright © 2021 by Suzanne Frank, Jessica Frank, and Patrick Huyghe
ISBN: 978-1-949501-19-3
Second Edition

The first part of this book was originally published as *The Big Splash* in 1990 by Birch Lane Press, an imprint of the Carol Publishing Group. *The Big Splash* is Copyright © 1990 by Louis A. Frank and Patrick Huyghe.

Cover image: The spectacular bright trail of the gases from the disruption of a small comet the size of a two-bedroom house took place 5,000 to 15,000 miles above the Atlantic Ocean on September 26, 1996. This unusually bright and long-lived trail was captured by a camera aboard NASA's Polar spacecraft. For this 54-second-long, time-lapse image, the object was in sunlight but the Earth below was in darkness, so a map of the Earth has been superposed to provide perspective for the view seen by the Polar spacecraft. Photo credit: NASA Goddard Space Flight Center and the University of Iowa. Map courtesy of Face of the Earth™ (Copyright © 1996, ARC Science Simulations, arcscience.com).

All rights reserved. No part of this book may be reproduced in any form, except by a newspaper or magazine reviewer who wishes to quote brief passages in connection with a review.

Book design by Seale Studios

Anomalist Books
3445 Seminole Trail #247
Charlottesville, VA 22911

Table of Contents

Foreword by Patrick Huyghe vii

Part 1: The Big Splash (1981-1990)

1	A Radical Departure	1
2	Cosmic Rain	5
3	The Black Spot Mystery	11
4	True Confessions	19
5	A Storm of Controversy	23
6	The Creation of the Oceans	29
7	Heat, Dust, and the Origin of Life	34
8	The Atmosphere and the Ice Ages	40
9	A Masquerade for Radar	47
10	How to Spot a Small Comet	52
11	Flying Saucers and Other Strange Events	56
12	A Scramble for Satellite Data	62
13	Extraordinary Evidence	68
14	Launch of the Artificial Comets	74
15	Some Cometary Competition	78
16	The Verdict	83
17	The Competition Fizzles	88
18	Why the Moon Doesn't Ring Like A Bell	93
19	Mars and the Outer Planets	103
20	Where the Small Comets Come From	108
21	Death of the Dinosaurs	114
22	The Debate Comes to an End	120
23	Astronomers Enter the Fray	126
24	Where Are You Now, Galileo?	130
25	The Great Search Begins	134
26	Double Trouble	140
27	The Turning Point	149

Part 2 Vindication (1990-2001)

28	The Rest of the Story	155
29	Polar Rising	162
30	Vindication	167
31	Grand Slam	176
32	Dessler and His Pawns	184
33	Steering Through Rough Waters	191
34	Contact Sport	198
35	Spacewatch Redux	209
36	Deep Waters	215

37	Last Dance	224
38	The Witch Hunt	228
39	Snowballs in Hell	233
40	Small Comets and the Future	238

Notes 241
Bibliography 271
Illustration Acknowledgments 273

Appendix 1: How to Search for Small Comets 275
Appendix 2: How This Book Came to Be 278

Name Index 280

From a drop of water, a logician could infer the possibility of an Atlantic or a Niagara without having seen or heard of one or the other.

— Arthur Conan Doyle

Foreword

Louis A. Frank, the Carver/James A. Van Allen Professor of Physics at the University of Iowa, died on May 16, 2014, at the age of 75. There was no obituary in *The New York Times* or the *Chicago Tribune*, though by all accounts there should have been. In the mid-1980s, before this mainstream scientist turned maverick, he was regarded as one of the most respected space physicists in the world; he was responsible for more scientific instruments aboard spacecraft than any other scientist, having been an experimenter, co-investigator, or principal investigator for instruments aboard 42 spacecraft.

A native of Chicago, Frank's first professional research activities began in 1958 when he assisted James Van Allen in the calibration of the first U.S. lunar probes, Pioneers 3 and 4, as an undergraduate student at the University of Iowa. Later, Frank was the principal investigator for the auroral imaging instruments for the Dynamics Explorer Mission, the plasma instrumentation for the Galileo Mission to Jupiter, the U.S. plasma instrumentation for the Japanese Geotail spacecraft, and the camera for visible wavelengths for the Polar spacecraft of the International Solar Terrestrial Physics (ISTP) Program. His scientific accomplishments were many: he discovered the theta aurora, the remarkable configuration of auroral and polar cap luminosities that looks like the Greek letter theta hovering above the polar cap; he made the first measurements of the plasma ring around Jupiter and Saturn; and he was the first to measure solar-wind plasma funneling directly into the Earth's polar atmosphere, as well as the belt of ions around the Earth known as the "ring current." Frank was a Fellow of the American Physical Society, a Fellow of the American Geophysical Union, and the recipient of the National Space Act Award.

But everything changed for Louis Frank in 1986. After five years of analyzing the puzzling small black spots that appeared in the images from NASA's Dynamics Explorer 1 spacecraft, which looked at the Earth in ultraviolet light, Frank and John Sigwarth, the graduate student who had been assigned the task of following this data wherever it led, concluded that the black spots were caused by water vapor remaining from the disintegration of small comets in the Earth's atmosphere. The black spots were not transmission errors, not paint flecks on the imager, nor any other artifact, Frank concluded, but "atmospheric holes," the remains of fluffy snowballs each weighing 20 to 40 tons and striking the Earth every few seconds. News of the discovery created a firestorm of controversy.

The story of his discovery of the small comets with images from the Dynamics Explorer 1 spacecraft and the uproar that followed is told in the first part of the book, which is essentially a straight reprint of *The Big Splash: The Scientific Discovery That Revolutionizes the Way We*

View the Origin of Life, the Water We Drink, the Death of the Dinosaurs, the Creation of the Ocean, the Nature of the Cosmos, and the Very Future of the Earth Itself, published in 1990. The second part of the book is the never-before-told follow-up, in Frank's own words, of the confirmation of the existence of small comets from images taken by instruments aboard the Polar spacecraft and the criticism he faced for upsetting so many scientific applecarts in the process.

It's now been 40 years since Frank first noticed the black spots in the images from Dynamics Explorer, the tale-tell sign of the disintegration of small comets in the Earth's atmosphere. This book now tells the story of the first 20 years of their discovery and controversy, as Frank's account ends in 2001. But much has happened in the past 20 years that lends support for the existence of these small comets, specifically news of the discovery of water on planets and moons throughout the solar system. It's difficult to account for such an abundance of water without a source such as the small comets. The small comets story has not come to an end. Much remains to be told.

Patrick Huyghe*
September 1, 2021

---

* My role in all of this is explained in "How This Book Came to Be" (Appendix 2)

Part 1: The Big Splash (1981-1990)

CHAPTER 1

A Radical Departure

I live on the plains of Iowa, drive a pick-up truck, and build instruments for satellites that study the Earth. I am conservative in the way I dress, in the way I lead my life, in the people I vote for, and in the way I do science. Science is my life.

But four years ago, pictures relayed from an orbiting satellite led me to a radical departure from the scientific mainstream. I proposed that the Earth was being showered by a vast number of previously undetected, small comets. I calculated that about twenty of these comets, each about the size of a small house and made essentially of water and ice, plunge through our atmosphere each and every minute of every day. If I am correct, then the water that we fish in, bathe in, drink, and are made of, is of extraterrestrial origin.

I did not make this up. The discovery of the small comets is based on a set of observations made at the limits of present-day technology. It so stretches our understanding of the world and impinges on so many areas of scientific study that the interest and ire of nearly the entire scientific community has been aroused. This new population of objects in the solar system has ignited a fierce scientific dispute. If I am right about the existence of the small comets, a whole generation of scientists in a dozen different fields will have been working from geophysical and astronomical assumptions that are simply not correct. The textbooks in a dozen sciences will have to be rewritten.

The proposition that a rain of cosmic ice is presently bombarding the Earth holds staggering implications for us, our planet, and the rest of the solar system. It would mean that our lakes, rivers, and oceans were not formed as we thought early in the Earth's history. It would mean that the substances necessary for the origin of life on this planet may well have arrived from space. It would mean that periodic increases in cometary showers could have caused the Ice Ages and been responsible for the death of many species, including the dinosaurs. It would mean that the planet Venus never had a primitive ocean, as many scientists suspect, and that the Martian landscape was formed by water, which may one day again flow on its surface. And it would mean that the volume of water on the Earth is slowly increasing and may one day submerge the planet entirely. It would mean, in sum, that our cherished notion regarding the Earth's isolation from the rest of the solar system and the universe will have to be discarded.

We once thought the world was small and flat. Columbus sailed across the Atlantic and enlarged it, Magellan sailed around the world and made it round. We thought that the world was at the center of the universe until Copernicus restored the Sun to its proper place and set the Earth in orbit around it. We thought the Earth and its Moon were unique until Galileo showed us that there were similar planets and moons elsewhere in the heavens. And we thought, until the turn of the nineteenth century, that not even a stone could fall from the sky, that nothing could shatter our isolation from the cosmos.

We now accept that the world is round, that it is one of many bodies that circle the Sun, and that, on occasion, a stone does fall from the sky. But science and history continue to maintain that the Earth is safe and secure from the turmoil of the cosmos. We may have bad winters, volcanic activity, and man-made catastrophes, but these are terrestrial affairs pure and simple. The cosmos has no effect on us. We are different. We are isolated.

Life is full of unwritten rules, and this is one that science holds dearest. Any tampering with the idea that we are isolated, any suggestion that places the Earth in a relationship with the solar system and universe that is different from the one we have always been taught to accept, will be strongly condemned. We are comfortable with the illusion that the Earth is isolated both in space and time. So if you propose—as I have—that something from out there is affecting us here and now, rather than millions of years in the past or millions of years in the future, beware the wrath of orthodox science.

Science guards our isolation well. Talk about a rock fifty miles wide that resembles a chocolate-covered cherry and lies at the other end of the solar system and scientists will say, "Oh, how fascinating." But talk about bodies close to the Earth and the reaction will be quite different. Say that there are dark objects falling from the sky, which we have never had the capability to see before, and that over the next ten-thousand years these objects will add another inch of water to our planet and scientists will object quite strenuously. Their safety will have been attacked. "Oh, no," they will say, "the water has been on the Earth since the beginning. Earth is here as it is."

Our isolation from the cosmos is staunchly maintained by the guardians of our culture. In 1985, for instance, an editorial appeared in *The New York Times* reflecting on a debate taking place among scientists about whether or not a distant companion star to the Sun could have set off a chain of events that led to the extinction of the dinosaurs. "Astronomers should leave to astrologers," the editorial said, "the task of seeking the cause of earthly events in the stars." The newspaper of record could only express ridicule at the notion that actions in the cosmos could have such a profound effect on the Earth, even though the event in question was said to have happened millions of years ago, far enough removed in time, you would think, to avoid any strenuous opposition. But such things are not possible, the newspaper editors were saying, because the Earth is isolated.

I, too, was taught that planet Earth spins its way through the void, essentially untouched by the chaos of the cosmos. But we know so very little about our solar system, about the interplanetary space that the Earth moves through, about the origins of its water, its carbon, or just about anything else. From elementary school on we read textbooks that make definitive statements, but we just do not know. These are just impressions we have. But I want to know. I would like to understand why we are here. What in the Earth's history, in the history of the solar system, has made it possible for us to be here? It is marvelous the way we inquire about our environment. It is wonderful how we are able to perceive and reflect on it. But just how do we perceive it, and do we perceive it correctly?

If you look at the Earth from space with the eyes that we put on satellites, you can see that this planet holds a unique place in the solar system. Even discounting the fact that there is life on it, it is the most exciting planet in terms of its composition and position relative to the Sun.

As occupants of this planet, we should probe further into what makes it so unique, how we got here, and what our relationship with it is. But we should awaken from the rosy dream that we exist separate from our past and future, from the solar system and the rest of the universe.

I think this is the final step in our education about the Earth. This is the next great debate. I do not believe the Earth is a fixed body, unchanged over billions of years. Changes are taking place continuously. The creation of the planets and other celestial bodies is not something that stopped billions of years ago. It is not something that only happens elsewhere in the universe. It is happening here, today. And the small comets are very likely a part of this process.

The comets I have found are much smaller, darker, and far more numerous than the large comets that are known to be wandering into our solar system. Their impact on our perceptions will also be infinitely greater. Yet for some reason it seems never to have occurred to anybody that there might be small comets. We know there are large rocks, large meteors that enter the Earth's atmosphere, as well as small ones. Everyone accepts that now. So if there are large comets, why not small ones? But there is next to nothing in the literature about small comets. The topic essentially does not exist. Until recently the capability to observe such objects has not been available, so without observations to establish their existence, the possibility of small comets in the solar system has been ignored.

In contrast, the known, large comets have been exhaustively studied. They are believed to be primitive remnants of the interstellar gas and dust that condensed to form the solar system. Their nuclei measure about a mile across and have the surface of a small city. Warmed by the heat as they approach the Sun, their surfaces vaporize and release long tresses of gas and dust known as comas. These comas extend for millions of miles from the cometary nuclei. The word comet, in fact, comes from the Greek word meaning hair. But not all known comets appear to us with these long flowing tresses. Most comets are no more than bright smudges in the sky and are only visible through telescopes. Less than a thousand different large comets have been recorded in all of human history.

You would think that astronomers looking for fainter and fainter objects would have found increasingly smaller comets, but this is not the case. Most astronomers suspect there is a real lack of comets smaller than a football field. They do not know why. They just say so. Small comets, they believe, would most likely be fragments produced by the splitting and breaking-up of large comets and would vaporize and dissipate very quickly. Such brief lifetimes, among other factors, would make small comets very difficult, if not impossible, to find. Unlike the large comets which are detected by their comas, small comets would only be detected by reflected sunlight. And because they are so faint, they would have to pass much closer to the Earth than the large comets in order to be seen. But if they do exist, a few astute astronomers have realized, there would certainly be many of them, and very likely they would be indistinguishable from the meteors we see entering our atmosphere.

This view is essentially correct, though no one would ever have thought that these objects would be as small or as numerous as I have proposed. As a consequence, the discovery of the small comets has turned into a broad scale intellectual battle among scientists at the forefront of science. There have been many brisk exchanges in public, but also much intrigue, back-biting, and small-mindedness behind the scenes. What I have experienced in trying to bring this discovery to light can best be described as a game of hardball science. Fairness has had nothing to do with it. Because losing has meant losing big, everyone has played to win.

So there are really two stories in this book. One is about the discovery of a new population of objects in the solar system. The other concerns how science works today, and in particular how controversial ideas are handled by the scientific community. Science is a mountain from which we can see far and wide into the surrounding terrain, the German physicist Max Planck once said, but the mountain itself, it is invisible. I hope that this story will shed some light on that mountain. It is important for all of us to understand just how science works, as it is one of the most influential institutions in our culture. Science shapes the way most of us think about the world.

The past four years have been a difficult time for me, but the truth, as I see it, is this: The infall of small comets into the Earth's atmosphere is taking place today. It is not an isolated event that took place billions of years ago or something that will take place billions of years in the future. It is not something occurring far away on another planet. This is a story about our perceptions of what is happening here and now, what it might have meant for our past, and what it might mean for our future. It has to do with the Earth. It has to do with the oceans, the water you drink, the atmosphere you breathe. It has to do with you. It is a very emotional topic.

CHAPTER 2

Cosmic Rain

Every year the Great Fool sets his trap. Without fail, one of the news stories of the day is a prank. No one expected this year to be any different. It was the 1st of April 1986.

On this day France removed its observer troops from Lebanon. The Central Intelligence Agency admitted to having long overestimated the yield of large Soviet nuclear tests. Wildfires continued to blaze across the southeastern portion of the United States. Navy salvage divers sixteen miles from Cape Canaveral continued to haul up more silt-covered wreckage from the space shuttle Challenger. And stock market analysts called the day's twenty-eight point drop a "plunge."

One story bore the mark. It read like something from a B-grade science fiction thriller. A scientist claimed to have evidence that the Earth was being pelted by twenty, hundred-ton comets per minute. These comets have been dumping water on us for more than four billion years, said the scientist, and in one swoop of mathematical magic, he pointed to the oceans and said "Voila!" I am that person. The English tabloids called me all day asking if this was an April Fool's joke. I told them it was no joke. But it did not seem to matter. The Fool had taken my story and made it his own.

I do not really know why the story broke on that date. I suspect that newspaper editors ran the story on April Fool's because it "fit." We had written up our findings in a pair of papers that would be published in the April issue of a refereed scientific journal called *Geophysical Research Letters*. Alex Dessler, the editor of the journal at the time, had sent early copies of our papers to the media at the end of March. Dessler was looking for publicity and he got it. I hate publicity. I do not like reporters and photographers interrupting my isolation and my work. But Dessler thrives on publicity. I think Dessler got more publicity out of this story than for anything else in his life. I certainly have.

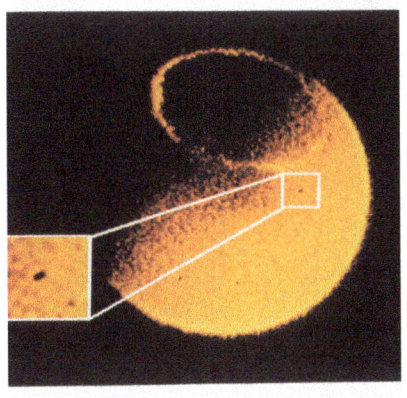

The cover of the scientific journal that announced the discovery of the small comets. The ring at the top of the ultraviolet image is the northern auroral oval. The inset is an expanded view of an "atmospheric hole" produced by a small comet. This black spot actually represents a cloud of water vapor about 30 miles in diameter at an altitude in the range of 200 to 600 miles above the Earth's surface. This is what remains of a small comet when the house-sized chunk of water-snow is vaporized above the Earth's atmosphere.

I was accused of making a mountain out of a molehill. Critics gleefully quoted Mark Twain's famous remark about science: "One gets such wholesale returns of conjecture out of such a trifling investment of fact." But what else could people say about someone who claimed to have discovered the existence of celestial objects never before seen by astronomers, yet which may be responsible for the origin of the oceans, the origin of life, the death of the dinosaurs, the unexplained falls of ice from the sky, as well as the water in Venus's atmosphere, the mysterious bright flashes sometimes seen on the Moon, the channel-marked surface of Mars, the ice moons of the outer planets, and a host of other phenomena throughout the solar system. My evidence, the so-called molehill, was a bunch of dark spots that had appeared on some satellite pictures of the Earth. It was, I admit, all quite hard to believe.

And very few people did. "That's as crazy as they come," said one astronomer. "I'm amazed," said an oceanographer, "I've never heard anything like it." One geologist dismissed it as simply a case of Halley's fever. But the long-awaited and much publicized flyby of Halley's Comet that year fizzled and my proposed comets had nothing to do with that celebrated non-event. My comets are not occasional visitors. They do not come by every seventy-six years without dropping in. No, these small comet-like objects slam into the Earth's atmosphere at the rate of ten million a year.

The evidence that led us to propose such a massive infall of objects into the Earth's atmosphere came from a high-altitude, polar-orbiting satellite called Dynamics Explorer. The satellite carries an ultraviolet camera which takes pictures of the Earth in the light which lies

The data that led us to propose the existence of the small comets and their present-day bombardment of the Earth came from an ultraviolet imager on this NASA satellite, Dynamics Explorer 1. It was launched on the 3rd of August 1981 from Vandenberg Air Force Base in California. The satellite continued to transmit evidence of the small comets until it died in the fall of 1990.

just beyond the violet end of the visible light range. These otherwise normal ultraviolet images were speckled with some very puzzling dark spots. We found more than 30,000 such spots in the images during some 2,000 hours of observing time.

At first I thought the spots might be due to faulty instrumentation. So with the help of John Sigwarth, one of my graduate students, and my colleague John Craven, a research scientist at the University of Iowa, we scrutinized the data for possible computer glitches, random flaws, radio transmission noise, failing sensors, even the possibility of paint flecks on the camera. But after years of checking and rechecking, testing and retesting, and having eliminated the possibility of spurious effects, we had to face the fact that the dark spots were real.

It was clear that the spots on the images represented something that lay between the satellite itself and the region of the upper atmosphere we were observing. Something was absorbing ultraviolet light in that region and creating a kind of dark "hole" in the uniform glow of the Earth. The signature, by all appearances, indicated the presence of water. But these dark spots were enormous, measuring about thirty miles in diameter. No amount of evaporation from the surface of the Earth could produce such large clouds of water vapor at that altitude. The water had to come from *out there*. The source clearly had to be extraterrestrial.

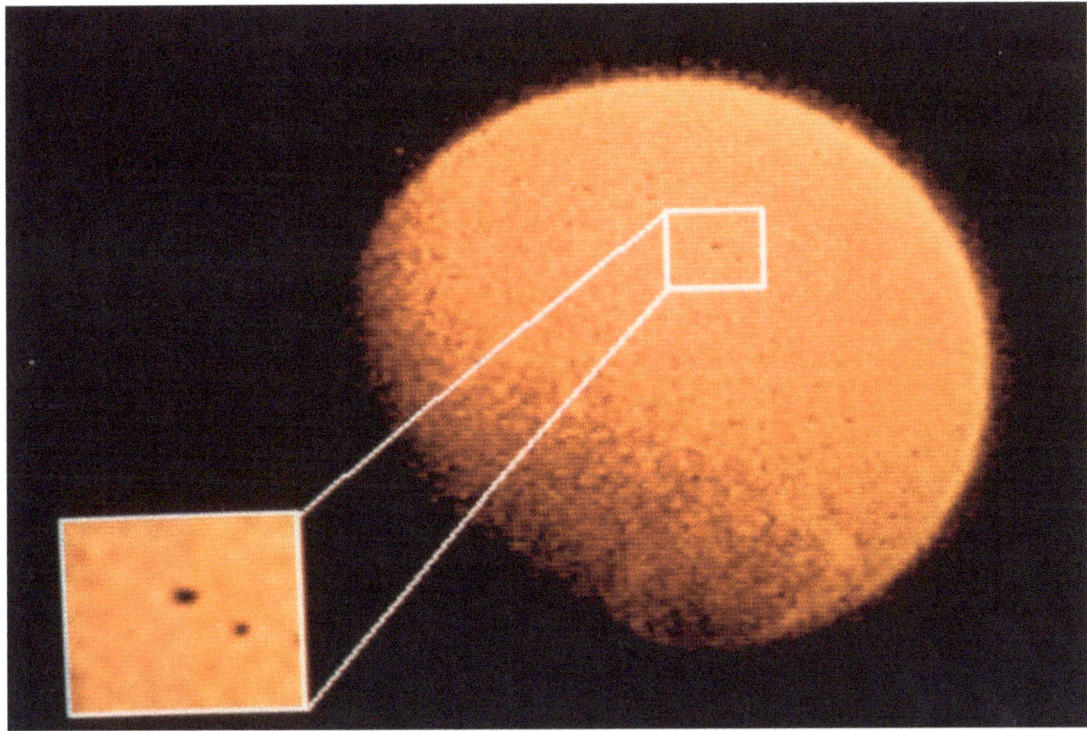

The images from the Dynamics Explorer satellite are not taken "at once" like a camera, but produced one line at a time. This means that when the satellite is at a high altitude, the black spots that appear in adjacent lines (see inset) will represent the motion of a single object as it plunges towards, and finally into, the Earth's atmosphere. An atmospheric hole that is detected in two consecutive scan lines is shown near the center of the inset, along with evidence of a bright feature in the next scan line, as the photometer's scanning mirror tracks the object. The other atmospheric hole in the inset is seen only in one scan line. A study of these "apparent motions" suggests that the objects—these water vapor clouds produced by the vaporization of the small comets—are travelling at about 10 miles per second just above the Earth's atmosphere.

But I knew that a cloud of water vapor in the interplanetary medium would disappear very quickly. So whatever had caused this large cloud in our images had to be in a form other than water vapor when it was well away from the Earth. I calculated it to be a relatively loosely-packed ball of water-snow the size of a small house. At one to two thousand miles above the Earth either the action of tidal forces, which are produced by the planet's gravity, or perhaps static electricity, would then break up this object and the Sun would vaporize its contents into a ball of gas. By the time it had plunged to within about 300 miles of the Earth, this hundred-ton spray of water and snow, traveling at about twenty times the speed of sound, would have expanded to a thin ball of gas some thirty miles across. That is not much water. It is thinner than a London fog. But it would be enough to produce the black spots in our images.

At about fifty miles above us in the atmosphere, the vapor cloud would slow to subsonic speeds, but continue to plummet until it reached a height of about thirty-five miles where it would get mixed up with the air in the upper atmosphere. Winds circulating into the stratosphere would then turn the water vapor into ice crystals. These ice crystals would fall and become part of the water vapor present at low altitudes, eventually turning into precipitation and falling on land and sea.

This, of course, is the scenario for just one comet-like object. At the observed rate of twenty per minute, over a year's time you would get one ten-thousandths of an inch of water on the Earth's surface. That is but a tiny fraction of the annual rainfall and quite indistinguishable from it. It is a mere drop in the bucket. But the icing on the cometary cake, so to speak, is that if you add up all the drops accumulated over the age of the Earth, you would have enough water to fill the oceans and more.

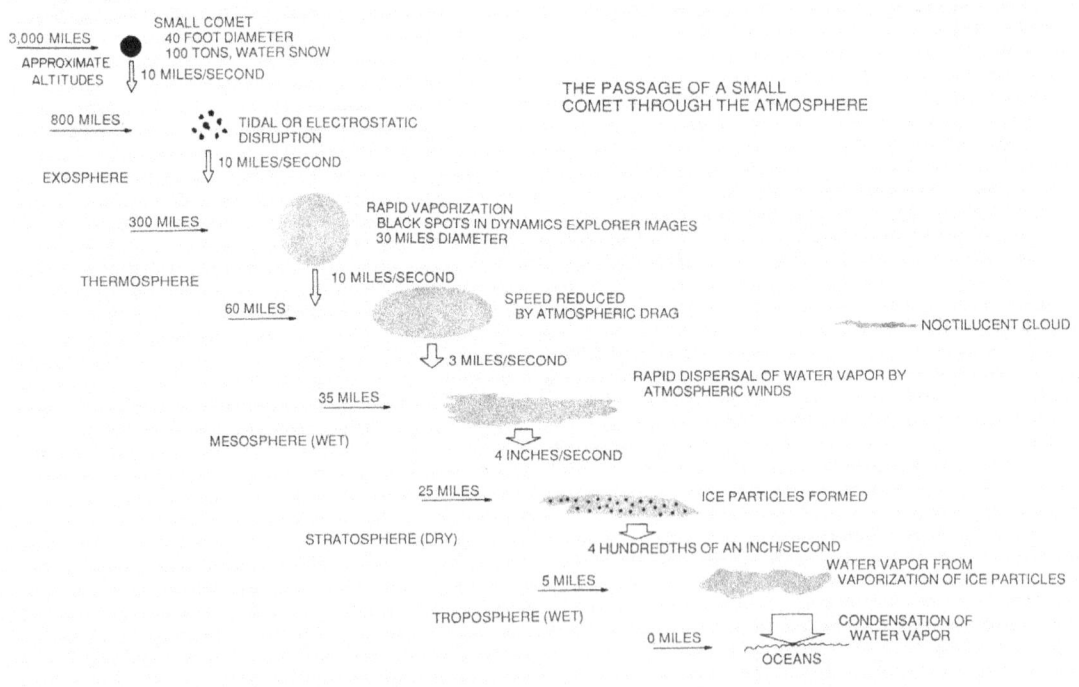

The sequence of events for the disruption of a small comet as it plunges into the atmosphere

That is how we came to the startling conclusion that the dark spots in the images were being formed by small, comet-like objects entering the Earth's atmosphere. The mind rebels. It is disturbing. But it is really quite reasonable. I can explain why the black spots in the satellite images cannot be anything else. I can explain why we are not seeing these comets all the time. I can explain just how the water from these comets manages to survive through the atmosphere. And I would have to do just that in the months and years to follow.

The first time I did so in public was in Baltimore a little more than a month after the publication of our two original papers. The occasion was the annual spring meeting of the American Geophysical Union. The AGU, as it is known, has 22,000 members worldwide "dedicated to the study of the Earth and its environment in space." My proposal had sparked a great debate among the community of physical scientists and many were here to apply the pressure, to crush this bothersome fly with a cinderblock if need be, before it created any further damage to the established notions of science.

The meeting room was packed. The seventy or so seats were filled and people were lined along the back wall. A television monitor was even provided so that the dozens outside the room could also witness the proceedings. It was hot in the room and rivers of sweat rolled down my face and soaked my white shirt. I removed my jacket and loosened my red tie. As I walked to the front of the room one of my colleagues told a reporter: "There's nothing like a good, bloody brawl in science."

It was clear why everyone was after me. I had committed two deadly sins. The first sin can only be made by an establishment scientist: I broke from the mold. I was working along with everyone else then suddenly seemed to reverse course. I did something that drastically disturbed what everyone else was doing and thinking. And while my proposal ran over everyone else's work—from planetary astronomy to oceanography—it had no real effect on my own work. That was my second deadly sin. It upset people greatly that this finding had so much of an effect on their work and none at all on mine. But I was caught between the proverbial rock and a hard place. People said I should not be wandering over all these fields. That was true. But I had no choice. I had to defend myself.

I could understand the reaction of my critics. Many of them, by the way, happen to be my friends. I would not have reacted any differently. A lot of people said it was impossible. That did not bother me. The existence of such a large number of small comets is difficult to grasp. It introduces a large, previously unknown factor, at least from an observational point of view, so it is upsetting for people in many fields. If it is correct it is going to change an awful lot of things. Its implications are widespread.

I am, as I have said, a very conservative person. One of the things that bothered people so much about my proposal was that this was just totally out-of-character for me. Of course, they did not consider how many years of effort I had spent on it. I knew this was a hot potato, and I did everything I could to get rid of it. Many people thought my career would be ruined. "What if you're wrong?" they said. But being wrong is probably not as great a risk for a scientist who has already established his reputation. I was not going to be destroyed by it. If this work had been done carelessly, certainly my career would have been ruined. But most people would agree that I have not been careless about this at all.

I have no complaints about the harsh treatment I have received. I expected it. Many of the arguments were clever and on the mark. But unfortunately much of it had nothing to do with the science at all but with belief. I harbor no bad feeling about it. I asked myself two questions before I published the papers. Was I old enough and did I know enough to be able to carry out this debate? I was forty-seven at the time, and I decided that I was old enough and knew enough, however little that was, to be able to do it. But I also wondered if I were physically able to go through with it. I was overweight at the time. I wondered if this debate would send me to my grave.

So I cut down on the weight and proceeded to work continuously for two and a half years. I slept four or five hours a night for months at a time. I took no weekends off, no vacations, no holidays. I developed a hernia at the time, but I had to postpone the operation for several months. I worked until two and three in the morning every day, Christmas and New Year's included. I thought that it was my responsibility. If I was going to cause this much fuss, this much apparent harm to science, then I must pay the price.

Nor did I wish to stop my work in plasma physics—the clumps of charged particles in the cosmos—during the four years that this controversy raged on over the small comets. I did not want to drop from the circuit, the endless rounds of meetings and seminars in my field which, like circuits for golf or tennis professionals, require that players be on the road for much of the year. The circuit is where the newest ideas and data are discussed long before they are published in the journals. If you just publish papers, but never travel, never give talks, and never attend meetings, you cannot be a major player in modern science. It is not a glamorous existence. You are on the road so much, living out of a suitcase, that you need your appointment book to know whether you are in Tokyo or Stockholm. And when you return, you are overwhelmed with paperwork, writing grant proposals to finance your work, and reviewing papers for scientific journals.

It has been an exhausting four years. But one of the things that made this effort possible is that I am isolated here at the University of Iowa. There is no name on my door, so there are no distractions. I do not have a heavy teaching load. The University of Iowa is not a well-known university. It is an intellectual outpost in a soybean field.

CHAPTER 3

The Black Spot Mystery

The black spots on the images were like flies walking across a television set. They were annoying. They were there from the start, on the very first images. They are still there today. There is no question about who saw them first. Everybody saw them. We would give talks and the black spots were there on the images for everyone to see.

It all began back in 1981. On the third day of August a Delta rocket lifted off from Vandenberg Air Force Base carrying a pair of NASA satellites, both known as Dynamics Explorer, into elliptical orbits. One of these satellites circles the poles of the Earth at an altitude ranging from 350 miles to 14,500 miles. I was responsible for three of the instruments aboard the satellite, one of which is an ultraviolet camera, built and operated by my colleague John Craven.

This satellite was designed to examine the Earth for certain light emissions that are invisible to the naked eye. I and other scientists hoped that these emissions would provide further insight into the nature of the auroral lights that occur in the Earth's polar atmosphere and detect any effects associated with them. I was particularly interested in getting the first global pictures of the Earth's aurora and I was not disappointed.

John Craven

The pictures sent back from the ultraviolet camera on the satellite were spectacular. The remarkable auroral crowns encircle the poles of the Earth, while the planet's dayside looks like a bright ball illuminated by a flashlight. This bright feature is known as the dayglow. Dayglow is produced by the interaction of sunlight with the atomic oxygen present in the Earth's upper atmosphere. The ultraviolet light emitted by this dayglow is not visible to the naked eye but is within the range of the satellite's specially-designed camera. The emissions it captures are transformed into a normal photograph.

But the images of the Earth we obtained beginning in late 1981 contained an unexpected feature. The blanket of dayglow was not uniform. It was speckled with dark spots. Strictly speaking, these spots were areas of greatly reduced brightness. In other words, there seemed to be holes in the dayglow. Everyone saw them. Everyone assumed they were noise—those random fluctuations in data that are due to chance.

Life went on. During the summer of 1982, an undergraduate student named John Sigwarth began working for me. Sigwarth had come to the University of Iowa in 1979 and had taken his first physics class with me. He was very interested in space science and was very bright. I wanted him to take enlargements of the satellite images and scan them for signs of gravity waves, small scale ripples in the upper atmosphere that sometimes follow the brightening of auroras. These waves sweep across the face of the Earth much like waves in the ocean. They have been detected on radar and I wanted to know whether these waves would show up as crests and valleys of light in the satellite images. I assigned to Sigwarth the task of processing the data to make the waves visible.

But he could not do it. He worked on the computer program, which was designed to extract subtle features from the original pictures and highlight them, but each time the computer stuttered and burped as it scanned those black holes in the atmosphere. The

John Sigwarth

black spots kept getting in the way. Then one day in the fall of 1982 Sigwarth, quite exasperated, walked into my office on the second floor of Van Allen Hall and said: "How can we get rid of this stuff?"

Sigwarth, Craven, and I pondered the question and concluded it must be some disturbance in the camera's electronics that would occasionally put these annoying little black spots in the image. Perhaps a transistor was fluky, or a computer was not functioning properly, or a problem arose as the pictures were being transmitted down to us from the satellite. It was possible. Each picture involves the transfer of an enormous amount of information. Like the image produced by a FAX machine or by the display on a home computer, each of the satellite pictures is constructed of a large number of little dots called pixels. Each of these pixels is described by eight "bits," actually a string of "0's" and "1's" that indicate the intensity or brightness of that pixel. So perhaps some of the image pixels were not being transmitted down properly from the spacecraft. Whatever the source of the noise, I did not relish spending my time tracking it down. Sigwarth got the job.

It was tempting to simply remove the spots from the images and get on with the search for gravity waves. But you cannot alter data on a mere assumption. You have to have a reason. We needed to show that the spots were either detector noise, or produced by electronics on the spacecraft, or generated by computers on the ground. Only once that was accomplished could we eliminate the spots from the processed images and get on with our work.

Sigwarth worked very hard trying to solve the mystery. From time to time he would come into my office and say he was not having much luck with it. The year 1982 drew to a close. Sigwarth kept working on it. But he was unable to trace how the holes appeared on the images. One possibility that we entertained was that one of the two light counters on the camera was failing. Every other pixel in the satellite images is produced by an entirely separate set of electronics. The counters, in other words, take turns producing the dots that comprise each image. So Sigwarth had separated the data produced by each counter. But when he examined the data he saw that both counters were observing spots in the same sequence and at the same rate. This told us the counters were not dropping bits, those meaningful strings of "O's" and "1's." We knew that it was nearly impossible for two counters to malfunction in exactly the same way.

We also eliminated the possibility that these annoying little black spots were caused by errors in radio transmission from the spacecraft. We checked the entire system from the time the data left the instrument, passed through the satellite itself, traveled down to the ground, and was relayed to us. Because the instrument regularly transmits fixed words, or fixed bit patterns, we could check to see if any transmission errors were occurring. We calculated that dark spots due to telemetry noise would appear once in every 200 images. But these spots appeared in the images almost a thousand times more often than expected.

All of the tests we performed on the data produced results that ran counter to our expectations. We really wanted to show that the holes were noise. But we were out of luck. So Sigwarth began to look at the pictures in great detail, trying to see if a spot that appeared in one frame could be seen in any of the subsequent frames. A very careful analysis showed that the spots could be followed in this way, although the spots in the subsequent exposures were not as dark as those in the initial exposures. This seemed to indicate that the black spots were moving and changing.

Sigwarth then programmed the camera so that instead of scanning the entire Earth, it would scan just a small portion of it. This allowed the camera to return to the same area more quickly. The series of pictures produced showed that a black spot would appear and disappear in a sequence of frames. The black spots seemed to be objects in motion. This was not characteristic of noise. Noise should appear at random all over the image. This indicated the presence of a real object. Sigwarth came down to the lab where I was working and showed me the data. He was very excited. I looked down at the pictures and congratulated him. I

The gradual appearance and disappearance of a single atmospheric hole (circled), which is caused by a small comet, in a sequence of images taken 72 seconds apart. These images reflect the transient nature of the water vapor clouds, or remains of the small comets, as they plunge into the Earth's atmosphere. A few other atmospheric holes can be seen in the images.

thought we were on to something.

Other clues convinced us that the spots were genuine. The spots, or holes in the atmosphere, appeared to move in the same direction across the face of the Earth. If these holes were random events, due to malfunctioning equipment, for instance, you would expect to find half the spots going in one direction, and half going in the other direction. But this was not the case. Most of the holes appeared to move in the same direction across the face of the Earth.

By February of 1983 we had come to the conclusion that something, some kind of object, was absorbing the ultraviolet radiation between the camera on the satellite and the Earth and producing the apparent holes in the atmosphere. The more we looked the more it seemed that our images were actually snapshots of the movement of these objects above the atmosphere. We began to suspect that these objects were meteors of some unusual sort.

So we decided to compare the motion of our elusive black holes to the passage of meteoric dust and debris in our atmosphere. Much of this meteoric dust tends to orbit the Sun more rapidly than the Earth. It essentially catches up to the Earth, in other words, and approaches the atmosphere from the local evening face of the planet. Such motion relative to the Earth is called prograde motion. Whatever the black spots represented, they showed the prograde motion that is characteristic of meteoric material. This not only implied that the spots were real but that the objects they represented were extraterrestrial. So we assumed, given the large number of spots that showed up in our images, that the holes were caused by a particularly large influx of meteors such as you would have with a meteor shower.

We went public with these results for the first time in May of 1983. Two weeks after receiving his undergraduate degree in physics, Sigwarth presented our findings in a paper entitled "Atmospheric Holes Possibly Associated with Meteors" at the spring meeting of the American Geophysical Union in Baltimore. I sponsored the paper and John Craven was listed as a coauthor. People seemed interested and curious, nothing more. But over the next two-and-a-half years we presented three more papers on the topic at meetings of the American Geophysical Union and each time the audience grew.

Over this period of time Sigwarth and I analyzed over 10,000 images and learned a good deal about the black spots in the process. Our interpretation of the events continued to involve meteor impacts into the Earth's upper atmosphere. By counting the spots in our images we were able to estimate the rate at which these objects appeared. This was the simplest measurement to do. We saw ten holes per minute on the daylight side of the Earth. So we doubled that figure to obtain the rate of these objects over the entire face of the Earth. There had to be about twenty such objects entering the atmosphere every minute. That was an alarming number of objects.

We still needed to explain just how these objects, which we assumed to be meteors, could cause holes in the atmosphere's screen of atomic oxygen. We entertained three possibilities. The first and simplest explanation had the meteors laying a blanket of material over the atmosphere, preventing the light from getting through, and creating a black spot in our images. Another possibility was that the atomic oxygen up there was being depleted by some special chemical process. But we could think of no chemical reaction that could get rid of the atomic oxygen quickly enough and none that allowed the atmosphere to restore itself as rapidly as we observed in our images. A third possibility involved a catalytic reaction of the sort that takes place in the catalytic converter in your car. Could some small amount of material, some catalyst, be converting the atomic oxygen in the atmosphere into molecular oxygen and producing the dark spots in our images? It was not likely, as we were never able to identify any such catalytic agent.

The knowledge that the spots actually moved across the face of the Earth strongly pointed to the existence of some kind of an object that prevented light from passing through it. Whatever it was had to be big and blackening out the ultraviolet light at a certain wavelength. It could not be an atom. It could not be a rock. It could not be anything thrown up there from down here. It had to be a common molecule in the solar system that absorbs at the right wavelength. The only common molecule is water and water just happens to absorb at the wavelengths we were observing with our camera. There was no reason to look for anything exotic. Water, in the form of water vapor, fit the bill perfectly.

This explanation posed certain difficulties, all of which were more psychological than physical. When we calculated how much water we would need up there to produce a spot in our images, we came up with a figure of about a hundred tons. Anyone would tend to back off from such a large figure and initially we did too. Then we figured out how many such objects we needed to account for the holes in the images we observed over the course of the year. And it was not one, not a hundred, but ten million. There was the problem. One per year would not have been a problem. But ten million per year? Unfortunately, there was not much leeway in our numbers.

The size of the holes presented another problem. They were easy enough to measure. We knew the size of the area each pixel covered in our pictures and we knew the altitude of the spacecraft. But what looked like little dark spots on the images turned out, in reality, to be about thirty miles across. They could not be rocks because such large rocks would just smash the surface of the Earth to pieces. These were clouds of water vapor.

It was the only reasonable explanation we could find. It was the only reasonable explanation that anybody has ever been able to offer. Early on some engineers suggested that the black spots might be due to satellite parts falling into the atmosphere. But that is not possible. First of all, there are not ten million satellite parts out there falling down on us each year. And secondly, they are not thirty miles across. Many people said there had to be other explanations for the dark spots. But no one has ever come out with one.

The numbers were shocking. Earlier we had found that the spots on our images varied with the frequency of meteors falling into the Earth. At first, we were elated. But when we sat down to think about it, we saw a disaster pending. The objects were just too big and too numerous to be just another nice little geophysical fact with no real impact on our thinking. But once we ruled out noise and ascertained that the spots were real, the next step, the interpretation, was trivial. The only way to interpret these events was in terms of ten million objects falling into the Earth's atmosphere every year. That is an infall of material that is about ten thousand times more than anyone had ever imagined. So psychologically, emotionally, the interpretation was difficult to accept. But intellectually, it was trivial. There was no other reasonable explanation.

By December of 1985 the press had picked up the story. Stephie Weisburd had written an article for *Science News* entitled "Atmospheric Footprints of Icy Meteors" following our fourth presentation on the topic, which took place at the fall meeting of the American Geophysical Union in San Francisco. We still called the objects meteors at that time, but that was not quite right. One thinks of a meteor as dust and rock, not water, although some meteors are thought to contain a substantial amount of water before hitting the atmosphere. But the objects we were talking about clearly had to be mostly water. That could only mean one thing. Sigwarth and I finally realized after the meeting that these objects could be nothing other than comets, small comets.

People at the meeting seemed to be quite interested in our presentation but we were careful not to threaten anyone's perception of the solar system. We never came out and said just how big these objects were or exactly how many were falling into the Earth's atmosphere. We just said there were a large number of unknown objects falling into the Earth's atmosphere that had not been detected before. I also began discussing our findings privately, in other meetings and hallways. No one got too upset over it.

Science is no stranger to odd data. Sometimes anomalous observations are just put aside for later examination. But often they are just stuffed into a drawer and forgotten. We scientists normally have our hands full examining the data we were seeking in the first place. So odd data has a tendency to fall through the cracks. There is a lot of tucking under in science. Everyone thought I would eventually bury my data on the mysterious black spots. No one thought I would publish them.

They were wrong. I began writing the first draft of the original small comet papers in December of 1985. Eventually I handed out a few copies to people for their comments, but not before spending many nights pacing the hallways of Van Allen Hall and pondering the consequences of such an action. The papers touched on many of the objections that such a large infall of comets would raise. What would their existence mean for the Earth and Venus and Mars and the other planets? What was the lifetime of such objects? What kind of mantle did they need to survive repeated passages through the inner solar system? Where did they come from?

One of the copies of our original papers had gone to Thomas Donahue, a professor of atmospheric sciences at the University of Michigan. You cannot do any better than Donahue when it comes to experts on planetary atmospheres. I respected his opinion. Donahue looked at our papers and said that if we were right these comets would have brought in enough water, over the age of the Earth, to produce the oceans. And of course he was right. I had not thought of it because I had always assumed that the oceans had always simply been here. So did everyone else.

I submitted my first papers on the small comets to Alex Dessler in February of 1986. Dessler had just become editor of the American Geophysical Union's *Geophysical Research Letters*. He was ambitious and wanted to make the journal controversial, exciting, and well-known. Unlike the quarterly *Reviews of Geophysics*, which Dessler had edited previously, this monthly publication provided a rapid way to communicate the latest research in the field of geophysics. Shortly after seeing our papers Dessler called to say that this was just the kind of material he wanted. He would have the papers reviewed in forty-eight hours.

Scientific journals rely on a rigorous review process to weed out inaccurate claims and research findings. Editors, if they are favorably disposed to manuscripts submitted for publication, send them to two or more scientists considered experts on the topic treated in the manuscript. The comments of these referees are returned to the editor, who then sends them on anonymously to the original author. This allows the referees to be candid and honest. Often, however, this power is abused. Those who are judging the merits of a manuscript are often in competition with the author for grants or recognition. Some scientists abhor this anonymity and those who know one another occasionally send their comments directly to the author of the submitted paper.

I had this kind of unspoken understanding with Donahue. He was one of the referees and he nearly exploded. He asked me not to publish the papers, the interpretation paper, in particular. The first paper was simply a description of the black spots and no one could deny that the black spots were there. The second paper was the interpretation. In it we spelled out that we were dealing with ten million, comet-like objects entering the Earth's atmosphere per year, each one the size and weight of a small house. We also touched upon many of the topics this interpretation would seem to contradict, such as the origin of the oceans, as well as the lack of water in the Earth's upper atmosphere, on Mars, on Venus and on the Moon, as much as we could cram, in other words, into four pages. This was the limit on the length of papers published in *Geophysical Research Letters*.

The other referee also recommended against publication. He said, Dessler told me, that "if this was correct, we would have to burn half the contents of the libraries in the physical sciences." It was a dicey situation for Dessler. The two people he had asked to review our papers had advised him not to publish. So Dessler called me and asked if I would withdraw the interpretation. He warned that its publication might destroy my scientific career. But I told him it was necessary. Without it people were likely to regard the black spots as a mere curiosity.

Donahue also begged me to come to my senses. He did not want the interpretation paper published. I think he was trying to be helpful. But he must also have been worried about the tremendous repercussions these findings would have for many fields of science, including planetary atmospheres, Donahue's own little niche. Yet despite two negative reviews and the "uncomfortable ramifications for the community at large," Dessler decided to go ahead with publication. The final decision always rests with the editor. Reviews are designed to guide, not bind.

"If you restrict the journal to publishing only what pleases the referees you end up publishing what is popular," Dessler explained in 1987. "And while it does make everyone feel more comfortable, you are guaranteed to miss the occasional breakthrough. There is a price to pay, of course, for doing it the way I think it should be done. Nine out of ten of these things will turn out to be rather worthless. But I think the price is worth it. Occasionally one is so important that it makes publishing all the previous ideas that didn't turn out quite right worthwhile."

Anytime you break new ground in science you get attacked. This is the basic conservatism of science. But it is necessary to put all new ideas on trial. I had some notion of what the response to these papers would be. I had been involved in many scientific debates before. But the reaction I received was like none I ever experienced. I was driving a bulldozer through dozens of the neatly planted fields of science and everyone was upset.

I had, of course, been tempted not to publish these results. I was enjoying my work. People respected me. I had all the opportunities in the world and all the funding I needed. I did not need the aggravation these findings would stir up. On the other hand, I thought of all the scientists from Copernicus to Einstein who had come forward with seemingly outlandish but revolutionary results. I realized that if I was right and did not publish my results—and no one did this work for another ten or twenty years—then I would be wasting thousands of man-years of scientific effort.

People had warned me repeatedly and I understood what they were saying. Had I heard someone talking the way I did, I would have warned them as well. But I had agonized over the decision to publish this material for a long time. The decision came long after it became quite apparent that our efforts to show that the black spots were not real had failed. There were no other tests to be done. I knew this would have a big impact on science. But I felt I had no choice. I could not bury this material.

The notion of so much water falling in from space flew against all current observations and beliefs. Most likely the notion was wrong, so I had to make certain that it was not. I had done everything I possibly could to find something that said these objects could not exist. I only needed one big piece of evidence. Just one. A hundred pieces of evidence would not prove that they exist. But it would only take one to show that they did not. So I sat in libraries and read about astronomy, about oceanography, about geology, one field after another. And finally I decided that our findings must be published, no matter what the consequences. I could not live with myself otherwise. It was just morally incorrect.

CHAPTER 4

True Confessions

I was born in Chicago in 1938, the year "You Must Have Been a Beautiful Baby" became a popular song. We lived in Berwyn, Illinois, until I reached first grade. But my parents were concerned about the violence around Chicago. They thought that a better place to raise children might be somewhere out on the plains, so we moved to Topeka, Kansas. We lived there with my grandparents for a short while and then moved to Chanute, Kansas. I went to grade school there.

I was the oldest child. I have a brother, who is younger by two years, and a sister, younger by six years. My father worked for the Santa Fe railroad. We did not have much money, but I never knew we were poor. My father never made it through high school because he had to support the family. At an early age I was trained to be a pianist. I took piano lessons and played in a very emotional way.

School bored me. I sat in the classroom and seldom paid any attention to what was going on. Teachers ostracized me. Whenever we had spelling bees or math contests I competed against the last person and won. So the students ostracized me as well.

Just before junior high, we moved to Fort Madison, Iowa, a rivertown on the Mississippi. I spent my time reading science fiction and building things like cannons. I also continued to play the piano. By the time I got to high school, I was known as a problem child. Much of the problem was my poor memory for dates and names. Once I was registering for classes and could not fill out the forms. I could not remember the date of my birthday. The principal sent me home to find out when I was born. He thought I was mentally retarded. When I returned, he had signed me up for non-academic courses, like machine-shop, because he thought I would never make it to college.

High school presented no challenges. The teachers sometimes asked me not to attend classes because I depressed the students by my performance. I actually taught the chemistry and physics classes when the instructor was not there. I got every award you could get in high school, everything from the typing award to the American history award. I would probably also have gotten the home economics award but they would not let me take the course. I was class valedictorian.

My mentor was a chemistry and physics teacher named William Heinberg. He came from industry to see if he could teach. He actually stayed on an extra year, before going back to industry, just so that he could teach me chemistry and physics. He had a profound effect on my life. He trusted me to work in the labs and I became fascinated with science.

I went to college at the cheapest place I could find and that was the University of Iowa. I thought I would never again have the same problems in class. But I was not there a month before my English teacher told me that if I did not come to class, he would give me an "A." He said I discouraged the other students. But I loved science. I changed from chemistry to a physics major because that appealed more to me.

I enjoyed my freshman year. That was 1956. I needed work and found a job in the physics department helping to plot points for someone's doctoral thesis on cosmic rays. I enjoyed reading the data from the printouts and constructing graphs. Then, before the start of my second year, Professor James Van Allen called me into his office and offered me a graduate teaching assistantship. I was a sophomore undergraduate at the time. I was a kid. But I did it. Everyone in the class was older than I was. You had to take ROTC at the time and I would come into class in my ROTC uniform. It made me look even more like a kid. But we all had a really good time.

By 1958 the space program was underway and the University of Iowa was in the thick of it. Van Allen's lab was in the basement of the physics building. Things were stored in dog cages and at one in the morning bats flew around the hallways. It was like no lab you see today. Van Allen was in charge of the Geiger counter that would be placed on Explorer 1, the first U.S. spacecraft. It weighed just over ten pounds and was about the size of two oatmeal boxes. Explorer 1 was launched on the last day of January and discovered the radiation belts that gird the Earth. They now bear Van Allen's name.

In my junior year Van Allen asked me to help him calibrate the instruments for two other spacecraft, Pioneers 3 and 4, which would be launched atop modified ballistic missiles. Pioneer 4 would become the first successful lunar probe. The Pioneers were the size of an oatmeal box and shaped like a cone. Each spacecraft contained a battery, transmitter, and Geiger counters. Our interest in these spacecraft was intense since the day in October 1957 when we heard the news that the Russians had launched something called Sputnik. We had strung an antenna up on top of the physics building and tuned in as Sputnik passed overhead. It made a "beep beep beep" sound that I will never forget. I was impressed. But I was also distressed that the United States would not be the first into space.

The Pioneers were launched during my undergraduate junior year. Pioneer 3 did not have enough "oomph" to make it to the Moon. It only got about halfway there but it did make some beautiful measurements of the far boundaries of the radiation zones that had been discovered by Explorer 1. I spent a lot of time plotting and analyzing the data. Van Allen wrote the papers. I was barely twenty years old and I co-authored two scientific papers with him. Van Allen and Frank. Those were very exciting times.

I received my bachelor's degree in 1960, but continued on as a graduate student, spending many hours in the lab. The space sciences fascinated me. I published a number of papers, both on my own and with others. I got my master's and Ph.D. in about four years. I was undecided about what to do next and was interviewed by several places. The day I went for the interview at Berkeley, President Kennedy was shot. But the opportunities at Iowa were so great at that time that I decided to stay. By the age of thirty-two I was a full professor.

I have had experiments on about forty spacecraft. People tell me I have had more instruments on more spacecraft than anyone else on this planet, but I have never bothered to check that out. The point is simply that I love experimental physics. I like to build things and use the best technology available to probe our environment and solar system and expand our knowledge. My area of expertise is plasma physics. Plasmas are the charged particles that comprise 99 percent of the universe. You look up in the sky and the stars are plasma. Our Sun is plasma. The Earth's aurora are caused by plasma. In order to understand the universe, you must understand plasma. That is what I try to do.

I have been in the forefront of many plasma discoveries. I made the first measurements of the plasma ring around Saturn. I was the first to measure solar wind plasmas funneling directly into the Earth's polar atmosphere. I was the first to observe with a satellite instrument the belt of ions around the Earth that is now known as the "ring current." And I discovered the theta aurora, a luminous phenomenon which, seen from space, looks like the Greek letter theta stamped across the polar cap.

If you know enough about experimental physics, you can work in any field you like. That is because physics deals with fundamentals. I have always hoped to be in a class of people known as tinkerers. These are people who see something in a field outside their own that so stimulates their imagination that they make the necessary measurements in the hopes of pressing forward the frontiers of knowledge. That to me is the final stage, the ultimate achievement, of an experimental physicist.

My whole life has been driven by my work. Unfortunately, it has come at the expense of my family. I have not been a particularly good father or husband. I have not spent a lot of time with my two daughters. I am thankful that they have a fine mother who has raised them well. During the past few years, my daughters have seen me more on TV and in the newspapers than in person. They talk to my secretary more than they talk to me. That is not good, but this is the way I am. I realize that this is selfish on my part, but I am just consumed by what I do.

There is a wonderful restlessness to science. It keeps wanting to move on, to explore new territories. Since the beginning of the space age we have moved rapidly to explore the Moon, the planets of the solar system, and the furthest objects at the very edge of the universe. But for most of my career I have looked back at the Earth. I have been particularly fascinated by how nature produces those beautiful little crowns, the auroras, on the north and south poles of the Earth. The Earth and its royal crowns are every bit as magnificent as Jupiter and its big red spot. These space-borne views of the Earth have provided many insights to similar phenomena observed elsewhere in the solar system.

I am presently involved as an experimenter on the Galileo mission to Jupiter. My instrument will measure the plasmas at Jupiter. I am building what is probably the most advanced dim-light imager in the world for the Polar spacecraft to be launched in 1993. I am one of two U.S. experimenters putting entire instruments on a Japanese spacecraft known as Geotail, due to be launched in 1992. I have a lot to do.

People generally do not equate the "L. Frank" who works in plasma physics with the "L. Frank" who discovered the small comets. One appears to be the most conservative of scientists, the other a maverick hellbent on destroying the very foundations of science. But they are one and the same person. The small comets are only one aspect of my life's work. In the past five years, I have managed to author or co-author about sixty papers, only some of which have been on the small comets. I did not want, and could not afford, to let the small comet controversy stop me from my other work.

There is no easy or clear-cut blueprint for scientific discovery, but the process, it seems to me, usually follows one of two paths. It can occur, in one case, when you are pressing forward the frontiers of a science. This requires that you devote your life to one very narrow scientific topic. Take solid state physics, for example. The current frontier in that area is superconductivity—the ability to conduct electricity without resistance. If you are working in this field, you are trying to find the right combination of metals and alloys that will conduct electricity at higher and higher temperatures. If you are interested in making a superconductor, you are not reading about geology, or astronomy, or even what is happening in high energy physics. You do not have the time. You are working in your little niche day and night, knowing that there are dozens of other people competing with you. There are ten, twenty, thirty other large laboratories, not counting the many universities, trying to do the same thing you are. You are going to feel the pressure, the excitement of it. But although there may be a temporary clash of ideas, a debate perhaps, there is no conflict.

Conflicts occur when your findings clash with accepted scientific dogma. This is the second path to scientific discovery, but it is generally not a pleasant experience. Those who question the prevailing point of view are usually regarded as outsiders, as mavericks, as renegades. Their behavior is often seen as dogmatic and pathological. And the scientific community responds by either turning on you or turning away from you. The history of science is riddled with examples.

Hannes Alfvén, the Swedish astrophysicist now in his eighties, is one of those who was essentially ignored. He developed the fundamental principles of plasma physics in the 1940s but was not recognized for his work until three decades later. He won the Nobel Prize in 1970. But at the beginning, no one paid any attention to him. His work opened up a whole new area of science. Yet his very sophisticated ideas about the aurora, sunspots, solar wind, the flow of plasma, and about the origins of the solar system were far enough removed from the scientific thinking of the time not to upset any applecarts. Even today, with such concepts as "Alfven waves" firmly entrenched in space physics, the man is still regarded with ambivalence. Scientists respect his past contributions but remain skeptical of his theories.

Alfred Wegener suffered the opposite fate. In 1912 he proposed the revolutionary hypothesis that the continents are not fixed but rather have been slowly wandering during the course of the Earth's history. This proved far too radical a notion for most of his contemporaries to accept. The reaction was heated. Everyone turned against him. Wegener, who was occupied most of his professional life as a meteorologist, had no time to respond adequately to the increasing opposition facing his theory of continental drift and was essentially drummed out of the scientific corps. Wegener died in shame because he thought he was wrong. Vindication did not come until many years later. But even today there are still people who do not believe in his theory of continental drift.

I think that I understand how Alfred Wegener felt when he proposed his theory. He told people that the ground they were standing on was moving under them. That hit close to home. It cut across a lot of areas of science. I am telling people that tens of thousands of small comets are falling into the sky above them each day. That also hits close to home. And it also cuts across dozens of areas of science. No one plows over everybody's neatly planted gardens without paying the consequences.

CHAPTER 5

A Storm of Controversy

Even before publication of our discovery, the small comet proposal became the target of vigorous and even vociferous attacks. The battle lines were drawn quickly and easily—it was us against everyone else. No blood would be shed, but in the months to come a sea of ink would cover the battleground, the pages of *Geophysical Research Letters*. We had to answer every question that was thrown at us regarding the small comets and we had to do it quickly.

Our discovery drew a flood of critical comments. The editor of the journal, Alex Dessler, had welcomed controversial topics to its pages and had established a specific procedure, resembling the pre-arranged rules of a duel, for handling any formal critical responses to the published articles. The attack of the challenger he called the Comment. Dessler gave it an initial reading, but it was not refereed. Comments judged to be valid were then forwarded to the defender.

Dessler required that the defender meet the opponent's attack with a counterattack, labeled the Reply, but he would allow no ducking or side-stepping of the issues raised by the Comment. This parrying continued—with as much skill, speed, and finesse as the players could muster—until one side or the other chose not to revise its last submission. Dessler then published the final drafts of the Comment and Reply together in a future issue of the journal—point pricks, shoulder jabs and all. Dessler hoped that this would present readers with an accurate summary of the match, "a terse distillation of what might have been many pages of private debate."

For *Geophysical Research Letters*, the interest stirred up by the small comets was unprecedented. The first Comment reached Dessler's desk before the original papers announcing our discovery had even been published. Less than a month later, Dessler had accepted four Comments from the critics and within three months he had accepted ten critical Comments for publication. It would take more than a year out of my life to answer them all, a total of eleven Comments in the end. It was like a paddling line. There was no rest. If I had been ten years older, I would have fallen apart physically. As it was, I barely made it.

Sigwarth and I did the best we could to answer this laundry list of objections in as orderly a fashion as possible. I would receive the Comments, since I was listed as first author, and would make a copy for Sigwarth. He would read it and we would discuss the various points in the Comment. We then decided what we needed to do in order to answer the question it posed. Sigwarth scurried around several libraries on campus, digging out material that often had never been checked out before—or since. We reviewed the material together, learning about the topic raised by the Comment. Sigwarth would do most of the computer work. We both did the theoretical calculations and discussed our results with Craven. Then I wrote the

Reply.

The Comments received by *Geophysical Research Letters* defined the scope of the small comet debate. They covered a very broad range of subjects from geology to planetary atmospheres and from lunar science to astronomy. The entire point of this exercise was to find inconsistencies with the small comet hypothesis. If one was found to contradict the existence of small comets, these objects would then cease to exist. If I faltered, if I failed to answer any of the Comments, if I could not come up with a half-way reasonable physical explanation for the criticisms that were flung at us, then it was over. In fact, Dessler called me on occasion and said that I did not have to answer the Comments. But everybody in the world knew that if I failed to answer them, then it was an admission that the debate had come to an end.

Dessler arbitrated the debate. Sometimes several Comments made the same point. Not all Comments were accepted. The author of the first critical Comment to reach Dessler's desk and the first to be published was Thomas Donahue, one of the two referees of our original papers. Donahue is a good physicist. We had known each other before all this happened, but we got to know each other a lot better in the course of this controversy. During this time he was chairman of the Space Science Board of the National Academy of Sciences. This is a very prestigious position and, as far as advisory boards go, it is the most powerful one outside of NASA.

Our findings upset Donahue from the very start. So did the handling of the subject by the press. *The New York Times* quoted Van Allen, the head of the Department of Physics and Astronomy at the University of Iowa, as saying that our cometary hypothesis was "pretty unavoidable." This prompted Donahue to send off an angry letter to Van Allen, thinking that he had played a role in my work. But I am not on a leash, even though some people wish I were. I greatly admire Van Allen, but he has no control over what I do and has not had any since I got my degree back in 1964.

Donahue was not the only one to contact Van Allen on the controversy I had caused. Thomas Gold, the noted Cornell astronomer, did so as well. Perhaps he, too, felt uneasy about voicing his comments to me directly and decided to do it through a close associate. The small comets upset Gold terribly. I have known Gold since I was in my early twenties and his response came as a surprise. Gold is a maverick himself. He believes, contrary to current opinion in geology, that most of the oil and gas in the Earth does not have a biological origin. To prove his hypothesis Gold is presently trying to mine oil from an area in Scandinavia that is thought to be the impact site of an asteroid. Current opinion holds that the area is lacking in the geological foundation necessary for gas and oil deposits.

But Donahue would play a much larger role in this controversy than Gold. Donahue is a very explosive personality. When he talked to reporters about me, he sometimes sounded like he was on the next horse out to get the small comets. One reporter for *The Ann Arbor News* characterized Donahue's attack on my efforts to bolster the small comets theory as "bitter." Donahue denied it. But it did not matter. I expected this. When *The New York Times* ran its first story on our findings, the reporter quoted Dessler as saying that the comet theory, if correct, was "probably the most important finding in space science of the last decade," and that it would "make or break" my reputation. Donahue wanted to hang Dessler by his thumbs. He thought it was improper for an editor of a scientific journal to use his position to promote public controversy and put the reputations of contributors to his journal at risk.

Donahue asked me to co-author a letter to *The New York Times* to straighten things out. I agreed. "We regret the public contention," our letter said in part, "that has had the effect of transforming a civilized discussion between scientists interested in arriving at a mutually agreed upon explanation for a puzzling discovery into a life or death struggle for our scientific reputations. We two, who have been portrayed as bitter adversaries in some of these stories, wish to assert that contrary to the impression that has been created, we have very high respect for each other as scientists and as human beings."

The letter was never published. That further angered Donahue. He was also upset because I had used the word "startling" in the final version of our paper. Originally, I had used the word "plausible" in describing our interpretation. But Donahue and others had kept calling it "startling." So I appropriated the word for the published version of our paper. Donahue was not happy.

Thomas Donahue is one of the leaders in the area of planetary atmospheres. Another is Donald Hunten, a professor in the Department of Planetary Sciences at the University of Arizona in Tucson. Hunten has spent a lifetime on the theory of planetary atmospheres and in particular on how water slowly evaporates from our planet. It is a very elegant, very beautiful theory.

This is how it works. A water molecule comes up off the ocean and rises into the atmosphere, where ultraviolet rays split it into hydrogen and oxygen. The hydrogen, because it is so light, eventually flies off into interplanetary space. The oxygen, meanwhile, comes back down to Earth and oxidizes something, like iron, to form rust. That is the way you lose one water molecule. This is a beautiful theory because it explains why we have this big envelope of hydrogen gas above our atmosphere and it explains how some small fraction of the Earth's oceans has evaporated over time.

What could be wrong with that? In fact, only one thing. All the equations in the theory are based on an important assumption. The assumption is that there is no source for the water other than the oceans. But the small comets provide a source of water that amounts to one thousand times the rate of evaporation calculated by Hunten. For every gallon of water he has evaporating each year the small comets bring in a thousand gallons of water. Which means that the detailed mathematical equations which Hunten has spent a substantial part of his life working on are wrong because he does not account for this additional source of water. He did not make any mistake in his arithmetic, but in his assumption of the supply of water.

A scientist toils through life writing paper after paper that culminates in a magnificent theory. Then one day someone comes along and finds that a fundamental assumption is wrong and a large part of his life's work collapses just a few years before retirement. What do you expect his reaction to be? I understand how he feels, but there is nothing I can do about it. I feel like a spoiler. Hunten and I get along very well despite the fact that we both have very strong personalities. We openly argue quite a bit and these heated exchanges make a lot of people rather uneasy. Hunten is very intelligent, very honest, and very influential.

So Donahue and Hunten joined forces to defend their model of planetary atmospheres against this intrusion from outer space, the small comets. Call it damage control. It was not the first time they had tried to fight off the suggestion that the Earth's atmosphere has an external water source. But this time, Hunten said to Donahue, in effect, I'll let you take care of it. So even before our papers were published in *Geophysical Research Letters*, Donahue was preparing his criticism of the small comets theory.

Donahue saw "a large number of problems" with our small comet hypothesis but there was no anger in the tone of his published Comment. Compared to some that would follow it was almost kind. One of the problems that bothered Donahue was Venus. How could Venus avoid this cometary bombardment? Where was all the water on Venus? Venus, of course, is one of Donahue's very favorite subjects. He had become well-known for his Venus work.

Compared to the Earth, Venus today is exceedingly dry. Very small amounts of water have been measured in its atmosphere. It is estimated that there is a hundred thousand times less water on all of Venus than there is in just the Earth's oceans. Donahue proposed that Venus once had an Earth-like ocean early in its history but subsequently lost it to evaporation. Donahue came to this conclusion after analyzing the data from an instrument on a releasable probe carried by the Pioneer Venus orbiter in 1981. Just as the probe plunged into the lower cloud layer about thirty miles above the surface of the planet the instrument got clogged up with what was almost surely sulfuric acid. During this time a measure of the ratio of heavy hydrogen, or deuterium, to ordinary hydrogen was taken. The value was quite high. It was about a hundred times greater than the deuterium-to-hydrogen ratio on Earth. This is the smoking gun that, for Donahue, raised the issue of Venus's wet past.

To understand why, you have to realize that Venus and Earth have long been regarded by scientists as nearly identical twins separated at birth. Both planets were formed from the same stellar material approximately 4.5 billion years ago and resemble one another in fundamental ways, in size, mass and gravity. So both planets are thought also to have been originally endowed with the same ratio of deuterium-to-hydrogen. But while Venus grew up close to the Sun and went dry, the Earth was brought up farther out and was able to maintain a sizable ocean.

But what happened to all the water on Venus? Donahue proposed that a once Earth-like Venus lost its oceans due to a runaway greenhouse effect that boiled the oceans, sending much of the water into the upper atmosphere where sunlight tore both hydrogen and deuterium atoms from water molecules. The ordinary hydrogen would have escaped into space. But the deuterium, being twice as heavy as ordinary hydrogen, was left behind as a residue. The loss of ordinary hydrogen would show up clearly as an elevated amount of deuterium in the Venus atmosphere. But there are problems with Donahue's proposal. He makes an assumption about the original ratio of deuterium-to-hydrogen on Venus, despite the fact that this ratio varies widely throughout the solar system. He also assumes that the water abundance on Venus has been declining throughout its lifetime.

My Reply to Donahue simply said that the small comets could also account for the presence of water vapor in the Venus atmosphere. An ancient ocean was not necessary. Of course, these comets could not possibly bring in as much water to Venus as they bring to Earth. The flux of the small comets at Venus cannot be the same as it is on Earth. The Sun vaporizes most of the small comets as they approach Venus. Maybe only one out of a thousand small comets actually reaches Venus. So I said there was no need to invoke an early ocean on Venus to explain the water vapor in its atmosphere. A few small comets entering its atmosphere would do just as well.

Others had already questioned Donahue's scenario. One was Sig Bauer at the Institute of Meteorology and Geophysics at the University of Graz in Austria. Back in 1983 he authored a paper for *Annales Geophysicae* entitled "Water on Venus: Lack or Loss?" In it he had suggested a possible cometary origin for the atmospheric water content of Venus. In a letter to me a few years ago, Bauer wrote: "Your observations and interpretation are of particular interest to me in connection with Venus since I always held the (minority) opinion that Venus was formed with little or no water and the present water content came from the outside." John Lewis had voiced the same opinion back in 1974 while at the Planetary Astronomy Laboratory at the Massachusetts Institute of Technology. In a paper published in *Earth and Planetary Science Letters*, Lewis concluded: "It is probable that Venus's present hydrogen loss rate is appreciably smaller than the rate of injection of hydrogen from meteoritic, meteoric, and cometary water."

Lewis, like Bauer, had looked at the constituents of the Venus atmosphere and concluded that the water there could have been supplied by an occasional comet. Comets, whether large or small, are a much more reasonable source for the water vapor in the Venus atmosphere than the evaporation from an ancient ocean. But the proposals of Bauer and Lewis were not popular in their time. Mine is not popular today.

The core of Donahue's Comment had nothing to do with Venus, however, but with the Earth. If the small comets were bringing all this water into the Earth's atmosphere at an altitude of one hundred to two hundred miles, Donahue said, then someone would have measured it. My Reply was that these comets were not laying all this water on top of the atmosphere. You cannot stop a hundred tons of water falling in at three miles per second or more at those altitudes in the atmosphere. And once that water vapor plunges into the lower atmosphere it gets so mixed up with what is already there that the infall of new water is beyond detection at the surface of the Earth.

I think my Reply stung a little. Our atmosphere is composed of two main layers. The first layer consists of the winds that circulate near the surface of the Earth. These winds travel not just from Nebraska across Iowa but up to about seventy miles high, to a region called the homopause. The winds here are part of one big circulation cell, rather than a bunch of little ones all piled up on one another. The homopause separates the lower zone of the atmosphere from the upper zone. It too has winds in it, but they are not locked closely into the winds in the lower atmosphere.

Donahue's point was that if the small comets come in and stop before they reach the general wind pattern in the lower atmosphere, then the water should stay up there and build up to the point that anybody should be able to detect it. He cited the microwave measurements of John Olivero, a meteorologist at Pennsylvania State University, indicating that there is indeed no large amount of water up there. But if the water did remain up there, it would be obvious. There would be a worldwide ice cloud. On the other hand, if the water vapor from these objects actually plummets down through the homopause it would simply mix in with the water and clouds already present in the lower atmosphere.

Our work showed that the cloud of water from a disrupted small comet enters the Earth's atmosphere like a piston of water vapor. When this piston reaches an altitude of fifty miles, well below the homopause, it is still moving at almost twice the speed of sound. It is simply not going to stop at the top of the atmosphere. Hunten found my description of small comets as pistons of water vapor punching through the atmosphere "ingenious." But it is really quite trivial, a few calculations done on a lunchroom napkin.

Donahue also thought that the supply of water brought in by our small comets was just too large. He insisted that the infall of small comets into the atmosphere could not be continuous. If it is, said Donahue, then we had too much water coming in over geological time. Over the four billion plus year history of the Earth this would be enough water to fill the ocean basins four or five times over. That seemed to bother everybody. But who would expect the infall of comets to be exactly constant over some four billion years? Besides, it is just incredible that we were even talking in terms of ocean units in the first place. We were not arguing over a thousandth of an ocean, or a millionth of an ocean, or even a thousand oceans or ten million oceans. I said it was one ocean. Donahue said it was four or five oceans. What difference did it make? We were in the same ballpark.

CHAPTER 6

The Creation of the Oceans

It is a strange planet we live on. There is water everywhere. There is so much water on this planet, in fact, that it is beyond our ability to comprehend it. We do have estimates of the total world volume of water, but the numbers are beyond our direct experience. Perhaps the best way to begin to grasp the quantity of water in the world is to portray it in terms of the number of bathtubs we would need to contain it. If we take a bathtub to contain about 35 gallons of water, then the world's supply of water would fill about 10 quintillion, 566 quadrillion, 293 trillion bathtubs to the overflow vent.

You would need most of those bathtubs—more than 97 percent of them—just to hold the contents of the oceans. The rest would hold the world's fresh water supply, of which more than three-fourths is locked up in the polar ice caps. Most of the remainder lies in groundwater and soil moisture. Lakes and swamps account for a little more than a third of one percent of the world's fresh water supply and the atmosphere contains less still—just a few hundredths of one percent. But small percentages can be just as deceiving as large numbers. The atmosphere actually holds enough water to fill hundreds of trillions of bathtubs.

Current wisdom favors the view that most of the Earth's water has been here since the birth of our planet, which is estimated to have occurred some 4.5 billion years ago. It is widely believed that the Earth was formed through the accumulation of cosmic debris left over from the giant nebula, essentially a dust and gas cloud, that formed the Sun itself. The debris particles, which are thought to have contained water, eventually collected together through the force of gravity to produce our planet. The water then remained trapped in the Earth until the initial crust solidified and the gases from the interior were released in volcanic events. This outgassing, or degassing, from the Earth's mantle formed an atmosphere rich in water vapor and other gases, known as volatiles. Then, as the Earth's temperature fell, water condensed out of the atmosphere. When the rains came, the original crustal minerals were eroded and weathered, forming sediments and an initial ocean.

But some scientists have begun to doubt this view. Not until the late stages of formation was the Earth either large enough or cool enough to support water. In the early stages of formation, the Earth was nothing but a large, hot cinder that had grown from collisions with many smaller cinders. Though it was eventually large enough for gravity to prevent the easy escape of hydrogen, helium, and water from its atmosphere, the Earth still could not support liquid water on its surface until it had cooled considerably. So it seems that any water present during the Earth's formation from the cinders would very likely have escaped back into space.

The alternate view is that water was brought to the still primitive Earth in a bombardment of large comets, which are conjectured to have been quite common in those first hundred million years or so of the Earth's history. Some of this comet water would have remained on the surface and some would have mixed with the crust of the Earth and its underlying mantle and have been brought to the surface by volcanoes. Though appealing to some scientists, this scenario for water delivery through an early comet swarm is still not well-defined.

But in either case—water here at the start or brought in by large comets—it is generally agreed that the Earth probably had most of its water quite early in its history. I, too, had always assumed that the oceans were here in the beginning. I thought it had long been accepted as proven fact. I had never given it a second thought. Then one day Donahue, who had seen the early drafts of our original papers, realized that the small comets, if they were real, could have brought in enough water, over time, to supply the world's oceans.

This was the first major implication of the small comet discovery and the one that drew the most attention from the public. Initially, I was more upset than satisfied by the idea. So I did the calculation myself. A small comet falls into the Earth's atmosphere about every three seconds. Each one contains about 100 tons of water. Some 25,000 such objects fall to Earth every day, or about ten million on average over the course of a year. Some years would have more, some less. But since the Earth was formed, 4.5 billion years ago, about 40 quadrillion small comets would have fallen into the atmosphere. The total volume of water deposited by such an influx of small comets makes these objects not only a reasonable, but a likely candidate for the source of the world's oceans.

This is no magic trick. It is a simple deduction from the known facts. I will be the first to admit that it seems almost inconceivable that the ocean waters should come from *out there*. It sounds silly. One skeptical oceanographer even dubbed it "the trickle down theory of the oceans." But that is exactly what it is. I have spent some time psychologically adjusting to this rather striking disparity with my previous education. Being a very conservative person, I probably change my ideas more slowly than most.

I searched the scientific literature looking for solid evidence to back up the notion that the oceans were here in the beginning. I thought that if I could find proof that the supply of water originated early in the Earth's history, I would have to reject the notion of a constant bombardment of small icy comets. Reporters who called me on the story actually were quite helpful in this respect. They, too, searched for evidence to back up what they had learned in high school and college. And they, too, were disillusioned.

There is a lot of conjecture about what went on in the early Earth, but very little data. Our beliefs are based almost entirely on the geological record left by sedimentary rocks. And the oldest sedimentary rocks, which were formed in the reaction of crustal minerals and water degassing from the planet, have been dated at close to four billion years old. But there is no agreement on how much water was needed to produce them.

Just how quickly the oceans formed is actually a subject of some controversy among geologists. Most geologists lean toward an early formation of sedimentary rock and ocean. They think that the sedimentary material accumulated very early in the Earth's history, meaning that the water volume of the early ocean would closely resemble today's ocean. They believe there has been a kind of cannibalism of the old rock to make new rock. So that what looks like new sedimentary rock is actually old rock that has been weathered, eroded, and redeposited. Others think the sedimentary mass has accumulated gradually over time and the oceans likewise. Indeed, the proof that the Earth had as much ocean water in the past as it does today lies largely in the lack of evidence to the contrary, according to Gustaf Arrhenius, the highly respected geochemist at the Scripps Institution of Oceanography whose specialty is the early history of the ocean and atmosphere.

The small comet scenario calls for a slow acquisition of the world's waters. Adding one inch of water to the Earth's oceans every ten thousand years is undetectable. There are simply no rain gauges that accurate. In addition, no one really knows what the water levels are on Earth. If the small comets are real and we go back far enough, say a billion years, then we should find one quarter less water in the oceans. But we cannot determine that this was the case with any accuracy because the time period is so remote. Some people ask me whether the rock record shows an increase in water levels. The answer is: you cannot tell. The change in water level is just too small and the Earth's crust shifts too much.

Even Donahue has admitted that the small comets could not be ruled out to explain the origin of the oceans. "Data concerning ocean volume and sea level," he said, "are too imprecise and limited in time ..." So it is all still quite problematic, this origin of the ocean waters. No one is too certain about where the Earth's waters came from in the first place or just how quickly the ocean basins were filled.

I know people do not look at scientists as innocent, but they really are innocent. And the suffering that I have endured in this affair is mainly the loss of this innocence. I assumed, like everyone else, that science was relatively secure in its knowledge of the origins of the oceans. But it turns out that most of what we know about the subject are assumptions based on our mistaken impression of a famous paper by William Rubey, published in the *Bulletin of the Geological Society of America* in September of 1951. Everyone cites Rubey's paper on the "Geologic History of Sea Water" as saying that the water must have been here in the beginning. But Rubey did not say that.

Rubey had asked himself if all the water on Earth could have come from all the leaching and weathering of rock that had taken place over the Earth's four billion plus year history. And the answer he came to was: no way. The water has not come out of these rocks. Rubey said that by normal evolution, by weathering and such, there should be far less water, carbon dioxide, nitrogen, and chlorine, or excess volatiles as these substances are known, on Earth than actually are here. Or else, he said, our understanding of the history of the Earth is quite wrong.

Rubey thought that the water must have arrived on the surface of the Earth over time. He actually said that in the paper. He even briefly considered the possibility that the excess volatiles might have been gradually supplied by an extraterrestrial source. But he dismissed it, given "the exceedingly slow rate at which meteoritic debris is reaching the Earth today." He then concluded that the most likely source of volatile elements was the outgassing from volcanoes.

Rubey clearly condemned the quick soaks, those who prefer to have all the water here at the beginning, and came down in favor of the slow soaks, those who increase the water content of the oceans "by small increments over a much longer period of time." I was amazed when I read the paper. It was not subtitled "The Solution" but "An Attempt to State the Problem." Rubey had wanted merely to define the problem, not resolve it. What a loss of innocence on my part. So this was the foundation on which all our statements about the origin of the ocean waters are based. It was mind boggling. Bright man. Famous paper. But somehow misinterpreted, perhaps because it is a long paper and merely cited, not read carefully.

Besides, not everyone thinks you can squeeze the oceans out of hot rocks. The outgassing of the Earth through geologic time does not really solve the problem of excess volatiles, according to oceanographer Pat Wilde, head of the marine sciences group at the University of California, Berkeley. It is very difficult to get volatiles out of igneous or volcanic rocks. Though the distribution of some fifty-eight elements in sediments and seawater can be explained by the weathering of igneous rocks—a concept called Geochemical Balance—it does not explain the volatile content of the ocean. The numbers simply do not add up. You need another source, says Wilde, and the source may well be these small comets.

Wilde believes that the composition of the oceans may be the crucial clue to the problem of the excess volatiles. If the water was here from the beginning then it should have changed over time. When you start with a reservoir of water, it is like a fresh pool. But if you let it sit, it will become increasingly stagnant. It deteriorates because there is no refreshing of the waters. On the other hand, if you have a source of fresh water coming in continuously, then eventually the ocean should come into equilibrium with that source.

In the case of the oceans, if you start with as much water as we have now, then you would expect the salinity, for example, and the amount of dissolved materials in it to change over time. But the geologic record of the past 500 million years indicates that the salinity of the oceans has been relatively constant. This conclusion is based in part on living marine fossils that have a restricted tolerance for salinity but have been around and unchanged for about half a billion years. One example of such a living fossil is the brachiopod *Lingula*, a small invertebrate creature with a brown, tongue-shaped shell that lives in mud or sand and has a preference for quiet waters.

Wilde realizes that the small comets may bring in more than enough water to fill the ocean basins over time but he has no problem accounting for any excess. Extra water may be used to put water into the Earth's mantle and grease the motions of the continental plates, all the while keeping the actual ocean volume relatively constant. "Some of the excess water might be one of the reasons for plate tectonics," ventures Wilde. "You have to have this type of lubricant otherwise everything would lock-up occasionally." It would be "a startling geological revelation," he adds, if the composition of the small comets should mimic the composition of the oceans.

There is now evidence that this may be the case—at least for comets in general. Studies of Halley's Comet conducted by the European Space Agency's tiny Giotto spacecraft in 1986 indicate that the comet's chemical fingerprint seems to match that of the Earth's oceans. In the planetary sciences, the relative amounts of the isotopes of hydrogen, oxygen, and sulfur present in a sample can serve as a fingerprint of its origins. Isotopes are the members in the family of a given element, such as hydrogen, for instance, which has an isotope called deuterium. Hydrogen differs from the heavier deuterium atom in the weight of its nucleus.

Analyses of the Giotto results carried out by Swiss physicist Peter Eberhardt and his colleagues show that the Earth's oceans contain the same ratio of deuterium-to-hydrogen as Halley's Comet. The relative concentrations of oxygen isotopes are also the same. And the same is true for sulfur. Eberhardt believes that these findings are in agreement with models proposing that the Earth may have acquired its water and other volatiles from comets. He is referring to those who believe that the Earth's waters may have been deposited in a storm of large comets early in the Earth's history, of course, and not to the current influx of small comets that I have proposed. But he certainly does not rule out this possibility either.

The bottom line is that the oceans could well have come from comets.

CHAPTER 7

Heat, Dust, and the Origin of Life

Alice Springs lies in the heart of Australia. Nearby is Ayers Rock, a 1,260-foot-high monolith located in the center of the red Aussie desert. This natural wonder draws visitors from around the world, but in the spring of 1986, Alice Springs had yet another natural wonder on its hands. Tens of thousands of comet watchers flocked to this tiny, clear-aired town in the outback to catch a good view of Halley's Comet. But as everyone knows, the celebrated, once-in-a-lifetime visitor simply fizzled out. The comet was fuzzy and feeble and everyone was disappointed. The comet gazers had come so far, and the comet, just 39 million miles away, had come so close.

The let-down put such a somber mood on April Fool's Day in Alice Springs that a group of soldiers decided to fill a nearby trench with dry ice and announce that a chunk of Halley's Comet had fallen to Earth. They added credibility to their story by telling passersby that a group of scientists from the United States were coming to retrieve it. Word of the event spread rapidly and before long local residents were in an uproar. Americans, they said, were stealing "our comet."

I, too, was accused of stealing from the great comet. The timing of the publication of our findings irritated cometary scientists. There is certainly irony in the fact that just as our theory of the small comets attracted the attention of the scientific community, Halley's Comet so disappointed everyone else. A lot of people believed I had comet fever. They said all this was brought on because Halley's Comet was coming and everybody was obsessed with comets. They thought I was hopping on the Halley bandwagon only to distract everyone's attention away from it.

This was certainly not our intent. It is true that much of our work deducing the probable nature of the small comets took place just as scientific studies were being conducted on Halley's during its latest little visit to the terrestrial neighborhood. But all along our work was really totally independent of the 1986 appearance of Halley's Comet.

Water is the primary ingredient of all comets. For decades the assumption has been that large comets like Halley's were one-third to two-thirds water and the rest dust. But I thought the small comets could not contain much dust. If they contained as much dust as Halley's Comet they would be creating these tremendous fireballs in the sky as they entered the Earth's atmosphere. But we do not see such fireballs very often, so the small comets could not be very dusty.

The question of the composition of the small comets was one of the first issues we had to address. But unfortunately, thanks to Alex Dessler, we had to address the topic repeatedly during the course of the debate. I think that Dessler tried to bury me with Comments related to the composition of the small comets. He accepted the first Comment on this topic on April 29, 1986. Two days later he accepted another Comment on composition. Less than two weeks after that Dessler accepted a third Comment on the topic and a month later he accepted the fourth and final Comment relating to the composition of the small comets.

The topic was, I will admit, an important one. The small comets had to be composed of something other than icy water. How else could they survive the heat from the Sun as they reached the vicinity of the Earth? These comets had to have a protective cover, a mantle, in other words. Of course, whatever this mantle was made of would, like the comet water, eventually end up on the Earth and perhaps be of significance for the planet and life upon it.

Our critics on this issue were a distinguished group. There was David Rubincam of the NASA Goddard Space Flight Center in Greenbelt, Maryland, and four Californians: Christopher McKay of the NASA Ames Research Center, Donald Morris of the Lawrence Berkeley Laboratory, and John Wasson and Frank Kyte, both of the Institute of Geophysics at the University of California in Los Angeles. Their Comments were variations on the same theme.

What Rubincam and the others said, essentially, is that if these small comets were to survive their journey to the Earth, these objects would have to be protected from the heat of the Sun. This meant that the comets would have to have a mantle and most of them chose to construct this protective cover out of dust. They simply assumed that the small comets would be like the large comets and contain a lot of dust. McKay, for instance, concluded that the dust cover on the small comets would have to be about fifty feet thick to survive the Sun's heat, but this was clearly at odds with my estimate of the size of these objects—about forty feet in diameter.

Everyone insisted, Morris included, that a minuscule amount of dust would certainly not be enough to form a protective mantle around a small comet and that any such small amount of dust was inconsistent with our knowledge of the high dust content of known large comets. But even assuming you could mantle these objects with dust, they said, where is all the extraterrestrial dust that would have come from all the small comets fallen to Earth in the past? If the small comets were real, said Wasson and Kyte, there should be from ten thousand to a hundred thousand times more extraterrestrial dust on Earth than scientists have been able to find.

My critics were essentially correct. The small comets did have to protect themselves from the Sun's heat. They did need a protective mantle. But the mantle need not be dust. My critics could only think in terms of the large, known comets. They assumed that the small comets would have to have the same dust content as the large comets.

But the small comets are different and contain only a fraction of the dust of the large, well-known comets. I do not think it unreasonable to expect such a difference. After all, in a comparison of masses, a small comet is to a large, known comet what a large comet is to the Earth. Large comets are a hundred million to a billion times more massive than the small comets. Several simple scenarios could explain the different dust compositions of the large and small comets. Perhaps the large comets accrete more cosmic dust than their tiny cousins because of their greater gravity. Or perhaps they were formed in a different, dustier region of the solar system.

But current thoughts on cometary evolution do not predict small comets, McKay pointed out. He is absolutely correct. No one ever thought there might be such objects. The whole scientific literature is devoid of any real discussion of the possible existence of small comets. But I think the reality of these objects is a clear indication that the current models of cometary evolution are quite incorrect. And when I say the small comets are about forty feet in diameter, I refer to an average size. Some will no doubt be smaller, and others larger. We are not talking about two completely distinct classes of objects, the known large comets and my own small comets. I fully expect there to be cometary objects of all different sizes and different dust content cruising through the oceans of space.

Regardless of the amount of dust in the small comets, however, a dust coating simply cannot be the dress around these objects. The comets have to live longer than just a few trips around the Sun, so mere dust will not do. But water and dust are not the only cometary constituents. Organic materials are also present in comets in small amounts. And laboratory experiments by Louis Lanzerotti at AT&T Bell Laboratories, Robert Johnson at the University of Virginia, and others, have shown that if you have organics, you can get a crust on a comet. In a series of experiments, they bombarded water-ice, containing such organics as methane gas, with radiations similar to those found in space. The result was a durable, thin, black crust. It was not a very thick crust perhaps, but one that should be sufficient to protect a small comet in its approach to the Sun.

With a little vaporization and a little cooking from the Sun, the mantle on the comets should resemble nature's version of Saran Wrap. If you start out with simple carbon compounds and expose them to millions of years of energetic charged particles and ultraviolet radiation from the Sun and distant places in the cosmos, it is quite possible to produce a black crust about half an inch thick on a comet. This carbon coating would bind the comet together. By carbon, I do not mean what you find in a pencil. It would be a carbon polymer; the carbon, in other words, would be mixed in with other elements, perhaps into chains of molecules. So it is going to be an amorphous, perhaps flakey, tar-like substance. Such a carbon mantle would give the small comets, coming no closer than the Earth's distance from the Sun, just the lifespan they need—a hundred thousand years to one million years.

I proposed this black crust well before we learned that Halley's Comet itself had a very dark surface. It annoys people when I mention this, but it was just a natural extension of our work. Everyone had expected Halley's to look like a dirty white snowball. But the probes found that large areas of its surface were extremely black. In fact, it was described as one of the blackest things ever seen in the solar system.

The comet's nucleus was also not spherical, but irregularly shaped, like a potato, and had holes in it. Everyone thought the entire sunlit side of the comet nucleus would be illuminated. It was not. The comet's illumination came from the jets of gas and dust that are blown out from the holes in its nucleus. And these holes, it just so happens, suggest an intimate link between the large, Halley-type comets and my own small comets.

Large comets travelling into the inner solar system are known to split and flare. No one knows why. They just break apart. I thought if the small comets are really there, then no large comet could exist very long without being hit by them. So I did some calculations. I began by trying to show that the small comets were too numerous and that any large comet out there would just be blown out of existence by them in a short time. But it turns out that the number of small comets is just about what you would expect to account for the random splitting and flaring of large comets. Not only that, but it is also the right number required to put several holes in the black surface of Halley's Comet.

The photographs taken by the Giotto spacecraft clearly show the holes in its nucleus. I think Halley's Comet has been hit by a number of small comets. I did not bother to publish this finding even though no one else has been able to explain the random splitting and flaring of comets in interplanetary space. I knew that our conclusion would irritate people to no end.

I should mention that Donald Morris, to his credit, did wrestle with the possibility that the mantles on the small comets might be made of carbon, a basic ingredient in all forms of life. But, Morris argued, there are limits on the amount of carbon that comes down to Earth. If these comets are coming in, he said, then the carbon has to dribble down through the atmosphere and end up on the ocean floor. And there are measurements of carbon on the ancient ocean floor, core samples by geologists such as Wendy Wolbach of the University of Chicago, that put an upper limit on how much carbon you can bring in with these comets. And it turns out, Morris said, that the amount of carbon down here is too small, so your comets are out of luck.

The point was a good one, but incorrect. If I have small comets slamming into the atmosphere at supersonic speeds, the carbon is going to ignite. There is oxygen up there. So the carbon is not simply going to fall down as carbon. It is going to burn like a furnace. It will not even get close to the Earth. It has to pass through atomic oxygen, which is highly reactive. It will be consumed by this oxygen as it falls through the atmosphere and turned into carbon monoxide, carbon dioxide, and other carbon compounds. Very little of it will float down to the Earth's surface as soot.

So the more appropriate question is how much new carbon exists on Earth? If additional carbon can be found on the Earth, what is the upper limit per year that is not identified as already being here? Heinrich Holland, a professor of geology at Harvard, has conducted an exhaustive examination of this issue. He calculated just how much carbon exists in the atmosphere, biosphere, and crust of the Earth. He also estimated that about three-quarters of the new carbon on Earth is actually recycled carbon from the weathering of sedimentary rocks. But Holland could not quite pin down where the remaining twenty-five percent of new carbon comes from. He suggested that it might be coming from the deep interior, that volcanoes are bringing up juvenile or new carbon from the Earth's own mantle. Of course, where but down below would a geologist look for a source of new carbon?

But the amount of carbon brought in through the influx of small comets each year can also account for Holland's remaining twenty-five percent. In fact, if you take the total amount of carbon on the Earth and divide it by the current rate of additional new carbon, you come up with about four billion years. In other words, if the comets have been coming in steadily over the age of the Earth, they would not only have brought us the oceans, but most of the carbon on this planet as well. And if you go back to Rubey's paper on what the excess volatiles are, water is number one and carbon is number two. This is incredible. Of course, if it is true, then the carbon in the trees around us and the carbon in our own bodies will have come from comets as well.

Life, and its origin, is a delicate subject. Conventional wisdom holds that the key molecular building blocks of life were formed on the Earth itself. Modern laboratory experiments suggest that gases in the Earth's primitive atmosphere—such as water vapor, methane and ammonia—were broken up, either by exposure to the Sun's ultraviolet light or to lightning discharges, and then recombined to form other molecules, such as hydrogen cyanide and formaldehyde, which are thought by many scientists to be chemical building blocks of carbon-based life. But whether such photochemical and electrical reactions are indeed responsible for the overall terrestrial inventory of complex organic molecules is not known with any confidence. This uncertainty can be traced directly to the composition of the ancient atmosphere, about which very little is known.

Scientists during the past three decades have considered an entirely different scenario for the origin of life. The idea is that comets may have supplied the Earth with the organic materials needed to create life. It was back in 1961, in fact, that John Oró at the University of Houston first suggested that the biochemical molecules from which life arose on the primitive Earth may have been brought here by comets. It is a distinct possibility, as nearly a fifth of the comet's nucleus may be composed of organic matter. But many critics objected to the idea on the grounds that an object large enough not to burn up in the Earth's atmosphere would hit the ground with such force as to destroy all the organic material it contained.

Today, however, the notion is no longer wildly controversial and is the subject of ongoing study. A group of scientists at Cornell University led by Paul Thomas, along with Leigh Brookshaw at Yale University, now believe that much of the organic molecules needed to create the first forms of life on Earth could well have been brought in by comets that bombarded the planet early in its history. Their study, actually a simulation of the impacts formed by comets of various sizes, took into account their suspicion that the Earth's atmosphere was probably ten to twenty times more dense three to four billion years ago. This thick veil could well have exerted an "aerobraking" force on incoming comets of a certain size.

They found that large comets a mile or so in diameter would plunge into the Earth without sufficient aerobraking. They thought that small comets, on the order of those that I have proposed, would certainly slow down sufficiently, but would not contribute enough mass to make a difference. They concluded that the most promising candidates were comets of medium size, those measuring about 400 feet across. Such comets would have slowed down enough during entry into the Earth's atmosphere for their organic components to survive the impact. The implication of their work is that a cometary bombardment could have brought in a hundred to a thousand times as much organic material as the Earth itself would have produced photochemically during the same period.

But the small comets with their constant flux, smaller size and lower speeds do not require a thick atmosphere to deposit the initial inventory of life producing organic matter on the Earth. In fact, the initial impacts of small comets on the barren cinder that would eventually become our home could well have been responsible for producing the early atmosphere itself. Later, though the surfaces of the small comets would burn up in their descent through this new atmosphere, all organic matter buried inside them might not be destroyed. Some matter could well have floated down to the Earth on a protective cushion created by the comet's own gases. So it is quite possible for small comets to have carried to the Earth the precursors of life.

The implications were piling up. We were not just talking about bringing water, and lots of it, down to the Earth but the very ingredients for life itself. And more besides—like a good portion of the atmosphere. It seems that the small comets also address the problem of how early life, which needed oxygen to protect itself from the Sun's ultraviolet rays, could have flourished on a primitive Earth thought to have been devoid of oxygen. Some of the water molecules in these comets would have broken apart from the heat of impact and formed oxygen and hydrogen gases. This would automatically have produced a shield of oxygen to protect life from damaging ultraviolet rays.

So it may be that these small comets provided not only the chemical seeds for life on Earth, but the oxygen to protect it from the Sun, as well as the marine incubator—the ocean—in which it could grow and thrive. That, in essence, would make us all the children of comets.

CHAPTER 8

The Atmosphere and the Ice Ages

A strange and quite beautiful cloud can be seen over the polar regions during the summer months. It is a thin cloud, wavy or banded, colored silver or bluish white. It forms at an altitude of about fifty-five miles, in the coldest part of the upper atmosphere, a relatively unexplored boundary known as the mesopause. No other cloud occurs so high in the sky. It can be distinguished from ordinary clouds only when seen against a dark sky and illuminated by the setting sun. For this reason such clouds are called noctilucent, meaning quite literally "night luminous."

Noctilucent clouds are remarkable for many reasons. But nothing is more striking than their very existence. These clouds require considerably more water vapor than can be expected from ocean evaporation. No one knows how the water gets up there. No one really understands why these clouds exist. And finding out why has not been an easy task.

Those who explore the upper atmosphere—aeronomers, as they are known—have made little headway into the mystery. Since the 1960s rocket-borne instruments have been sent up to probe noctilucent clouds. Results indicate that the clouds are composed of ice crystals

These high-altitude noctilucent clouds are something of a mystery to atmospheric scientists. Not enough water is thought to evaporate from the Earth's surface and reach altitudes of 55 miles to form these clouds. But if the small comets are bringing the water in and depositing a little of it in the upper atmosphere, then the mystery has a very simple solution.

formed around meteoric dust particles. But some doubts have been cast as to whether the particles are meteoric or not. Still puzzling is their sporadic appearance. Sometimes the clouds are there and sometimes not, even though the temperature in the region is the same.

It would seem that the source of water vapor in the upper atmosphere varies over time. And indeed, if the small comets do exist, if they are responsible for depositing the water in our oceans and for sowing the seeds of life on this planet, then it would be no surprise that on their way down they should be leaving a trace of their passage—a little more water than we would normally expect—high up in the atmosphere.

But aeronomers do not like the notion of a wet upper atmosphere. They generally assume that the amount of water vapor up there falls sharply above forty miles because air carried from lower altitudes has been dried by frigid temperatures. On the other hand, there are a number of measurements in the literature that do not show this expected dryness. These higher water content measurements cannot be explained by simple upward movement of water vapor. Aeronomers are quite correct. The water is not rising from below. It is falling down from above.

George Reid and Susan Solomon are aeronomers at the National Oceanic and Atmospheric Administration in Boulder, Colorado. They have a model which computes how much water should be present at each level of the atmosphere. They used it to show that the small comets would bring far too much water into the atmosphere, even though this whole topic of water in the upper atmosphere is a kind of frontier of knowledge. We know less about the upper atmosphere, this area right above our heads, than most people realize. This is a no man's land. It begins at about thirty miles and extends another 100 miles up. We cannot reach it with airplanes. Even balloons will not soar through this region. And it is too low for satellites; the atmosphere would quickly pull them down to a fiery death. But rockets can do the job and so can certain types of radar and laser beams.

So Reid and Solomon took their model, injected into it the water coming in from the small comets, and concluded that the resulting water concentrations in the atmosphere would "greatly exceed those that have been observed." While most measurements of water vapor in the air at about forty miles and above are on the order of just a few parts per million, when the water deposited by the small comets was taken into account, their model gave a result that far exceeded this small trace of water at these altitudes.

Reid and Solomon did not want to throw the comets out with the comet water, however. They proposed an alternate scenario. They suggested that the water molecules brought in by the small comets would not survive the impact into the atmosphere intact. The molecules would instead be broken up into an H atom and an OH molecule, known as a radical. These OH radicals would then destroy the atomic oxygen in the atmosphere and produce the atmospheric holes in our images. Never mind that the mechanism for producing atmospheric holes cannot be related to a chemical reaction with oxygen atoms. This could only happen if the objects in our images were located at altitudes where atomic oxygen is dense, somewhere under 200 miles. But the fact is that the cometary water clouds are located above the atmosphere when they produce the holes by absorbing the light from the atmosphere.

Never mind. With such a scenario, Reid and Solomon continued, you could bring a good deal more water into the atmosphere than previously thought. The flux of the comets could then be thirty times less than what we had originally proposed. The small comets would only have to measure a mere seven feet in diameter instead of forty and weigh three tons instead of a hundred. This scenario, they said, would be consistent with current observations of the amount of water vapor in the atmosphere. Reid and Solomon were like everyone else. They were bothered by the numbers.

No one would have complained if my proposal concerned just one small comet coming in every year. Even two or ten comets would not raise many eyebrows. But at a hundred, people start thinking, that's one every three days. Still, they might concede that. At a thousand—three a day—the conservatives turn their eyes toward you and say, I don't know about this. At ten thousand, you have half the people in the world really upset. At a hundred thousand, they look at you like a creature from another planet. At a million, you are from another galaxy. And at the ten million I have proposed, well, you are not even sane.

But there is a problem with Reid and Solomon's model. It is based on a certain assumption. They assumed that the water from the comets comes to a dead stop on top of the atmosphere. It is as if some magical mechanism stops it up there. If the water did stop there, I agree that we would have too much water in the upper atmosphere. But they are wrong. This is a small problem. There is nothing up there to stop a hundred tons of water traveling at many times the speed of sound dead in its tracks. It is going to plummet deep down into the atmosphere. So their work is invalid.

What is most surprising is that Susan Solomon pointed out back in 1982 that you could not accurately calculate the movement of water vapor in the atmosphere by considering altitude only. What you needed to do, she concluded, was to consider also the motions of the atmosphere in latitude and longitude as well. This was ignored in their calculations of cometary water in the atmosphere. But what is valid one day seems to be invalid or not needed the next. There was no point in arguing.

If it is done correctly, I think there is much to be learned from running the small comet scenario through a model of the atmosphere. It could help us understand what effect the small comets might have on our climate, on the greenhouse effect, and the hole in the ozone layer over Antarctica. We know that water has a profound effect on the chemistry of the atmosphere. We know that water, like carbon dioxide, can also produce a greenhouse effect. And we believe that water ice may play an important role in the chemistry of the ozone hole.

It is also fairly obvious that an increase in the number of small comets could cause clouds to form about fifty miles up, reflecting sunlight and cooling the Earth. Comet storms could create a global ice cloud and may have been responsible for the Ice Ages. The last one, known as the Little Ice Age, was a reign of cold that began in the 13th century and reached its severest temperatures around 1750. Scientists are still not sure what caused the Little Ice Age, but it came as a result of a decline in global temperatures that averaged about one degree. Even small changes in the Earth's climate can have profound environmental effects.

Could small comet storms be responsible for the Ice Ages? It is impossible to tell with our present knowledge. We will not know just what effect the small comets have on our atmosphere until the influx of water they bring in is inserted into one of the computer models available to us. I have been trying to do that for several years. There is a place called the National Center for Atmospheric Research (NCAR) in Boulder, Colorado, where Raymond Roble has designed and runs on a large computer one of the world's best models of the upper atmosphere. I have been there several times trying to convince him to see what happens when you inject all that comet water into the atmosphere. I would do the problem myself if I could, but without access to this model it would take a lifetime's worth of work for several people.

Roble could think I am daffy. He tells me he will do it, but four years have passed and I still have been given no results. NCAR is supposed to be of service to scientists. The government puts a lot of money into it. But perhaps Roble is feeling the pressure. Such pressures are not uncommon in the scientific community. The word gets around quickly and certain factions in this community can make life very difficult for you. All that attention can be disconcerting and dangerous, especially if you end up with positive results and are foolish enough to report them.

The small comets proposal was not the first to challenge aeronomers. The problem of a wet upper atmosphere has come up again and again. Back in the mid-1970s a group of East German scientists from the Institute of Electronics at the Academy of Sciences took a great number of density profiles of the upper atmosphere using instruments aboard the Intercosmos satellites. Their results were puzzling. Sometimes, at altitudes above 60 miles, they found a very effective absorber of ultraviolet light. They concluded that this could only be due to the presence of nitrogen oxide or water vapor. Most arguments spoke in favor of water vapor, although such a conclusion led to much higher concentrations of water vapor than are permitted by models of the upper atmosphere. Some concentrations were surprisingly as high as a thousand parts per million. They were never happy assuming that all this water had to come from down below.

More recently, Klaus Grossmann, a physicist at the University of Wuppertal in West Germany, conducted an impressive series of rocket-borne measurements of water vapor in the upper atmosphere using an infrared light detector. Most efforts to observe water concentrations at these altitudes by rocket-borne instruments are stymied by problems of various kinds. The most frequent problem is contamination. Rockets carrying experiments designed to measure high altitude water content often end up inadvertently carrying the water up with them, making their results useless.

But apparently Grossmann had designed his experiment in such a way that contamination was not possible. It was a very carefully done rocket experiment. Three times he had flown his instrument aboard sounding rockets launched from northern Sweden and three times his samples indicated there was more water up there than could be accounted for by evaporation from the oceans. The launches all took place in the winter and in the winter you expect the water vapor to remain fairly constant with altitude. The water is supplied from below and there is no loss because there is no sunlight to fragment the water molecules.

Each time Grossmann found an increase of water vapor with an increase in altitude. There was, in fact, two to three times more water in the upper atmosphere than expected. So either the water was coming up from below in a way that current evaporation theory cannot account for, or the water was coming in from outside, from an extraterrestrial source. There was no way for Grossmann to tell if small comets might be responsible for his results, but I certainly think they are.

I believe there is more water up there than most aeronomers would care to admit. But my critics were mistaken in thinking that the water from the small comets would just lie up at the top of the atmosphere. Certainly, a small amount of the water gets left up there but most of it barrels down through shear momentum and mixes in with the water present much lower in the atmosphere. If I am correct about this, then the last chance to trace the hot water vapor cloud as an object would be in the middle atmosphere, at altitudes about forty miles above us.

Luckily, it is possible to monitor the presence of water vapor in the middle atmosphere. Such measurements began in the 1950s using aircraft, rockets, and balloons. But the results were inconsistent in part because water samples were contaminated by water brought up by the instruments themselves. Later, remote sensing techniques were developed that had the distinct advantage of removing the uncertainty of what was instrument water and what was atmospheric water.

Since the late 1970s, ground-based microwave radiometers have been used to determine the amounts of water in the middle atmosphere. A radiometer is essentially a telescope that is sensitive to microwaves instead of visible light. These microwaves are basically high frequency radio waves much like the ones used by your home microwave oven. Because water vapor absorbs and emits at these frequencies, you can cook food in a microwave oven. When your oven pumps in microwaves, the water that occurs naturally in food absorbs that radiation and heats it up.

What microwave radiometers have found is that water concentrations in the atmosphere reach a peak at an altitude of about forty miles and decrease very rapidly above this height. But, of course, if cometary water vapor clouds do exist, these instruments should be capable of detecting large, momentary bursts of hot water as the cometary material descends and encounters an increasingly dense atmosphere.

Virtually the only people in the world who have spent long periods of time looking continuously at water vapor concentrations in the middle atmosphere are John Olivero and his group at Pennsylvania State University. They are the best in the business. Olivero is a professor in the Department of Meteorology and the instrument he uses is the Penn State radiometer, a horn antenna which can collect microwave radiation. It is a fixed instrument, meaning that it always looks at the same spot in the atmosphere. A single reading for the presence of atmospheric water vapor takes twenty minutes to collect. But the signal from a single reading is so weak that thirty-six of them are grouped together to give a single measurement for water vapor in each twelve-hour period.

Early in the game Tom Donahue cited the work of Olivero. Microwave radiometers used to monitor water variations where your objects are supposed to come to rest, Donahue said, do not see anything. They should see an increase in water vapor concentrations at least during the time it takes the objects to pass within view of the radiometer. Donahue checked with Olivero to see if these increases occurred. They did not. But Olivero recognized that if these things were plunging into the middle atmosphere, the event would not last long, just a matter of minutes. Olivero, however, was averaging his measurements over long periods of time, hours and days. If a quick event did occur, it would be lost in these averages.

So Olivero decided to go back and look at his data again. He would break the data apart, look at each twenty-minute reading separately, and compare them to the long-term averages of which they were a part. Any momentary increase should show up in the individual measurements. He did not expect to find anything. He did it, he said, for one reason. He wanted to "bury" the small comets. He was going to resolve this issue once and for all. "We thought Frank's hypothesis was very interesting," Olivero said at the time. "And regardless of whether it was true or not, it made us all reexamine our assumptions about physical processes going on at these levels of the atmosphere. That's a very healthy thing to do."

It would take a long time to look through so many observations. I did not know it at the time but Olivero had assigned the search to a graduate student named Dennis Adams. It would be his master's thesis. Adams, being in the Air Force, was the right choice. The Air Force, and a lot of defense agencies, have an interest in scientific projects of this kind. Rapid fluctuations in atmospheric water content at high altitudes could potentially affect the detection of airborne intruders, as some defense systems are designed to spot aircraft and rockets by looking for water vapor from their exhausts.

Olivero intended to dispose of the small comets. That was good. I appreciated that. That is exactly what is supposed to be done in science. Olivero is a good scientist and a solid experimenter. He had a lot of courage. I have a lot of respect for him. Any result he came up with, I would believe.

I heard nothing more about the project until I was thumbing through the summaries for talks being presented at the May 1987 American Geophysical Union meeting in Baltimore. I saw that Olivero had a paper with someone named Dennis Adams. A third co-author was listed as Charles Croskey from the Department of Electrical Engineering at Pennsylvania State University. The paper was on their search for rapid variations of water vapor in the upper atmosphere. No results were given. The abstracts for the schedule book need to be submitted so far in advance that the results of a particular study are often not known when the abstracts are written up.

When Sigwarth and I arrived, there were a lot of people in the meeting room. I knew who Olivero was, but he did not know me. He was getting some water to drink and accidentally spilled it all over me. There seemed to be a lot of nervous people around. I began to think there would be more to this than someone getting up and saying they did not see anything. I took a seat in the front row. When the next paper was announced I saw this young man walking up to the front who I thought might still be in high school. It was Adams. People were surprised.

Adams shuffled his viewgraphs around for a while and began with a description of how their instrument worked, how they received the microwaves, and how they analyzed their data. There was no hint of the conclusion to come. He then explained that they had examined 390 hours worth of microwave measurements one by one for fast variations. It was a lot of work. Adams then paused and said, "Now I'd like to talk about the results." But before he put up his next viewgraph he said, "We found something."

I thought half the people in the front row were going to have a heart attack. I have never seen anything like it. Adams then showed us the events themselves. There were not many, but each was clearly very real and not noise. Each of these twenty-minute events showed a large increase from its adjacent twenty-minute periods. That seemed to indicate that sometime during this twenty minutes the water vapor suddenly appeared then disappeared from that part of the atmosphere. The short duration of the events was their most telling feature.

Adams then walked off and sat down. I noticed Olivero was sweating a bit. It took a lot of courage on their part. If you do not have much funding to begin with, and Olivero did not, you run the risk of people ridiculing you for doing something like that. A reporter once came up to me and said, "You seem to be outnumbered ten thousand to one." I replied, "I did not know the odds were that good."

Adams found four cases of "extreme variability" in the 390 hours of data collected between February 22 and March 26, 1986. At a later American Geophysical Union meeting he presented another paper in which these four events had been thoroughly scrutinized for any signs of error. They were not sporadic events due to instrument failure or errors in data analysis. These were real blobs of water coming into the middle atmosphere. They could not be impulsive injections from the surface, from the oceans, moving up. They were not due to volcanic events spewing water vapor into the atmosphere. They had to come from the top down and the only way to do that is with an extraterrestrial object.

All that remained to be done was to calculate how often these events occurred and how much water each one contained. These numbers would then have to be compared with the known frequency and the estimated water content of the small comets. Adams wrote his master's thesis on this subject. He calculated that the average time between these impulsive events was a little over four days. The expected frequency of comets for the area of the atmosphere viewed by their radiometer was about one every day and a half. The average rate we had calculated for small comets based on the appearance of atmospheric holes in the Dynamics Explorer pictures was about twice as much. This difference fell well within the variability of cometary events themselves, which we had observed to change by a factor of ten over a three-month period.

Adams also estimated the number of water molecules seen in these impulsive bursts. He figured that the smallest size for these water bursts had to be about ten tons. So the amount of water present in these events was in the same league as the small comets themselves. Adams was convinced there was a causal relationship. "Observed, short-term perturbations in microwave spectra," he concluded, "are most likely the result of a temporary increase in the water vapor concentration of the upper atmosphere from extraterrestrial 'events' ..."

John Olivero was less certain. But I have no doubt, knowing the kind of scientist Olivero is, that Adams's thesis can be counted on. The examining committee of faculty members must have thought so too or Adams would not have gotten his degree. Adams has since returned to the Air Force. His thesis, unfortunately, is buried in the library of Pennsylvania State University and no part of it has been published in a scientific journal.

But Olivero would not let me down. Two years later I would hear from him again. I have some results, he wrote in March of 1990. Another graduate student on leave from the Air Force had analyzed the four remaining years of data and had found another 111 "water events." After eliminating "all other reasonable sources for the observed signals," they had concluded that these water events were most likely the traces of the passage of small comets through the atmosphere.

CHAPTER 9

A Masquerade for Radar

Fred Whipple is widely regarded as the patriarch of cometary science today. He was born in Iowa in 1906 and was the first to suggest, back in the 1950s, that the large comets might resemble dirty snowballs. But when the subject of my small comets came up in 1986, it infuriated him.

"If his comets have all this dust in them that real comets have," Whipple said at the time, "then our radars should be picking them up like mad. Imagine, things of that size! My God, our radars can pick up things the size of basketballs most of the way to the Moon. I don't see how they could be missed. He must have them made of pure ice."

Whipple is, of course, quite correct. The small comets cannot have a lot of dust in them, as we have already noted, otherwise we would see these tremendous fireballs across the sky. The small comets are largely water snow with, most probably, a thin carbon mantle. Can our defense radars pick up such objects before they hit the upper atmosphere? It was a good question. A lot of people had been asking the same thing. Even if these small dark objects escaped visual detection, it was thought that radar, at least, should be able to spot them.

But the question is not easily answered. To begin with, the cores and mantles of the small comets are not made of the highly reflective material that radar depends on to track airplanes. The radar signal from a small comet must be weak and must mimic that of a much smaller metallic object. This is entirely within the realm of possibility. It is this type of confusion that has enabled the Pentagon to design a large bomber, known as the B-2, which some say resembles a mere insect on a radar scope.

The early warning system used to defend the United States from airborne attack consists, in part, of large arrays of over-the-horizon radars. It is one of the best radar systems in the world. But the data this system produces is all classified. From time to time we hear of reports of strange objects in their field of view. But even scientists are not able to penetrate the security around such events. I have tried. Supplying us with that kind of information would reveal the characteristics of our defense system and would compromise national security. I believe, nonetheless, that our new radars are sensitive enough that an occasional small comet might be buried within all their data. I would also expect the defense establishment to have little interest in such objects, however, as their paths are from deep space and not from a launch pad in another part of the world.

Reporters trying to pry some information from the North American Aerospace Defense Command (NORAD) found the agency closemouthed about its ability to pick up a forty-foot chunk of snow and ice. "Our radar is infrared," Kay Cormier explained in 1986. Cormier was working for NORAD Media Relations at Peterson Air Force Base in Colorado. "It intercepts the heat from missile launches. That's why our satellites are infrared and pick up heat. We can pick up forest fires, even garbage-can fires, but not ice. Ours are not geared for that. Unless it's a man-made object we wouldn't pick it up. Our satellite sensors are geared solely for man-made objects. I've never heard of a radar that picks up ice. Our tracking system is solely for missile detection and for satellite tracking."

Wes Bodin, of NASA's Goddard Space Flight Center in Greenbelt, Maryland, is familiar with the capabilities of the NORAD system and was able to provide a few more details. "NORAD has a grouping of radars sort of like a fence," he explained. "They are broad in one direction but not in the other, so that things that cross it will show up. If an object doesn't hit in that area, then nothing will show up. Now, NORAD radar is very sensitive: it could get a signal off one of these comet-like objects. And they probably get stuff that probably looks like noise."

The ability to track a small comet-like object depends on the sensitivity of the radar. Low-gain, broadband radars, like those used by airports, track objects that are close by. Such radars do not have to be precisely aimed at a target to pick it up. On the other hand, the higher the gain of the radar system, the smaller the target it can track, but the greater the need for the radar to be pointed precisely at the target. So a high powered radar has a fairly narrow beam that requires precise pointing information—or a lot of luck—to find what it is looking for.

Snow and ice can reflect radar waves but the radar has to be able to point to it. Comets move fast. To track them you have to be looking for them. Large comets can be tracked on radar. But their trajectory is known. "If you know an object is coming and you know its trajectory, you ought to be able to track it with a high gain radar," Bodin said. In other words, comets might occasionally cross a radar's path, but no one would be likely to know what they were unless they were specifically looking for comets in the first place.

Perhaps small comets are being picked up by radar disguised as some other objects—such as meteors. It is well known that at least several billion meteors streak through the Earth's atmosphere every day of the year. No one argues this point. Most meteors are mere specks of dust, are largely devoid of water, and weigh but a tiny fraction of an ounce. Only a few of the larger ones are bright enough to be seen at night with the naked eye.

With the introduction of radar in the 1940s scientists were able to greatly expand our knowledge of these celestial visitors. When a meteor rips through the upper atmosphere, its molecules, and the molecules of the air it passes through, are torn apart to form a short-lived trail of electrically charged ions that will reflect radar signals. Radar has a big advantage over visual observation when it comes to the study of meteors. Unlike a human being, a radar can track meteors during daylight hours. And at night, an observer might expect to see only a few of the objects that a radar viewing the same sky can sense.

I now believe that a tiny fraction of these radar meteors, as they are known, may actually be small comets. Because most meteor surveys conducted by radar simply tally the observations of ionized gases left by the impact of extraterrestrial objects with the atmosphere, I would think that a small rock is probably indistinguishable from a hundred-ton snowball. We cannot interpret radar signals very well. More often than not, when we get a reflection on radar, all we know for certain is that something has passed through the atmosphere.

I thought that if a link did exist between the influx of small comets into the atmosphere and the appearance of meteors on radar, then the rate of atmospheric holes in our satellite images should show the same variations in time as radar meteors do. It turns out that meteors—like showers in April, flowers in May, and weddings in June—do show seasonal as well as other variations in time. But did the small comets show such variations? Finding out was not an easy task. It is extremely difficult to observe the same area of the Earth's sunlit hemisphere for an extended period of time with Dynamics Explorer. But we found that one such period of observation did occur soon after the satellite was operational. This period took place between the 1st of November 1981 and the 20th of January 1982.

At first we thought that the rate at which these holes appeared in our images would increase along with the number of bright meteors in the sky. So we counted up the number of holes per day over those three months and compared it to the dates of the meteor showers that occurred at these times—the Northern Taurids, the Leonids, the Geminids, the Ursids, and the Quadrantids. Much to our surprise, we found that the occurrence of the holes had nothing to do with meteor showers.

But there are two components to the meteor rate. One is the background rate, which varies slowly over days and weeks and is continually present. On top of this rate are the showers which usually last just a few days to a week. These showers occur when the Earth passes through a stream of debris created by the breakup of a large comet or an asteroid. Because the meteor rate is thought to be similar from year to year, we searched for a study of radar meteors that had occurred over the same months as the data we had collected on atmospheric holes with Dynamics Explorer.

We found a study of radar meteor rates for June 1955 through August of 1956 conducted by two Canadian radiophysicists. When we eliminated meteor showers from their data and looked only at their background meteor rates for the months of November through January, we discovered that the rate declines after the first week of November, falls to a minimum in the middle of January, and then rises again afterwards. We then plotted the rate of the holes occurring in our satellite images and found the same variation. There was a maximum at the beginning of November, a decline to the end of the month, followed by a fall to a minimum in mid-January, and an increase afterwards. The similarity between the background meteor rate and the rate for atmospheric holes was stunning.

This would suggest that the small comets are distributed spatially about the Sun in the same manner as a certain population of meteors tracked by radar. The problem is that there are far more radar meteors than small comets. During the November 1955 through January 1956 time period there were about a hundred times more meteor events as there were atmospheric holes or small comets during November 1981 through January 1982. So while small comets are likely to be related to these radar meteors, they are not one and the same. But many radar meteors could well be the debris from the break-up of some of the small comets in interplanetary space. Little pieces from the mantles of the small comets would not produce sufficiently large clouds of water vapor to be picked up as atmospheric holes by the satellite but they would certainly be large enough to be picked up by radar.

I felt that these temporal variations in the occurrence of the small comets should be widely known. Anyone attempting a search for these objects should know that the comets were ten times more likely to be seen at the beginning of November than in the middle of January. So in June of 1986 I prepared a paper on the topic and submitted it to *Geophysical Research Letters*.

The reports that came back from the referees were very critical. I think it was mostly an emotional outlet for them. This can happen when things are done anonymously. Two of the referees never understood what was going on in my paper on time variations. They were talking about one thing. I was talking about another. Only one referee stuck to the point. He insisted that observations of geophysical phenomena—whether atmospheric holes or meteors—were all modulated by the Earth's rotation on its axis and its revolution around the Sun. So any similarity of daily or seasonal behavior was unlikely to imply any deep relationship between such phenomena. Even so, he was of the opinion that neither the daily nor the seasonal variations in the occurrence of the atmospheric holes were consistent with meteor observations.

I could have answered the referee, revised the manuscript, and pushed for its publication. But Dessler was inflexible. In August of 1986 he informed me that the small comets hypothesis was not gathering support. He also felt that the journal was already providing an adequate forum to defend our findings. This debate would have to be fully aired before he would allow us to publish anything else on the subject that was not in the form of a Reply to someone's Comment. He had already been burned because he had published my first two papers. He did not feel that it was appropriate to go against the advice of the referees once again. This was as far as he would go. So Dessler rejected my paper.

Ultimately it did not matter to me. Some time later I was able to insert these findings in a Reply that Dessler would publish anyway. More importantly, I now knew the answer. I understood how the occurrence of small comets in the Earth's atmosphere varied over time.

The billions of trails left by meteors as they flash through our atmosphere are now being enlisted by the Pentagon for relaying messages in case of a nuclear war. The Defense Communications Agency and the Air Force recently completed an emergency communications network that, unlike conventional communications links, would not be disabled by the electromagnetic pulse created by a nuclear explosion. This network exploits the impact of individual meteors on the atmosphere, or the continuous background of many overlapping impacts, to reflect radio signals from one part of the Earth to another. The technique, called meteor burst communications, uses high-speed equipment controlled by computers to first sense a meteor trail and then transmit a "burst" of highly compressed information during its split-second existence.

In 1985 Billy Ficklin had been conducting experiments for the United States government on meteor burst communications using a radar installation in Alaska. Ficklin is a senior research engineer in the Special Communications Systems Laboratory at SRI International in Menlo Park, California. Certain anomalies had cropped up in his data, and after hearing about the small comets in 1986, Ficklin came to believe that these anomalies might have been caused by the presence of small comets or some other unidentified objects.

Between September 21st and October 25th, 1985, Ficklin had observed sixty large meteor scatter signals that lasted more than thirty seconds. These were larger and of longer duration than the meteor bursts that commonly occur. He suspected that these events could be created by the type of comet we had described. He may be right. The disturbances Ficklin reports are at about the right rate and at the right sort of scale you would expect from the small comets. His findings suggested a favorable prospect for conducting a radar search for these icy intruders in our atmosphere.

It took more than two years after the discovery of the small comets for someone to initiate a search dedicated to finding these objects with a radar. It took place during the first week of November 1988. It was not done by an American, but by a Frenchman named Gérard Caudal working at the Centre de Recherche en Physique de l'Environnement. I had met this young man at an international scientific meeting in Helsinki, Finland, the previous summer and we had discussed his future search for small comets. He was quite eager and I was under the impression that he was looking for a quick way to make a name for himself.

Early in 1989 he submitted a paper on his radar search to *Geophysical Research Letters*. The paper was entitled "EISCAT Finds No Small Comet." EISCAT, which stands for European Incoherent Scatter radar, is a powerful instrument whose transmitter is located near Tromsø, Norway. Caudal had sent up radar pulses, looking for echos reflected back from the small comets at a distance above their break-up altitude. He assumed that anything moving faster than three miles per second would be a comet rather than a satellite. Caudal expected to see six comets during his search but had found none.

I was asked to review his paper. At first glance, Caudal appeared to have done a fantastic job of eliminating the small comets with his radar search. But on closer inspection, I found that he had made a couple of crucial mistakes. The most important one was that he had looked in the wrong direction. It was difficult to believe that he had made such an error because his paper cited the three papers in the literature that told him where to look. But he made the assumption that the distribution of the small comets was isotropic, in other words, coming into the atmosphere from all directions. He did this despite the fact that the literature specifically stated that the objects were streaming past the Earth in a preferred direction. This mistake had allowed him to make the claim that he had not found the small comets. Caudal had failed to correctly cite the literature and that is something you just do not do.

My review of his paper was not a pretty sight. Not only had Caudal looked in the wrong direction, but he had also changed the character of the small comets, making them more reflective than they actually are, so that his radar would have a greater chance of seeing them. He did not know how much signal the small comets would reflect. He assumed that their surfaces were bare water snow, though the literature plainly states that these comets are mantled to suppress vaporization. When he revised his paper, he again incorrectly referred to our published work, assuming that the composition of the small comets was the same as for the large comets. I began to wonder if he had even bothered to read the journal articles on the topic.

Caudal never admitted that he had looked in the wrong direction. But he should have said so in his revision, despite the embarrassment. He claimed instead that he had looked in the right direction about half the time. I eventually had to draw the simplest of diagrams showing he never did at all. Caudal was overeager.

When you are young, science can be a very risky business. You have no established reputation. And failing to analyze your results properly can be a deadly sin. I feel sorry for young people who jumped into this fray thinking it was a child's game.

The importance of looking in the right direction cannot be underestimated. It would also be of considerable help to anyone looking for these objects with a telescope. But first I would have to clear up all the misconceptions people had about just what the small comets would look like in the night sky to the naked eye.

CHAPTER 10

How to Spot a Small Comet

No one thought for a minute that the impact of a hundred-ton comet would be invisible. Meteor observers can see the arrival of an object no larger than a grain of sand. So everyone thought that you should be able to walk out in your yard at night, look up, and see the small comets blazing across the sky like burning chariots. Alex Dessler believed that too. He thought we should be able to see many of these objects every night. A few astronomers wrote to me saying that if there are ten million of these things a year the sky should be afire with them. These thoughts were not based on any calculations, but on intuition. How could we miss these objects? How could we not see them? But if anyone is to blame for these mistaken impressions, it is I. I said there were ten million house-sized objects falling into the atmosphere each year and somehow that just incited people to riot.

No one had the right perspective. No one ever bothered to compute how often you would see these objects, how long you would have to sit outside in the dark until one was within your field of view as it crashed into the atmosphere. The proper calculations show that you can indeed see them with your naked eye but you might have to stand out there a good long while to see just one. If I were to go in my backyard, it would take at least a hundred hours of clear night viewing before I would see one of them. The problem is that most of us do not go out and stare at a clear dark sky for a hundred hours in a whole lifetime. And even then, what you would see is a peculiar little streak, no sky afire.

We do not see the small comets all the time because the twenty or so that hit the Earth every minute are scattered around the planet and an individual's view of the sky is quite small. A camera on a distant satellite orbiting the Earth has the advantage of seeing an entire face of the planet at once. With such a large field of view it is able to record almost half of all the comets falling into the atmosphere at any given moment. And our satellite sees them all the time. Their entry into the Earth's atmosphere has been more or less the same since 1982. We have tens of thousands of them. Practically every image shows the vapor clouds from the comets, even in January when they are much less frequent.

For a ground observer, not only is the field of view greatly limited, but so is the viewing time. To begin with, these small comets are not visible from the ground during the day. Sunlight vaporizes these objects at high altitudes and transforms them into clouds of water vapor. Though incoming comets may generate an increase in the visible light the atmosphere normally emits, the difference is too small to notice. You simply will not see these objects glowing up there during the day.

At night, the situation is quite different. The comet may not necessarily dissipate into a cloud of water vapor before it enters the Earth's shadow. Without solar heat to vaporize it, the comet remains a compact object much longer than it would during the day. Most of the slowing down at night occurs in its last dozen miles, where the atmosphere thickens considerably. So at night you are not going to have this large cloud of dimly glowing gas but a much smaller object plowing through the atmosphere. It will come in and hit the atmosphere the way a meteor does. But because it is made of water snow it will neither glow as brightly, nor penetrate the atmosphere as deeply, as a rock.

All studies done on cosmic matter falling into the Earth's atmosphere involve dust particles and iron meteorites only. (Meteorites are meteors that you find on the ground.) An incoming particle hits the atmosphere and produces a flash. You see these things all the time. Most of them are iron and dust particles. These objects are little. They may weigh just a fraction of an ounce, but they glow. The metallic ions in these objects are very efficient at producing light seen by the human eye. What happens is that the electrons in these metallic ions are easily excited by collisions with atmospheric molecules. Then, when they relax, which they do rapidly, they emit light. Anything that contains a little bit of metallic ions glows brightly. But there are light bulbs and there are light bulbs.

Just how much a rocky or metallic object glows when it falls from the sky was worked out in the 1920s by an Estonian astronomer named Ernst Öpik. Öpik devised the double count method of tallying meteors, in which two observers view the sky at the same time. Öpik spent a good part of his life looking up at the night sky, estimating how fast an object was moving, how bright it was, and how long you could see it. These observations led to his Tables of Meteor Luminosities, which tell you its brightness, given the weight of the object, the speed when it impacts, and the source of the light—whether it is stone or iron. Öpik's understanding of the effects of atmospheric resistance and heating on meteors would later prove to be of great importance in connection with the design of nose cones and heat shields for ballistic missiles and spaceships.

Öpik's tables make it clear that just a fraction of an ounce of iron is pretty visible in the night sky. So if you scale it up to the size of the small comets, to three million ounces, let's say, then these objects should be tremendously brighter than that little chunk of iron. Those are the arguments the critics were using against us: We should be seeing extremely brilliant fireballs every thirty minutes or so. But a hundred-ton rock hitting the ground would be an unforgettable event, even if you were miles away from the impact. If that is the interaction with the atmosphere we can expect from a hundred-ton comet, then we are history. We probably would not be around if there were that many incoming, hundred-ton, rock-like objects falling down on us from the sky. At best, we would be living in caves.

But these comets are not rocks and do not behave like rocks. They contain little dust and lack the iron and other metals necessary to glow like a neon sign. And because they fall apart and vaporize so much faster than a rock plunging through the atmosphere, they can be much more massive. So I could not use Öpik's tables to calculate the brightness of incoming small comets. I looked for a Table of Snowball Luminosities but found that none exists. No one had ever worked out what happens when a soggy waterball falls through the atmosphere at very supersonic speeds, how it would glow, how bright it would be, or for how long. I would have to work it out myself.

I began looking for these answers late in 1986. A comet, I thought, would have to glow. It slams into the atmosphere at supersonic speeds, pushing and heating the air in front of it. At an altitude of about seventy-five miles, the pressure created by ramming the atmosphere equals the strength of the comet's mantle and the object is disrupted. But there is no explosive burst of visual radiation just yet. The water snow then vaporizes rapidly. This is followed by an expansion and deceleration of this blob of hot water vapor. Momentum carries it down to an altitude of about fifty miles, at which point its speed is reduced to less than one mile per second and the gas begins to cool. It is at altitudes of sixty-five to fifty miles where the gas glows brightest, though only a fraction of its energy is converted into visible light. What we see is thermal radiation, like the glow that comes off a hot stove.

But just how much glow do the small comets produce? We wanted to know whether they would be too bright. Because if they were as bright as the Sun or the Moon, we could just write them off. We needed to know the answer. The purpose of this exercise was to find something inconsistent. It is much more decisive to find one thing wrong, than to find a hundred things right. All along it has been my stated intention that if I could find something wrong with the small comet proposal, I would be the first to say what was wrong with the proposal and withdraw it.

The nose of a supersonic jet slamming through the atmosphere glows by compressing and heating the gas in front of it. This is good to know if you want to create a defense system. But it is also good to know if you want to calculate the glow from a soggy comet. In these days of ICBM missile heat shields, there is considerable experimental data on how much glow an object produces as it plows through the atmosphere. There are tables on this phenomenon. In Öpik's time it was not possible to even discuss how bright a small comet would be because nobody knew how bright atmospheric gases would glow when they are very hot and compressed. The numbers are now available to us, but calculating the glow is an immensely difficult task. Even to get a rough estimate required the use of a computer. So I asked Kent Ackerson to tackle this problem. He is a member of our research group at Iowa and was then working on an instrument that would be flown to Jupiter on the spacecraft Galileo. Ackerson is a very competent scientist and brilliant with computers.

By early 1987 we had our answer. Ackerson had come up with a surprising result. The glow produced by a small comet would be equivalent to the glow in front of a plane going about thirty times the speed of sound. That glow is roughly equivalent to the brightness of the planet Venus.

But the small comets are strange objects. They appear as bright as Venus for only about two seconds. That is how long it takes them to fall from an altitude of about sixty-five miles to fifty miles, when they slow down and snuff out. People think the small comets will hit the ground because they start out bright. At first, they do look like a big rock. But they do not go anywhere near the ground. They break up and vaporize far up in the atmosphere. It is strange for an object from interplanetary space to stop glowing that high up in the sky. A rock should glow all the way to the ground. But this is not a rock. It is a snowball.

So every hundred hours or so, you should see something in the night sky that is about the brightness of Venus for about two seconds before it vanishes. The question is, are there such objects? The answer is yes. They are objects that were never explained before and are known as Öpik's dustballs. The dustballs were never explained because they do not behave like rocks. A rock should burn its way down through the atmosphere. Dustballs start out like a very bright rock, but travel only a short distance, then snuff out.

Öpik's dustballs were known about long before the small comets. He thought the dustballs were just a big conglomeration of dust particles, big enough to cause a large glow but so weak as to break apart and not continue down into the atmosphere. Some people called the dustballs "dirt clods." But no one could ever capture one because they blinked out long before they got to the ground and nothing was ever found at the place they were expected to hit.

Öpik never could figure out how all the dust particles in a dustball would manage to stay together. But I think the answer is water-snow. A small comet of water-snow impregnated with minor amounts of cosmic dust should emulate the visual signature of these dustballs. I think the small comets look just like Öpik's dustballs. They should have the signature of this very peculiar kind of meteor.

Öpik believed that many visual meteors were in fact dustballs. To confirm this notion, we had to establish that some ten million objects as bright as Venus were indeed being seen in the Earth's atmosphere each year. We went to a book of lists for how often visual meteors appear and looked up the number of objects with the brightness of Venus that appear annually in our skies. The number is close to the predicted number of small comets. It is not one, one hundred, or a hundred million. It is on the order of a million to ten million.

So it seems we must discard the conventional wisdom which holds that all meteors are due to stony or iron objects. Some of them are small comets. One person was quick enough to catch this. His name is Fraser Fanale. He is a planetary scientist at the University of Hawaii in Manoa. He told me once that the small comets should have the signature of Öpik's dustballs. He was right.

To observe a small comet, you must stand out on a clear dark night until you see a short bright streak that quickly snuffs out. It looks like it might be a big rock at the beginning but then winks out like a candle. You will have to be out there for a hundred hours or so to see one. A hundred hours of clear night viewing does not happen often in the average lifetime. It depends on whether you like the outdoors or not. Most of us spend our time indoors, drinking a beer or two, not watching the night sky. But even if you do, you are only going to see a few of these in your lifetime.

I think I may have seen one or two myself.

CHAPTER 11

Flying Saucers and Other Strange Events

The small comets hit a responsive chord. We received a flood of mail on the subject from doctors and businessmen, writers and children, teachers and preachers. Some wanted additional information, others wanted to share their own wild ideas. One geologist, for example, thought the spots in our satellite images might be produced by the impacts of large meteorites in the ocean, splashing water into orbit. But most people seemed to think we were on the right track and wanted to know more about the possible effects and consequences of such an infall of cosmic snowballs. An eleven-year-old boy wondered if germs in the comets might have killed off the dinosaurs. And a letter-writer who claimed to be "illegally confined" asked if my proposal had been responsible for the University's loss of a twenty-one-million-dollar contract with NASA. It was not.

A number of people wrote to tell us that they had actually seen the small comets. Many of these letters came from amateur astronomers, a fact which lent considerable credence to their stories. No one spends more time looking at the sky than amateur astronomers. They probably know the night sky better than anyone else in the world. They are the million eyes looking at the sky for uncountable hours, and they do see strange things.

John West of Texas is one of these amateur astronomers. He has kept a keen eye out for meteors during the past two decades and wrote about a couple he had seen that were clearly "out-of-theordinary." One he labels a "fuzzy" meteor. "This meteor," he said, "is not characteristic of the majority seen, in deceleration, quality of light envelope, lack of distinctive 'edge' to the path, a very brief afterglow and no color." His other shaggy dog he calls dark meteors, "a brief streak blacker than the sky." He notes that others had seen these along with him and he insists they were not tricks of the eyes.

Eric Kurczewski of New Jersey also believes that "dark" or "nebulous" meteors may be evidence of the demise of small comets. "Unlike a trailed meteor," he said, "a nebulous meteor appears as a blur and is of very low surface brightness, only slightly above the light level of the background sky. The length of travel of this type of object ... is five to ten degrees across the background sky. Very dark sky conditions are required to see this event..." He claims to have seen three such meteors, not associated with any meteor showers "in 200 hours of observing far from city lights."

The number of references to nebulous meteors was striking. These people were obviously sincere. But I had never heard of nebulous meteors. So I got in touch with Edward Ney, a topnotch astronomer now retired from the University of Minnesota. He had never heard of nebulous meteors either. But I later learned that most professional astronomers pay little attention to the topic. It seems that a century ago in Great Britain the reality of nebulous meteors was a hotly debated topic. There are apparently many solid observations of this phenomenon from amateur astronomers. Nebulous meteors are generally thought to be loose clumps of fine debris of cometary origin. But unfortunately there is no organized information on these observations. There is nothing that allows us to document what is happening here. So I cannot say with any certainty that nebulous meteors have any bearing on the topic of small comets.

Then there was the letter from Keith Bigg, a physicist from Sydney, Australia. He had, he said, "direct evidence," a picture of what might be a small comet entering the atmosphere back in November of 1968. His account had appeared in *Nature* the following year. Bigg had been studying dust layers in the stratosphere with a camera mounted on a high-altitude balloon when the camera happened to snap a picture of a group of meteor trails with a red glow beneath them. The brightness of the trails suggested the object was "very large" before fragmentation. Bigg had been puzzled by "their short paths" and by the lack of "bright flares typical of large meteors." He now believes this could be evidence for our small comets. It is, of course, quite possible for a large object or comet to split in the upper atmosphere and leave all sorts of trails. So Bigg may be right.

A mystery object photographed by a DSMP (Defense Satellite Meteorological Program) satellite in the late 1970s may be yet another portrait of a small comet. One infrared image from this weather satellite shows a strange object above the cloud cover. The object is very white, indicating the "maximum cold" the satellite could register, or less than minus 76°F. Its shape appears almost rectangular but may not be because the DSMP imager does not photograph a whole area at once the way a camera does. Instead, like Dynamics Explorer, it creates a picture one line at a time. So any relative motion between the object and the satellite while these scan lines were collected could have distorted the shape of the object in the image.

Bruce Maccabee, a Maryland physicist who works for the Navy, performed a thorough analysis of the image on his own. He determined that the size of the object could be anywhere from tens of nautical miles, if the object was located near the Earth's surface, to as little as tens of feet, if the object was hovering within several hundred feet of the satellite. Unfortunately, there was no way to determine the object's altitude. But Maccabee showed quite clearly that the object was not created by random electronic or mechanical noise. Nor could it be explained as a cloud or aircraft flying near the Earth itself. And he thought it unlikely that the object was another satellite or piece of space junk traveling in an orbit beneath the DSMP satellite. Maccabee, as director of the Fund for UFO Studies, obviously suspected a UFO.

I estimated the size and speed of a cometary water cloud at a few hundred miles altitude and concluded that the DSMP picture could be of such an object. But in order to prove that this object is a cometary water cloud, other sightings would have to be made with this satellite in order to establish their frequency and hence the number of such objects. Because the speed of the DSMP spacecraft must almost match that of the cometary water cloud, such events are expected to be rare, perhaps as infrequent as once every several thousand orbits or more. Other kinds of features, not as spectacular or easily identified, such as short white lines in single or consecutive scan lines could also be expected in the images. What this makes clear, however, is that unidentified phenomena are observed with spacecraft. But I cannot say with any certainty that this particular observation relates to a small comet.

I avoided many topics in the course of the debate on the small comets and UFOs was one of them. Our hypothesis was already controversial enough. We had angered thousands of scientists. We did not need to bring such a taboo subject into the fray. But there is no longer any need to avoid the subject.

A number of people suggested to me that perhaps some of these objects entering the atmosphere could produce the illusion of a flying saucer. Alvin Lawson was one of those who had noticed that the small comets share many ephemeral qualities with UFOs. Lawson, a professor of English at California State University in Long Beach, has investigated UFOs for over a decade and is a skeptic on the issue. He wrote to me saying that if UFO sightings have a cause, then the small comets might be "the most likely explanation yet."

"Might one assume," he asked, "that dust, released gases, and varying sizes and speeds of the comets would result in occasional visibility? I am aware that most of the numerous cometary events occur at high altitudes; but if only one percent of them incinerate there would be potentially 288 'weird lights in the skies' each 24 hours worldwide."

Such an explanation, Lawson believes, could well account for the vast number of UFO reports, as well as their widespread origin, long history, sudden appearances and disappearances, odd behavior, and virtual silence. And if small comets are the cause of UFO sightings, he mused, how ironic it would be: UFOs would actually have turned out to be extraterrestrial.

I think Lawson may be right. In our computer modeling of the comets and how they plunge through the atmosphere, we came across a result that we did not advertise. We learned that water clouds produced by the occasionally larger small comets, those that penetrate more deeply into the atmosphere, would have a peculiar shape. If you look at how these water clouds are deformed, how they might become frozen, then you can see how someone might go out in full daylight and look at the sky and see a saucer-shaped object that looks like it is moving very, very rapidly.

Normally, of course, a small comet should not be visible during the day. And at night you should simply see a short streak across the sky. The average small comet does not penetrate deeply enough into the atmosphere. But not all the comets are expected to be the same size. Some will be large. Of course, the larger the size, the fewer such comets we would expect to see. But also the larger the size, the deeper into the atmosphere they would fall. If the comet is large, but not large enough to hit the Earth, then as it descends into the cold parts of the atmosphere ten to twenty miles above us, the hot steam would slow down and perhaps cool sufficiently to condense into ice crystals and become visible. The underside of the water cloud would be shaped like an ablation cone. It would look just like a curved saucer.

Human perception is such that someone seeing this sort of an object would have considerable difficulty judging whether it was located a hundred yards away or twenty miles away. Then, as the ice crystals on the surface of the cloud disappeared, the object would appear to diminish, and the observer might get the impression that the object was receding at tremendous speeds. This shrinking cloud could then be picked up by the winds, producing the illusion that the object had made a sudden sharp turn. My scenario, of course, will not account for the reports of tiny aliens snatching frightened humans from their homes and taking off from pumpkin patches in flying saucers.

I tread across this territory carefully. Many others have explored this murky, no-man's land of unexplained phenomena before, though most not so cautiously perhaps. One forerunner was a Bronx-born clerk named Charles Fort. In 1919, he published *The Book of the*

Damned, his first work of odd data culled from the scientific journals and newspapers of his day. Fort had a fondness for tales of "aerial icebergs" and "shafts of water" from "blue skies," and he poked fun at the efforts of science to explain them. To Fort, the reports of falling ice suggested "that there is water—oceans or lakes and ponds, or rivers of it—that there is water away from, and yet not far remote from, this Earth's atmosphere and gravitation ... Sometimes fields of ice pass between the Sun and Earth." Fort almost got it right. I think the small comets may well account for at least some of the many unexplained reports we have of falling ice.

There is a long history of ice chunks of all sizes falling down from the sky, sometimes even crashing through homes. Such tales of falling ice dot the popular press and meteorological journals worldwide. Some of the pieces are said to measure as much as twenty feet in diameter and weigh hundreds of pounds. Though most such hydrometeors, as they are known, are considerably smaller, they are still too large to be explained by the prevailing theory of hail formation in storms. Consequently, hydrometeors are usually blamed on passing aircraft, even though reports of the phenomenon predate the Wright brothers. A study conducted some time ago by the late James McDonald, an atmospheric physicist, found that only two out of thirty ice falls in the United States in the 1950s could be attributed to aircraft.

One of the best documented falls, according to Arthur C. Clarke, who also seems to have an interest in such things, occurred over a tree-lined street on the outskirts of Manchester, England, on the 2nd of April 1973. It was just after 8 PM when a Dr. Richard Griffiths saw a large object hit the road just outside a liquor shop. The ice block was quite large, according to Griffiths, who picked it up and stuck it in his freezer overnight. The next day he took the four-and-a-half-pound chunk of ice to his lab at the Manchester Institute of Science and Technology for analysis.

Griffiths tested it the way a hailstone is tested to determine its history. He sliced it into thin sections and found that the ice was made up of fifty-one layers separated by some thinner layers of trapped air bubbles. The regularity of the ice layers and the size of the crystals made it unlike any hailstone he had ever seen. It was also considerably larger than the largest scientifically verified hailstone, the 1.67-pound Coffeyville Hailstone, found in the town of the same name in 1970. Griffiths then made inquiries at the local airport. He discovered that one plane had landed just after the ice block had fallen, but engineers informed him that there had been no icing on the aircraft. Clarke believes that this type of ice fall "ties in beautifully" with my small comet hypothesis.

He is not alone. In 1989 I received a letter from Hu Zhang-Wei, an astronomer at Nanjing University in the People's Republic of China. He told me that on three occasions—the 29th of December 1982, the 11th of April 1983, and the 17th of November 1984—large chunks of "natural" ice had fallen from the sky near the city of Wuxi. He performed a detailed chemical analysis of a fragment from the latest fall, found that it differed from hail, and concluded that perhaps the ice had come from a comet.

I take the possibility of such falls seriously. If the comets are large enough to begin with, chunks of ice could reach the ground. Under certain conditions, perhaps the center of the cloud does not vaporize at all, or turns to ice in the lower atmosphere, and then crashes down to the Earth's surface. We expect the small comets to vary in size from time to time. Not all of them will be a hundred tons. A few will be smaller, a few will be larger. So for every thousand normal-size small comets that come in, maybe there will be a larger one. Some may be as large as 10,000 tons.

The very largest of these comets are certain to leave their calling cards on the Earth. One certainly was left at Tunguska in Siberia in 1908. What caused the giant explosion there has been a puzzle for some time. But I think it was a comet. It formed a shallow crater and felled trees for several hundred square miles around it. That is exactly what we now expect to happen when a large water comet plows into the Earth. The impact created only a shallow crater, which is not surprising since the impact pressures for a water comet are ten to fifty times less than for a solid rock. But the heat from the incoming hot piston of gas charred the area around the crater. And the atmospheric pressure wave it created left behind an eerie landscape of felled trees, all pointing away from center of the impact.

There may be other Tunguskas around. The flat-floored crater at Flynn Creek, Tennessee, may be one. These events will happen once in every 100 or 1,000 years; for every billion small comets you are going to have a large one that measures hundreds, perhaps thousands of feet in diameter. No one can deny the existence of these larger comets.

If the small comets, these icy interlopers, are real, then there probably exists a good deal more evidence of their presence within all the reports of strange and unexplained phenomena buried in the scientific literature. Oceanographic and meteorological journals have long published reports of bright flashes of light lasting from one to several seconds and covering large portions of the sky. Bright luminous patches have also been seen close to the horizon or moving steadily and deliberately across the sky. And large amounts of water vapor high in the atmosphere might account for a wide variety of optical effects that have been reported, including odd halos and strange rainbows.

The small comets may also be responsible for a phenomenon described by pilots and others who have seen it as "an expanding ball of light." A study of fifteen such cases, taking place between 1959 and 1985, was conducted by psychologist Richard Haines and published in the *Journal of Scientific Exploration*. Pilot reports indicate that the initial ball of light forms at a relatively high altitude. The light, usually white or yellowish, then gradually expands until it becomes increasingly transparent to stars in the background. Eventually it fades completely from sight. Some reports also tell of a towering cumulus-like cloud which appears to rise out of the stratiform layer. I think these clouds might be a comet's vapor trail at lower altitudes. But the phenomenon would have to be related to the atmospheric impacts of somewhat larger comets than those responsible for the atmospheric holes. And we expect that the number of such large comets, penetrating deeply in the atmosphere, to be much less.

Some pilots may also on occasion experience air turbulence caused by the infall of these larger comets. One pilot, Harold Blake of Florida, wrote to me pointing out the existence of a phenomenon called Clear Air Turbulence, or CATs. These, he said, are "elusive and destructive clear air turbulences that most military pilots fear and commercial and private pilots avoid 'like the plague'." They happen without warning. I think that a piston of gas coming down at a hundred miles an hour from one of these comets could cause such turbulence. Near the ground, Blake believes, CATs might also account for wind shear. But there is very little research in this field. Some of these CATs are probably associated with the jet stream in the atmosphere, but are there CATs that are not? To be certain, we would need to see airline records on CATs, if such exist, then calculate their rates and see if they might match a reasonably expected rate of infall of the occasional larger comets.

One pilot who regularly flies the Pacific wrote to me saying he has sometimes seen round holes through clouds. Others have seen them as well. These holes may be as large as several miles in diameter, and appear in otherwise dense, uniform cloud decks. There are also some credible observations of precipitation from cloudless skies. And sometimes layers of unexplained haze are detected by balloon-borne instruments at altitudes above fifteen miles. Could all these be effects due to the small comets?

In 1976, Frank Gibson of the Air Force Geophysics Laboratory in Bedford, Massachusetts, published a letter in *Nature* entitled "A Rare Event in the Stratosphere." On the 20th of June 1974, a high-altitude balloon was sent up in the clear skies over New Mexico to measure the scattering of light by aerosols, or tiny solid particles, in the atmosphere. Upon reaching an altitude of about ten miles, a strong signal was recorded indicating increased aerosol concentrations, a cloud in other words. This haze extended at least another seven miles up and had a reach of some fifty miles across, but it was not visible from the ground. A momentary injection of aerosols would be necessary to form such a cloud. But there was no obvious source for it. Whether it might be related to a small comet is hard to tell.

There is no end to such curiosities. Recently scientists at the University of Illinois claimed to have found clouds of sodium atoms that rapidly appear and disappear at an altitude of about fifty miles. They observed these sodium clouds with an instrument called a lidar. The lidar sends up pulses of laser light into the atmosphere and measures the intensity and wavelength of the light that returns. The scientists, headed by Timothy Beatty, suggest that the clouds are vaporized meteor material. Or perhaps, once again, cometary material. It would certainly complete the recipe for a salty ocean.

Quite obviously there is a whole folklore of flashes and dimmings and strange things going on in the sky. It is tempting to see the small comets as an explanation for many of these puzzling phenomena. But whether these things are related to the small comets is not certain. The phenomena are interesting, but they are often lacking in documentation and so are difficult to interpret. Some could be due to the passage of a small comet, who knows? At the moment, all that we know for certain is just how little we know.

CHAPTER 12

A Scramble for Satellite Data

Meanwhile, the debate raged on. My discovery of the small comets with Dynamics Explorer had sent a number of scientists off on a frantic search through the data libraries of other satellites. It was a sensible thing to do. After all, hundreds of satellites had been placed in orbit during the past quarter century or so since the launch of Sputnik and probably less than ten percent of all the data received from these spacecraft has ever been closely looked at. So it was reasonable to assume that, if the small comets were real, some other satellite might have already gathered evidence of the passage of these objects through the Earth's atmosphere. Most people were, of course, quite convinced that the reason no satellite had ever reported such objects was simply because the small comets did not exist.

Of those hundreds of satellites, however, only a few were likely to have recorded the signs of the impact of cometary water clouds into the Earth's atmosphere. One of these was the Orbiting Geophysical Observatory, OGO-4. The Naval Research Laboratory placed an instrument on OGO-4 that provided continuous observations of ultraviolet light for a period of about eighteen months during 1967-1968. Nearly twenty years later Robert Meier and Talbot Chubb, physicists at the Naval Research Laboratory's E.O. Hulburt Center for Space Research, located in Washington, D.C., decided to reexamine the measurements from 2,000 of the satellite's orbits for evidence of atmospheric holes. Shortly afterwards they submitted a paper to *Geophysical Research Letters* saying they had failed to find any such evidence.

I refereed the paper at Dessler's request. The paper had one glaring problem. Their data did not support their conclusion. The other referee picked up on this as well. There was an abundance of unidentified fluctuations in their dayglow observations that might be ascribed to the presence of atmospheric holes. A casual examination of even the very few orbits in the published literature revealed a number of possible events. In fact, there appeared to be one or two events in every orbit. In a revised version of their paper, Meier and Chubb presented all kinds of excuses for the dark spots in their measurements. I think they went a little out of their way to avoid the possibility of atmospheric holes.

They claimed that a draftsperson had added a few bumps which were not actually present in the data. Other blips, they said, were caused by sunlight, tropical arcs, particle precipitation, periods of lost spacecraft orientation, and auroras. You name it and it was there. It was clear that the authors had decided from the start that all the fluctuations in their observations were due to something other than cometary water clouds, even though no definitive identification of every isolated dark spot was possible. In fact, very few of the events in their dayglow observations were likely to be caused by cometary water clouds—just one in every 100 to 1,000 orbits. But the authors had failed to convincingly identify each dark spot in their data, so they could not convincingly exclude the presence of cometary water clouds with the OGO-4 observations. Their paper was rejected.

Our Dynamics Explorer spacecraft had a twin called Dynamics Explorer 2. Both were carried aloft, piggyback-style, on the same rocket back in 1981. Scientists hoped that this pair of satellites would give them a ringside seat on the Earth's royal crowns, its auroral lights. We wanted the big picture, in other words, not just the worm's eye view aurora chasers get from down here on Earth. Our Dynamics Explorer was stationed at a high altitude to study how this natural wonder is produced, how a beam of electrons from space hits the Earth's atmosphere as if it were a giant television screen and produces these auroral "pictures." Dynamics Explorer 2 was supposed to cover the same territory but from a lower altitude, skimming over the upper atmosphere, just above the "screen."

Every few months there were working group meetings on the Dynamics Explorer mission. They were usually held at the Goddard Space Flight Center in Greenbelt, Maryland, and each one lasted a couple of days. From time to time I would present our findings on the atmospheric holes at these working group meetings. Some people not involved in the Dynamics Explorer mission would even come in and listen. It created considerable interest. One of the people who was kept up to date on our findings and interpretation of the holes was Bill Hanson, the director of the Center for Space Sciences at the University of Texas in Dallas. I took great pains to let him know exactly what we were doing on the atmospheric holes, but he got very emotional about it and was to remain so throughout the controversy.

Hanson had an instrument aboard Dynamics Explorer 2. His instrument examined the contents of the ionosphere, the Earth's radio-reflective covering that consists of ionized gases. It is here that the Sun's ultraviolet radiation tears neutral gases in our atmosphere apart, producing ions and electrons. The result is a veritable shroud of ionization, or charged particles, around the Earth. Hanson was certain that if the small comets were real they would be causing drastic effects to the Earth's ionosphere. So he speculated on how the ionosphere would be affected by the passage of a cometary water cloud. He believed that the water vapor from an impacting comet would cause a mammoth hole in the ionosphere that would only return to normal about ten minutes after the passage of a small comet. He thought that a satellite moving through the ionosphere, such as Dynamics Explorer 2, should encounter such holes about once in every two hours of satellite flight.

Unfortunately, Dynamics Explorer 2 did not have a long life. But it did manage to collect data for a year and a half before falling out of its orbit. Hanson ended up examining only five hours of measurements from Dynamics Explorer 2 and another eighty hours of data from two older spacecraft, Atmospheric Explorer C and Atmospheric Explorer E. Hanson, however, found absolutely no unusual perturbations of the ionosphere. He concluded, in his Comment published in *Geophysical Research Letters*, that either we had overestimated the number of small comets impacting the Earth's atmosphere or that the ionospheric effects of these objects were "more subtle" than he had assumed.

We also examined Hanson's Dynamics Explorer 2 data. But unlike Hanson's, our search did reveal some evidence of the passage of small comet water clouds through the ionosphere. We found it by looking first at those times when the Dynamics Explorer 2 had passed near an atmospheric hole that had been spotted by our Dynamics Explorer. We expected near simultaneous sightings by both spacecraft to be rare and in fact we ended up eliminating all but one or two of our 26,900 candidate atmospheric holes.

One of these atmospheric holes occurred near a disturbance in the ionosphere seen by Dynamics Explorer 2. The event happened at low altitudes and was quite brief. This suggested to us that the interaction taking place between a cometary water cloud and the ionosphere was not a large one. Rather than a severe loss of the number of ionospheric ions, the comet clouds appear only to create a small region of turbulence in the ionosphere. Actually, the passage of the cometary water cloud through this region resembles the disturbance caused by a large, fast object, such as the space shuttle.

We then went back and searched for the presence of other similar disturbances in a limited portion of ionospheric data from Dynamics Explorer 2 and found three more such events. Of course, these disturbances may or may not be due to small comets. But their size and frequency are about what you would expect for water vapor clouds produced by small comets as they come plummeting through the ionosphere. That is what we published in our Reply. Hanson had looked into the basket and did not see any apples. When we looked, we saw apples. Hanson seemed taken aback and angry. I think we had taken him by surprise.

On the 22nd of February 1986, just six weeks before news of the small comets first hit the newspapers, the Swedes launched a satellite called Viking. Like our Dynamics Explorer, their Viking satellite was equipped with an ultraviolet camera designed to take pictures of the aurora. It had two imagers actually, both provided by the Canadian government. This made Viking, of all the satellites in the sky, the most likely candidate to confirm or deny the existence of the cometary water clouds above the atmosphere that had first been seen by our Dynamics Explorer.

So, not long after Viking began sending back its images of the Earth's royal crowns, the press began to inquire about whether or not the satellite had seen any "atmospheric holes." The job of fielding these questions fell to Sandy Murphree, a physicist at the University of Calgary who was then the principal investigator for the Canadian imagers. Murphree, it so happens, had obtained his degree at Rice University while Dessler was chairman of the Space Physics and Astronomy Department. It was a small world.

The situation proved to be tremendously distracting for Murphree and his small team of Viking scientists. Few were asking questions about their beautiful pictures of the aurora and this, I am sure, they found distressing. I never realized that the publication of our original papers would cause such a problem for them. If the roles had been reversed, I also would have been tremendously annoyed. Besides, scientists often spend years studying the data from their satellite instruments. The question about atmospheric holes was not an easy one to answer. You have to look very carefully at your images. You really have to do your homework. So I understand how the Viking people felt.

Unfortunately, the press wanted a quick yes or no answer. But scientists and journalists usually march to the beat of different drummers. Science is slow and cautious. The press is always in a rush and not always cautious. They forced Murphree and his Viking team to issue a quick statement. And what quick statement could they make other than to say, "We don't see dark spots in our images"? Of course, they had not done an extensive search for these holes and never claimed that they had.

Early in 1987, the Calgary team published a paper on their Viking imager, which briefly mentioned this negative result. They had inspected a "small sample" of the images but had found no cometary holes of the sort we had reported with Dynamics Explorer. A more "extensive analysis" would be needed, they said, to confirm or deny the existence of the small comets. Once again, Dessler had asked me to referee the paper for *Geophysical Research Letters*.

I suggested a small qualification. I thought the investigators should clarify whether their instrument could have picked up such holes, if they did exist, and under what conditions it could do so. I wanted the Calgary team to say that they had not completed the analysis of their search for the atmospheric holes, and that by just looking at the pictures they could not tell if a dark spot in their images was an atmospheric hole or not. Dessler told me that he would take care of it, but he never did. This precipitated a very bad situation. Everyone got the impression that the Viking people had searched their images very carefully. But you cannot do a thorough search in such a short time and be sure it is correct.

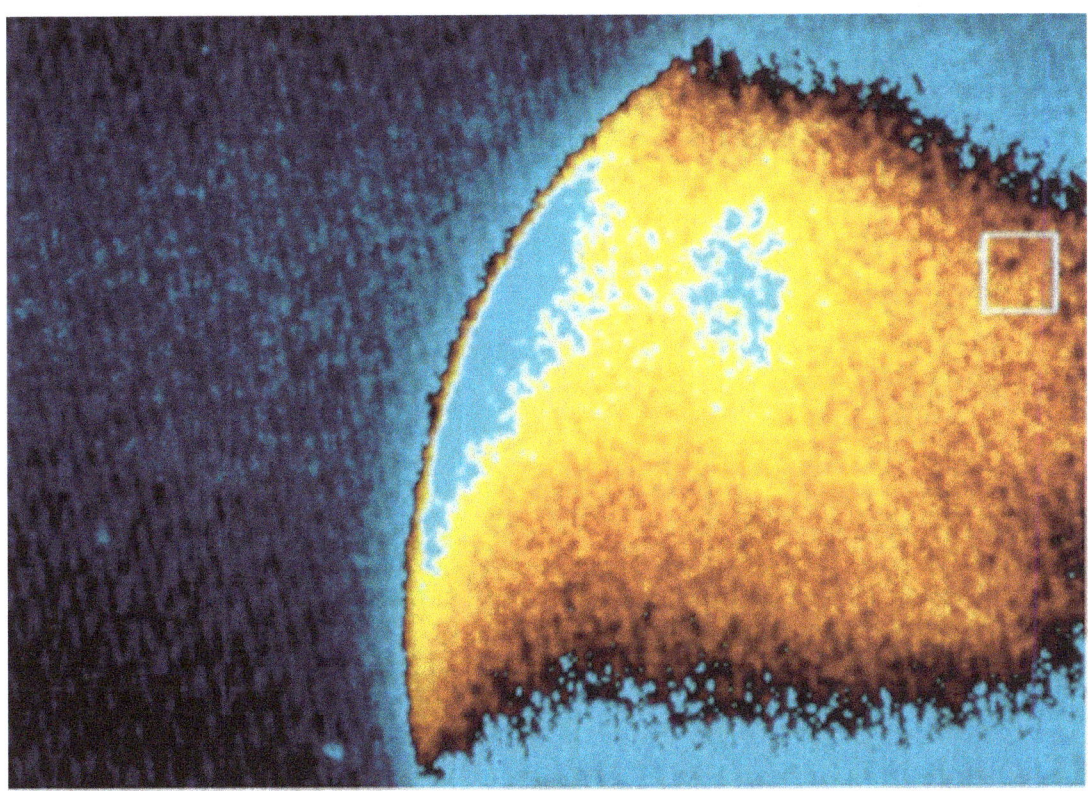

An ultraviolet imager on another Earth-orbiting satellite, the Swedish Viking satellite launched in February of 1986, has also shown evidence of atmospheric holes caused by the small comets.

Sandy Murphree and his team said they had looked for, but not found, any trace of the small comets in their images. But when we first saw their images during a presentation of their findings at a meeting of the American Geophysical Union in the spring of 1987, we could see things that looked suspiciously like atmospheric holes. They did not appear as frequently in the Viking data because the satellite was in a lower orbit than Dynamics Explorer. This lower vantage point offered Viking a smaller view of the Earth's atmosphere and consequently fewer small comet atmospheric holes would appear in their images. Besides, the

Viking had a different imager on board than the Dynamics Explorer and therefore a different analysis was required.

We suggested a collaboration. We had already agreed to work together on the aurora. The beauty of the Viking measurements of the aurora is that they could look at the North Pole while Dynamics Explorer looked at the South Pole. It was a fantastic opportunity to find out whether the aurora looked the same over both poles. We were not about to sacrifice this sort of opportunity over a petty argument over atmospheric holes. But they agreed to collaborate on them as well.

So in January of 1988, Sigwarth, Craven, and I travelled up to Calgary to start our collaboration on the aurora and examine certain Viking images for the possible presence of atmospheric holes. They had two cameras aboard Viking. Unfortunately, the camera that provided them with the most images was not suited for spotting atmospheric holes. It had a noisy sensor and the camera appeared to be picking up stray light reflected from atmospheric clouds. But it was still well suited for auroral work.

The other Viking imager had a coated mirror that provided protection against cloud reflections and we saw no reason why it would not be able to detect holes in the atmosphere produced by the small comets. Unfortunately, the camera had died on the 5th of May, just a couple of months after launch, so there were not many images. In addition, most of the images from the failed camera had been taken from the Earth's nightside to see the aurora, rather than from the dayside, where atmospheric holes should be visible.

So we began to examine the few existing dayside pictures. Murphree was not quite sure what atmospheric holes would look like in his images. He pointed to some dark spots and asked us if they were blackened due to some electronic malfunction. Then we changed the color scales on the television set to get better contrast and the spots began to look very much like the holes in the Dynamics Explorer images. It was immediately apparent to everybody—including Murphree—that if you looked carefully at the pictures from the Viking camera that should see the holes, you saw things that indeed looked like holes. Of that there was no doubt.

So in May of 1988 we presented a joint paper at the American Geophysical Union meeting in Baltimore. I gave the talk in Murphree's absence. I had told him what I was going to say. I would show that there were holes in the Viking images. The Viking images even showed the atmospheric holes in greater detail than in the Dynamics Explorer images. The satellite was at a lower altitude and therefore closer to the holes and its camera had a better resolution than ours had. So the Viking images showed these holes not in one or two pixels, or picture elements, but in clusters of six or seven. They were globs of pixels. Yet the holes were the same size as those seen by Dynamics Explorer. They were also of the same blackness and, even more importantly, they occurred at the same rate. The probability that a half dozen pixels could randomly come together in an image to form such a dark spot was extremely small. I knew the results would cause a furor.

Reporters quickly called Murphree. It is unfortunate that this debate has been so public. You could not sneeze without having your handkerchief exhibited for all to see in the newspapers. Afterwards the press reported that the Calgary group was withdrawing from the paper. But that was not true. We agreed to work side by side until a mutually acceptable version of the paper could be found. No one realized at the time that the task would take a year and a half. And I did not realize that this cooperation eventually would fail.

Problem number one was that the Viking images were noisier than those from Dynamics

Explorer. In particular they have an amplifier noise that produces diagonal lines across the image. So when you do a detailed investigation, it is very hard to show convincingly that the atmospheric holes could not be due to noise. We first examined the laboratory calibrations of the camera and then decided that an even more rigorous in-orbit calibration would be necessary. If the holes were noise they should appear randomly throughout an image. If the holes were real, they should only appear where we expected them to appear.

Problem number two was that the Viking imager had produced few usable pictures for such analysis. But there seemed to be just enough images to get a satisfying check on the camera to see if the holes were real or just artifacts. To understand how the calibration works you must realize, first of all, that most of the glow appearing in these ultraviolet images is produced by atomic oxygen when the camera is looking nearly straight down into the atmosphere. This is called nadir viewing.

But it is a property of the atmosphere that as you look further and further toward the edge of the Earth, which is called limb viewing, there is a greater intensity coming from nitrogen molecules. Since the water vapor from the small comets does not absorb the light from nitrogen very well, you should not often see holes near the limb. We found this to be true in the Dynamics Explorer pictures. When we examined the Viking images we found seven dark events when nadir viewing and only one for limb viewing in the same set of images. This was a profound confirmation that the darkened spots in the images were indeed not noise but actual atmospheric holes.

We exhausted all possible avenues for analysis. It took a tremendous amount of work, both on our part and by the Calgary group. But in the end the Calgary group was not completely convinced that the holes were not noise, although they were never able to identify a specific electronics failure or malfunction that could have produced such an effect. Perhaps they still felt bound by their previous hasty statement. In any case, they would not reverse themselves and in June of 1989 the Calgary group finally decided to withdraw their names from the paper. There was no bitterness between us. But we were disappointed. We felt that the findings of Dynamics Explorer had been confirmed.

One last feature of the Viking images is worth mentioning. Next to the darkened pixels representing the atmospheric holes in the images were adjacent pixels that showed a bright light. The images from Dynamics Explorer sometimes showed a bright region around an atmospheric hole but most often this brightening was not seen, probably because of the camera's poor resolution. But the Viking camera had a better resolution than Dynamics Explorer's. If the dark spots in the images really were holes, then bright rims should appear around their edges as the clouds of water vapor came crashing through the Earth's atmosphere. The face of the cometary water cloud should be hot due to the collision with the thin atmosphere at these altitudes. The Viking images showed these bright rims on its atmospheric holes.

The Viking scientists, which included the Calgary team, had working group meetings just as we did on Dynamics Explorer. Apparently some people at these meetings would point to those bright rims around the dark holes and ask what they were. They are noise, the Calgary people replied. Murphree had faced the same problem we had. When we first presented our Dynamics Explorer images at working group meetings, people would see the pretty aurora and ask, but what are those black spots? The Viking scientists had a similar question for the Calgary group: "What are those bagels?" they asked. Everybody referred to the black spots with bright rims as bagels.

CHAPTER 13

Extraordinary Evidence

Extraordinary claims excite the media, but their staying power is often fleeting. The recent flurry of excitement over cold fusion is a good example. The media trumpeted the claim in its usual way—as a fantastic new discovery. They had done the same with the small comets. But there was a difference. The chemists who announced the startling results of their cold fusion experiments did so without having firm evidence in hand. There was no published paper. (An informal paper was circulated privately and I received my first copy from a Los Angeles stockbroker who was once a student of mine.) Those who tried to reproduce the cold fusion experiments failed.

Our claim also was extraordinary. But our results had been published in a pair of refereed papers and the Viking spacecraft had, in effect, confirmed our findings with Dynamics Explorer. The evidence we had on hand was nearly overwhelming: tens of thousands of pictures showing the holes in the atmosphere caused by a previously undetected population of small comets. Each one of these pictures was worth much more than the proverbial thousand words. They are worth more than ten thousand words, actually, since each standard image is composed of 18,000 pixels.

Each of these pixels, or picture elements, has a lot to tell us about what we are looking at. But like words, pixels can sometimes mean different things to different people. Some of our critics interpreted the darkened pixels in our images as noise in our instrument rather than as clouds of cometary water vapor in the atmosphere. But the simple fact was that instrumental errors, telemetry problems, and computer malfunctions occurred far too infrequently to account for atmospheric holes. Our opponents challenged us on this point repeatedly during the year and a half that the debate took place in *Geophysical Research Letters* and we were forced to tread this ground over and over again. But rather than simply deny their claims, we utilized these opportunities to publish an in-depth analysis of the properties of these atmospheric holes. What we had found confirmed that the phenomenon was indeed very real.

Our opponents used several strategies to blame the appearance of atmospheric holes on instrumental noise. One concerned the size of the holes with respect to the satellite's altitude. If these holes were for real, they argued, then they should appear much larger—and consequently cover many more pixels in the images—when the satellite was at its closest approach to the Earth. Talbot Chubb of the Naval Research Laboratory in Washington, D.C., was the first of many to make this argument. At its closest point, or perigee, Dynamics Explorer flew at about 350 miles above the Earth. At its farthest, or apogee, the satellite orbited at an altitude of about 14,350 miles. Chubb claimed that the holes were not any larger at the lower altitudes. We found that they were. And we had the pictures to prove it.

To anyone with a clear understanding of how our satellite acquires its images, these pictures and our Reply to Chubb were devastating. Our ultraviolet camera works much like a FAX machine in that it produces its images line-by-line. The camera obtains its lines one picture element at a time, pixel by pixel, in other words. Each pixel is a measure of the light intensity of a small portion of the Earth's dayglow and is recorded during 3.9 thousandths of a second by one of the camera's two count registers. The registers alternate in producing the pixels in each line. A line of pixels is recorded as the satellite makes one rotation, which takes six seconds. After each rotation, the scanning mirror which feeds the light intensities into the count registers is tilted another "step" to obtain the next line of the image. Each image consists of 120 lines of pixels and takes twelve minutes to transmit to the ground.

A careful examination of our images showed quite clearly that the majority of atmospheric holes were no larger than a single pixel when the satellite was at high altitudes. Only a few holes were two pixels wide at high altitudes and when this was the case the additional pixel occurred in the next scan line and next to the first dark pixel. It was as if we were tracking the movement of the cloud from one scan line to the next, like pointing a rifle at a distant moving object. This situation occurred ten times more often than what you would expect by chance. Sometimes the cloud could even be traced for three or four scan lines. This is just what you would expect if the water vapor clouds were far away and had an apparent speed of about 20,000 mph.

We also found two occasions in which the holes blackened three consecutive pixels in the same scan line. Both times the satellite was at low altitudes. These events occurred at a rate that was about forty times greater than chance. But this was the rate expected for such events, because the probability of sighting an atmospheric hole diminishes very rapidly at lower altitudes where a much smaller patch of the atmosphere is viewed than at higher altitudes. The rate of detection was also diminished at low altitudes by the fact that the spacecraft itself spent very little time there. Yet when a cometary cloud was spotted while the satellite was low, it appeared larger than when the spacecraft was high. This is exactly what you would expect to happen. So instead of appearing as a single pixel in a scan line as it did when the satellite was high, when the satellite was low, the cloud was seen as a string of dark pixels in a single scan line of the image.

But the clincher was this. A satellite at low altitudes is closer to the cometary cloud than at high altitudes. So the cloud would move too rapidly to appear in the next scan line when the satellite was low. By the time the camera records the next scan line six seconds later, the cloud would be far beyond the view of the camera. That is just what happened. We have no images of holes appearing next to one another in adjacent scan lines while the satellite was at low altitudes.

So Chubb's attempt to dismiss the small comets on the basis of an altitude dependence failed. But he covered himself with other criticisms. He argued that even though our instrument sampled the dayglow sequentially by two counters, both counters shared some of the same electronics. A failure in one of these shared parts, caused perhaps by a cosmic ray, could create an atmospheric hole, said Chubb. But we had long ago considered this possibility and rejected it. Our instrument was relatively immune to cosmic rays.

None of these criticisms upset me. But I was angered when Chubb said we had treated some of the atmospheric hole images in such a way as to make the holes look larger than they actually were. Not true—all the images had been processed the same way—the way the images from many satellites are processed, in fact. Chubb had also consistently misspelled Sigwarth's name in his Comment, putting an "o" in place of the "a."

I wrote to Dessler asking that these items be corrected. They were not matters of interpretation. They were straight factual errors. Dessler should have forwarded my comments to Chubb, but I do not know if he ever did. Perhaps I should have communicated with Chubb directly. I am sure that he would have corrected the errors. In any case, Chubb's Comment, errors and all, was published in the October 1986 issue of *Geophysical Research Letters*. It was very frustrating.

Bill Hanson was also convinced that our instrument was in error and that the barrage of hundred-ton comets hitting the Earth was a fantasy we had conceived. Earlier, he tried to show that a large population of small comets was inconsistent with the results he had obtained with our sister satellite, Dynamics Explorer 2. Then, just a couple of months later, he seemed to be after us again. In June he asked to have some of our images. He did not tell us what he was going to do with them. But we very promptly sent the magnetic tapes for the images down to him at the University of Texas at Dallas.

A few weeks later one of Hanson's colleagues, Bruce Cragin, showed up at the Holiday Inn in Iowa City for a meeting on another subject, Jupiter. But he had brought with him a paper saying that the atmospheric holes in our images were instrumental noise. I like Cragin. He is a good physicist, but he is heavily influenced by Hanson. He handed me the paper in my office and I could tell he felt a little nervous. I looked through it and saw that it dealt with an old problem.

We had spent three years trying to find a way for the instrument to take the blame for the black spots in our images and failed. Hanson, who was first author, along with Cragin and two of their colleagues, worked on the problem for just a few weeks. They found that "the dark pixels are primarily single pixel events at all altitudes" and concluded that the atmospheric holes were an instrumental artifact. This, despite their admission that they had been "unable to identify a design problem or possible failure mode in the counting circuit that could account for the dark pixels."

Dessler rejected their Comment on the grounds that Talbot Chubb had brought out essentially the same point earlier. But Hanson, like the others, was out to get us. He had lost the earlier round and now he was striking back. He seemed totally consumed by his efforts to find something wrong with our work. So he and his colleagues revised their paper. This time Dessler accepted it. I never understood why Hanson was allowed to publish two Comments. The usual ground rule is one Comment from any single individual or group. But then again, Hanson and Dessler are good friends. They had been at Lockheed together and then at the University of Texas at Dallas.

Their revised Comment was thinly disguised. It listed Cragin as first author and Hanson as one of the three other authors. The conclusion was much the same and so was much of the material. They still harped on what they perceived to be our altitude problem. They claimed that the frequency at which the black spots was observed remained the same regardless of the spacecraft's altitude. But if you look carefully at the results in their paper, the frequency of the atmospheric holes does vary with altitude. I really think they wrote the conclusion before they even saw the data. I sent them a letter directly pointing out that they had found an altitude dependence. But they ignored my letter.

Their argument did have one new thrust, however. They were aware that the pixels that comprised each image overlapped somewhat. So, they argued, if the camera is taking a string of pixels and one of them is dark due to an object of some kind, then the pixels on either side should show some signs of darkening as well. They were, of course, quite correct. They calculated that these adjacent pixels should be darkened by about nine percent but found that they were not darkened beyond about one percent. Since there was not enough darkening of the adjacent pixels, they concluded that the holes were simply noise.

Though their argument was essentially correct and presented with great force and conviction, their arithmetic and interpretation of the images were all wrong. When we did the same analysis, we found that there was in fact some darkening of the adjacent pixels. But it was not the same amount of darkening Cragin and Hanson had expected to find. It was not the amount of darkening expected if the cometary cloud was all dark. Granted, if the cloud was all dark, then the pixels did not fit. But Cragin and Hanson had forgotten that an object slamming into even the thinnest of atmospheres would also have to glow.

A careful examination of the pictures from Dynamics Explorer shows that the clouds of cometary water vapor are not merely dark. The black spots are sometimes accompanied by bright features. When such bright features appeared around the dark spots in the Viking images, they had been dubbed "bagels." Even though the resolution on our imager did not compare with Viking's, the bright features were still visible in some Dynamics Explorer images. We had mentioned these bright features at several of our talks. These cometary clouds would have to glow a little as they plowed through the upper atmosphere at twenty times the speed of sound. But Cragin and Hanson failed to take this feature into account and had assumed that the holes were strictly dark objects.

Their model did not accurately describe what was seen in the pictures. Obviously, if the cloud also has a bright feature to it, the darkening of the adjacent pixels would not be as large as Cragin and Hanson had assumed. We showed that if you took these bright edges into account, you would get exactly the amount of darkening in the adjacent pixels that you see in our images. Rather than trying to impose our conclusions on the images, we let the images draw the conclusions for us. And in this case it was obvious that the instrument had provided substantial evidence for the observation of small, opaque objects that glowed along the edges that took the impact of the atmospheric gases.

We had begun the computer programs involved to reach this conclusion long before we had received the Cragin Comment. Sigwarth had been working on it as part of his master's thesis. Much of the work that he did went into our Reply to Cragin. It would be, at four pages, one of our longest replies, and for Sigwarth one of the most challenging. The problem was straightforward, but it required the use of many images and much computer time.

We knew there was a luminous edge to these infalling clouds of cometary water vapor but we did not know their shape or form. We first assumed that each dark cloud would appear to be disk-shaped and surrounded by a bright ring. But such an object did not fit the actual darkening of the adjacent pixels in our images. The model that best fit the observed holes was of a dark disk with a bright half-limb. Sigwarth's thesis went into detail on how any change in the shape, brightness, or thickness of this limb would not produce what we were seeing in our images.

Sigwarth did a fine job. He worked out the problem correctly. He did it rigorously and without approximation. It was a time-consuming task that often ran into the wee hours of the morning. But by looking at enough objects, he was eventually able to build up a clearer picture of them. Sigwarth showed that the side of the water cloud ramming into the Earth's atmosphere would have a bright face, while the side facing the camera would be mostly dark with only a little light showing at its edge. The bright face of the cloud could either be due to the high-speed collision itself, or by a handshaking exchange of charges between the cometary water molecules and the oxygen ions in the upper atmosphere.

But Sigwarth's thesis went one step further. He made a prediction. If the bright spots are there and manage to show up against the relatively bright atmospheric screen when the camera is pointing Earthward, then when the camera is looking above the limb of the Earth into the darkness of the sky we should be able to see the bright face of these incoming cometary water clouds. The cloud's dark side would not be seen above the Earth's limb, because the sky itself is too dark. But the bright leading edge of the clouds should be visible as they come burning in through the atmosphere. So he went searching for these bright spots and he found them. The images are stunning.

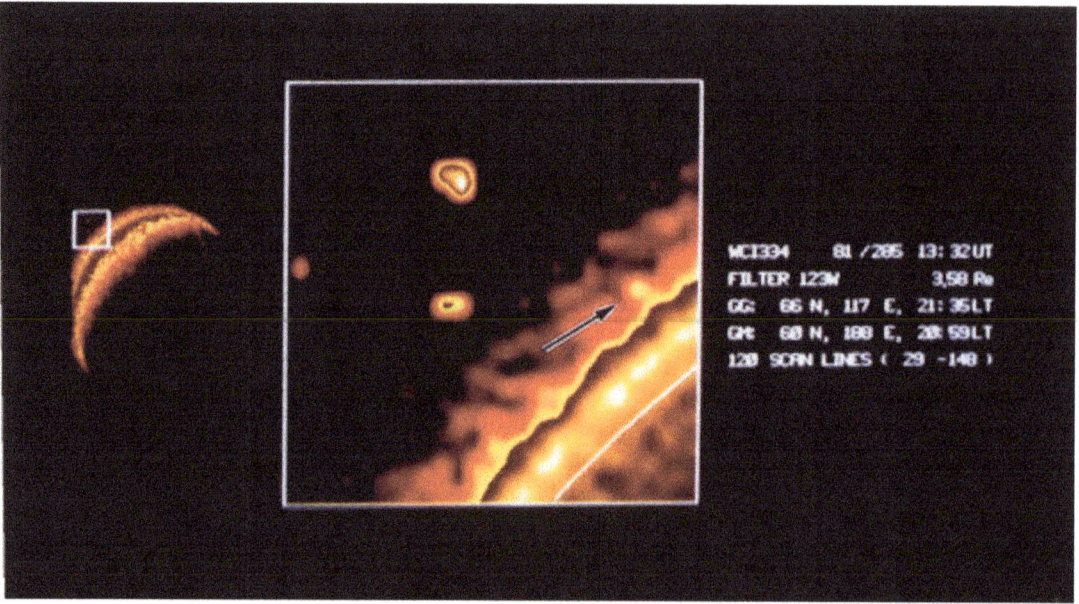

The cometary water clouds should have a bright leading edge from plowing through the thin gases in the upper regions of the atmosphere. The global image (left) shows the bright front-half of these objects as they collide with the Earth's tenuous atmosphere at high altitudes. The inset in the center is enlarged six times the original size. The left half of the inset shows a pattern of three stars. The arrow in the inset indicates the bright leading edge of a cometary water cloud.

The master's thesis by Sigwarth is a beautiful piece of work. It is the equivalent of a Ph.D. thesis. When he presented it at the American Geophysical Union meeting in San Francisco in December of 1988, there was standing room only. Despite all the negative comments, our small comet presentations were still drawing SRO crowds. And sometimes even standing room was hard to find. I was sitting a little way behind Hanson and he had a large grin on his face before the presentation began. He was unaware that I was sitting two rows behind him. But a minute into Sigwarth's talk, not even Hanson was grinning. It was tremendous. A lot of people came up afterwards to congratulate Sigwarth on a very impressive piece of work. I think Bill Hanson was very unhappy. I do not know if he had ever lost a public argument before. But here he had lost twice.

CHAPTER 14

Launch of the Artificial Comets

Not everyone was out to get me. Michael Mendillo was one of those rare other individuals who would conduct an experiment to shed some light on the small comet mystery. Mendillo is a professor of astronomy at Boston University and a space scientist for its Center for Space Physics. His work would show that our proposal was, at the very least, quite plausible.

On September 10, 1988, a Terrier/Black Brant sounding rocket carrying a special payload of explosive chemicals was launched from the NASA Wallops Island Flight Facility in Virginia, 150 miles from the nation's capital. Shortly after noon, the payload was detonated by ground command, releasing 584 pounds of water vapor, ice, carbon dioxide, and debris 185 miles up into the atmosphere. Almost four thousand miles above, Dynamics Explorer was watching. The satellite, which had been reprogrammed to take a series of one-minute images of the narrow region surrounding the release of the rocket's payload, managed to spot the event.

This was the third and final launch in a series of experiments that Mendillo had designed. The experiments were named ERIC (Environmental Reactions Induced by Comets), after his son. They were meant to simulate the deposition of a cloud of cometary water vapor in the atmosphere. It could not be done exactly, but you could release a water-producing bomb in the upper atmosphere, explode it, and see if you could pick up the event with the camera on a satellite.

You could not, of course, put as much water up there as we thought the comets were dropping down on us. You could take up a few hundred pounds, but versus a comet's one hundred tons, that was not much. Besides, Mendillo was simply depositing the water cloud at an altitude of less than 200 miles. The water clouds from the small comets pass by this altitude at great speeds and come to rest at an altitude of about thirty-five miles. But despite such differences, Mendillo's cloud should have produced the same kind of effect on the upper atmosphere as a small comet, and it did.

We first met with Michael Mendillo in San Francisco in December of 1983. John Sigwarth had presented our second talk on the mysterious black spots. Some people were starting to get excited about the subject. Mendillo was one of them. He approached us and suggested an experiment that might confirm our suspicion that the black spots in our images actually represented water vapor. What if you fired a rocket that would release a payload of water into the atmosphere? he wondered. How would the imager on Dynamics Explorer record such an event? Would it show up as a black spot also? It was an excellent suggestion. We were interested. But at the time we were too caught up in other projects to devote ourselves to it.

We were to meet again soon. Mendillo and I both had experiments aboard one of the last flights of the space shuttle Challenger, the one that carried up Spacelab 2 at the end of July 1985. Mendillo had an interest in the effects of water on the upper atmosphere and on this flight he and his co-workers used a ground-based radar to track holes in the ionosphere created by the water vapor exhaust from the space shuttle's engines. I knew then that he was determined to carry on his little experiment with us.

Mendillo is strong, ambitious and energetic. He is also a good physicist and conservative. I have a certain requirement in people. You have to be conservative. You have to know physics. And you have to work hard. Mendillo is one of those people. He wanted to know how small an amount of comet-like material would be detectable in the Earth's atmosphere and no one was about to stop him.

It never ceased to amaze me how quickly developments on the small comets front would spread. The nature of the development, be it positive, negative or neutral, mattered not. Every time anyone had anything to do with small comets, everyone seemed to know about it. Mendillo, with a smile on his face, told me about a strange little episode that occurred shortly after he had decided to go ahead with ERIC. During a conference someone had come up to him and began ranting and raving while he was standing at a urinal in the men's room. "Can you imagine believing enough in the small comets to go out and do something like this?" the man had told Mendillo.

But NASA was receptive to ERIC and funded Mendillo on three separate shots. All three were launched from NASA's Wallops Island Flight Facility. The first took place on the 7th of August 1987, and deposited 300 pounds of water, carbon dioxide, ice, and other gases high up in the atmosphere. Two satellites had been positioned to observe the event. One was the Dynamics Explorer satellite. The other was Polar BEAR (Polar Beacon Experiment and Auroral Research). Polar BEAR was orbiting at a lower altitude than our satellite, so we expected to get a better signature with it. But this little satellite was full of surprises.

Polar BEAR has had an unusual history. It was originally part of a series of navigation satellites for the U.S. Navy. Built by the Applied Physics Laboratory at Johns Hopkins University, these satellites proved to be so reliable and unlikely to need replacement that a "spare" was donated to the Smithsonian Institution's National Air and Space Museum in 1976. It spent the next eight years on display in the museum's Gallery of Satellites.

But in 1984 it was called back into service. The Applied Physics Laboratory made this unusual request after learning that it would no longer have Navy and Air Force satellites available for modifications. The museum was offered, and gladly accepted in exchange, a similar satellite of an even earlier design. The old navigation satellite was then refitted to monitor auroras, magnetic-field changes, and other ionospheric effects due to disturbances such as solar flares. It was placed back into service in November of 1986.

Nearly a year later, there was a failure aboard the satellite—just one day before Mendillo's first ERIC experiment. The mirror on the satellite's ultraviolet camera locked up. It could not scan the entire area in the atmosphere where the artificial comet was released. All it produced was a single line across the picture.

Mendillo and Sigwarth had worked very hard to find the right time to launch the rocket. They had to take into account not only the Earth's rotation but also the orbits and motions of the two satellites. A tremendous amount of coordination was necessary. The pictures had to be taken on the fly. Dynamics Explorer had to be as close to overhead as possible. From its high-altitude position, the satellite had no chance of picking up such a small event if an entire picture was taken. So in order to get as many images of the water release as possible during the few minutes it was up there, Sigwarth had to limit the scan of the mirror used by the Dynamics Explorer's camera. The cameras on both satellites were programmed to scan over the section of the east coast of the United States where the water would be released. The launch and water release had to be timed perfectly. Everything had to be done just right. It was a fabulous job. I really had little to do with it, other than biting my nails over how much they were limiting the viewing area of the camera. They could so easily have missed the target area entirely.

We saw a little signature with our satellite on the first ERIC shot and so did Polar BEAR, miraculously enough. There was not a lot of water in the payload so we expected the signature would be small, and indeed the black spot that resulted was quite small. Unfortunately, this decrease we observed in the ultraviolet light was within the limits of normal variability. So perhaps that little black spot was due to chance, but then again, perhaps not. The results were tantalizing.

With Polar BEAR out of the action, Mendillo searched for a backup for Dynamics Explorer before the second ERIC launch. He decided to use the Millstone Hill radar operated by the MIT Haystack Observatory in Westford, Massachusetts. Mendillo thought that this radar installation, operated by John Holt and David Tetenbaum during the payload's release, might be able to track the disturbance caused by an influx of comet-like matter in the ionosphere.

The second ERIC launch deposited some 430 pounds of water vapor and carbon dioxide about 180 miles up in the atmosphere. Again it was observed by our satellite and again the signature was a small one. But the cloud also produced atmospheric effects that were picked up by the Millstone radar. Within minutes of the release, the cloud caused a well-defined ionospheric depletion, or hole, that was monitored by the radar for about one hour. The location of this ionospheric hole coincided with the coordinates of the darkened spots on the Dynamics Explorer images.

I think Sigwarth and Mendillo were lucky. Each time the rocket drifted off one way or the other, almost enough for the camera to miss the region of the payload's release entirely. Each time I told Sigwarth to allow a little leeway and each time he said: "Oh yes." The first signature appeared at the edge of the image. Before the second launch I asked Sigwarth if he had put in more pad. "Oh, yes," he assured me. But that made me a little nervous. The second shot, on the 29th of January 1988, produced a darkened spot right at the edge of the picture. I asked Sigwarth if on the next attempt he would make the picture a little wider. Again we were lucky. The third shot, which took place the following September, also produced a weak dark spot in the images. But again the Millstone radar observed a well-formed decrease in the ionospheric density that coincided with the location of the darkened pixels on the Dynamics Explorer images.

Project ERIC produced good, if not spectacular, results. We observed the faint dark spots produced with all three injections of water into the atmosphere. And Mendillo found that if you scaled up the size of the atmospheric holes, or the darkening produced by these artificial comets, to those produced by the real comets, the amount of water required would be about a hundred tons. So the weak signature was about all we could expect with these small payloads.

This experiment did not prove our hypothesis. It was a calibration. It was enough to establish the fact that if you have water vapor up there, you are going to have a black spot on the satellite images. The radar results were also encouraging. Mendillo hoped to get an indication of how to take a radar and look in the ionosphere and get a unique signature for cometary material. He now thinks he might be able to use such a radar to pick up the intrusion of small comets into the atmosphere. But I believe the dynamics of the two situations are quite different. Mendillo was just laying water vapor up in the atmosphere, but there is no way of stopping the water vapor clouds injected at high speed by the small comets at those altitudes. Only a trace of the cometary water will be left at these altitudes.

It really did not matter, however, because no one paid any attention to the results. No one cared that during the third ERIC experiment, the camera aboard the Dynamics Explorer satellite captured the signature of one of the real water clouds from the small comets, just some fifty miles away from the area of the darkened pixels produced by Mendillo's artificial comet. Reality, it seems, had reserved itself the right to have the last laugh.

CHAPTER 15

Some Cometary Competition

I am getting a little ahead of my story because so many things happened at the same time. Just as the debate taking place in *Geophysical Research Letters* was coming to an end, Tom Donahue, like a jack-in-the-box, popped up with a surprise for everyone in the summer of 1987. I was in Vancouver at the time presenting a paper on the Earth's auroral lights at an international conference of space scientists. It was a nice sunny day and I was walking across the lawn going from one meeting to another. Suddenly someone rushed up to me. "Did you hear about the unscheduled paper Tom Donahue just presented?" I had not. "Everyone has been looking for you."

Donahue, I was told, now claimed to have found a new population of baby comets in interplanetary space. I was flabbergasted. This is the same Tom Donahue who spent more than a year trying to destroy the small comet hypothesis. Donahue had been fiercely critical of the small comets even before our original papers had been submitted for publication. Then, when he later refereed these papers, he recommended that they be rejected. And once the papers were published, he was also the first to submit a formal criticism of the small comets for publication. In fact, Donahue had spent more time than anyone else trying to debunk the small comets. And now, to everyone's great surprise, he had discovered a population of such objects on his own.

That was the good news. The bad news was that Donahue had said that his newly discovered objects had nothing to do with my small comets. I was taken aback. How could this be? This was all the more confusing because the data Donahue was now using as evidence for his new population of celestial objects, he had used just a few months earlier to show that no large population of small comets could possibly exist in interplanetary space.

Donahue fired this last shot in our direction in December of 1986. He argued, in a paper submitted to *Geophysical Research Letters*, that if the small comets were as numerous as I had proposed, no amount of protective mantle would be sufficient to halt the vaporization of the comets and prevent the flooding of the solar system with water. The action of sunlight on these watery comets, Donahue insisted, would release large amounts of hydrogen atoms in interplanetary space in the vicinity of the Earth. And because hydrogen emits ultraviolet light at a wavelength called Lyman alpha, he looked to measures of Lyman alpha taken by interplanetary probes for a sign of the presence of small comets.

Donahue reasoned that any brightness in the Lyman alpha greater than that normally produced by hydrogen atoms coming into the solar system from interstellar space could be attributed to the small comets and would be easily seen. But, he pointed out, spacecraft such as Pioneer 10 and Voyagers 1 and 2 failed to find even a small increase above the sky background rate as they left the vicinity of the Earth on their way to rendezvous with the outer planets. Donahue proposed that the only way to stifle the large production of hydrogen from the small comets would be to cover them with thick dust mantles. But the problem with thick dust mantles, as we already know, is that they would bring, said Donahue, a "forbiddingly large" influx of dust into the Earth's atmosphere.

My principal criticism, as a referee on the paper, was that the Lyman alpha glow due to the presence of the small comets would not be easily detectable. Lyman alpha is very difficult to measure. It is not at a wavelength that we can see with our eyes and it is hard to make the instruments that are sensitive to it very accurate. Whenever we measure the Lyman alpha it is against an ever-present background glow. This light results when the Sun illuminates the interstellar hydrogen that surrounds us. It is very bright relative to the dim glows we expect to see from the small comets. Add that to the fact that the Sun's illumination of this hydrogen constantly changes and you have a very difficult measurement.

Lyman alpha is a tease and I had long ago learned to stay away from her. The work of Ralph Bohlin, one of my former undergraduate honors students, had warned me. Bohlin was a bright young guy. He had gone on to obtain his Ph.D. from Princeton and then joined the Laboratory of Atmospheric and Space Physics at the University of Colorado. Bohlin wrote a paper pointing out some of the discrepancies in the measurement of Lyman alpha in a comparison of data from three spacecraft, Mariner 6, Mariner 9, and OGO-5. When I looked at his paper and those of others on the subject and saw those discrepancies, I realized I would be in trouble if I based my small comet figures on Lyman alpha.

Bohlin's work had warned me. If I was going to propose that water was coming off these comets into the interstellar medium, then I ought to stay a long way away from Lyman alpha measurements. I would have to get the information some other way. So I did an end run around Lyman alpha. I went to my own specialty for an answer to the amount of hydrogen in interplanetary space. I realized that if the hydrogen from the small comets was out there, it would not remain a neutral atom very long. Either a solar wind proton or the ultraviolet light from the Sun would strip off its electron. Once it was ionized this hydrogen atom would get picked up in the wind of protons and electrons from the Sun and eventually be swept out beyond the planets. So I looked to the concentration of stripped-down hydrogen in the solar wind to determine the upper limit for how much hydrogen the small comets could produce in interplanetary space. I used this upper limit to estimate the outgassing of water from the small comets and found there should be some excess hydrogen light. But at most it should be only two to four times greater than the upper limit given by Donahue from the Voyager spacecraft.

Then in the spring of 1987 Donahue went back to the Voyager data to assure himself that no excess hydrogen light had been detected by Voyager. He hoped to deal the small comets their final blow. But instead he found this great cloud of hydrogen around the Sun in the vicinity of the Earth. It was this discovery that had led him to propose a new population of comet-like objects in interplanetary space. The proposal shocked everyone, I think, but no one more than Donahue himself.

When I finally ran into Donahue at the conference in Vancouver he let me see the paper that had caused the uproar. He called his objects cometesimals. These are distinctly different from small comets. Donahue's cometesimals are the rocky cores of comets in the making. They are baby comets. My small comets are comets in their own right. They are snowy on the inside and surrounded by a protective crust like Halley's, only much smaller.

I had been very careful in naming these objects. I could have called them cometoids, but that sounded too much like hemorrhoids. Some people have called them mini-comets or CLOs, meaning Comet-Like Objects. But others, aware that comets normally take their names from their discoverers, insisted on calling them FSCs, after the three names on the original papers, Frank, Sigwarth, and Craven. Still others have jokingly referred to them as FSCs because they are Frank's Small Comets. I think the term small comets is best.

Donahue's cometesimals were remarkable creatures. They were ingeniously designed so that, unlike my own small comets, it was their first trip through the solar system. The cometesimals were new bodies with a nice, pure-ice coating. Just enough, in fact, that by the time they reached the Earth's orbit the Sun would have melted all the ice away. I guess it was possible to do that. With a little imagination one can construct almost anything. But there must have been good quality control somewhere out there while his cometesimals were being painted with a uniform coating of ice, the way peanut M&Ms are evenly coated with chocolate and sugar. Donahue and his colleague at the University of Michigan, Tamas Gombosi, and Bill Sandel of the University of Arizona, wrote a paper on the subject for *Nature*.

The ultraviolet spectrometer aboard Voyager 2 had taken a series of seventeen observations of interstellar Lyman alpha from hydrogen as the spacecraft sped away from Earth in 1977. The figures showed a rapid decline in Lyman alpha as the spacecraft travelled from a point about 7 million miles away from the Earth to one that was about 140 million miles away. The initial measurements showed an excess brightness that ran about one-third greater than the light due to interstellar hydrogen but eventually dwindled to nothing as the spacecraft receded from the Earth. Clearly this indicated the presence of a cloud of hydrogen near the Earth's orbit. The most reasonable explanation for it, Donahue concluded, is that a large number of comets were losing their ice as they passed through the inner solar system.

Donahue's cometesimals bore a striking resemblance to my own small comets. His had a diameter of 26 to 164 feet. Mine measured 40 feet on average. His weighed 30 to 30,000 tons. Mine weighed about 100 tons. His comets were small enough to have escaped detection by ordinary telescopes. So were mine. But his were boulders with thin ice jackets. Mine were snowballs with thin carbon mantles.

Where we differed the most, however, was in the number of objects. His proposal called for a hundred million fewer comets than mine. His rate, Donahue explained, was consistent with the number of craters on Mercury, Mars, and the Moon. Donahue had, in fact, limited the number of his cometesimals through a count of the number of lunar craters that such objects were likely to produce. He found that his cometesimals would account perfectly for every lunar crater measuring up to about 5,000 feet in diameter. These craters were previously assumed to have been formed by stone or iron meteoroids.

Donahue's cometesimals would have less of an impact on Earth than my small comets. His baby comets would hit our planet only once in every eight years, while the largest ones would impact about once in every 10,000 years. My small comets are hitting the Earth's atmosphere at a rate of about twenty per minute and have deposited enough water here over time to have filled the ocean basins. But the total cosmic rainfall from Donahue's cometesimals would be insignificant. They would not have brought the ocean waters to the Earth.

His discovery had not caused Donahue to change his mind about my small comets. His was the simpler hypothesis and therefore the best, he concluded. In a far more detailed exposition of his proposal that he prepared for *Icarus*, Donahue actually invoked the rule of Ockham's razor in support of his cometesimals. William of Ockham was a 14th century scholar who said that "entities must not needlessly be multiplied."

Modern science has since interpreted these words to mean that if two theories fit equally well all the observed facts, the theory requiring the fewest or simplest assumptions was more likely to be valid. Most scientists do not use this razor because its blade often cuts the wrong way.

Donahue's call for Ockham's famous razor to decide between our two proposals puzzled me. But it infuriated George Wetherill, the world-famous scientist from the Carnegie Institution of Washington who also refereed Donahue's paper for *Icarus*. "Yuk!" wrote Wetherill. "This is an argument that is useful only to an uninformed audience whose poor educations left them with the impression that a medieval scholastic understood inversion theory better than modern scientists. Usually it is used as an implicit admission of a lack of a good reason for rejection of someone else's hypothesis, but to have to use it against Frank's holes???!" Obviously, Wetherill was not fond of my small comets either.

It seemed to me that Donahue had completely reversed his previous position. He might be denying the existence of my small comets, but at the same time he had entered the cometary competition with an icy celestial population of his own. I was no longer alone. I was being faulted for numbers that were too high. Donahue was being faulted for numbers that were too low. Some people were beginning to think that the truth might lie somewhere in between. "I think that Donahue's work has had the effect of giving great credence to the existence of this class of objects," Dessler said at the time. "Now they are only dickering over the numbers." Never mind that our numbers were a hundred million apart.

Despite my excitement, I was wary of the tease of Lyman alpha and I wondered why the measure of hydrogen light from Voyager failed to agree with the measurements from other spacecraft. When Halley's Comet came rolling through the solar system, various satellites were turned towards it. One that could look at the hydrogen gas around the comet was Pioneer Venus. Ian Stewart of the University of Colorado headed that effort. We also looked at Halley's with Dynamics Explorer. My colleague John Craven did the analysis. What we learned from these observations was that the hydrogen envelope around Halley's is enormous.

The measurements from Dynamics Explorer and Pioneer Venus agreed. This meant that the two instruments reported back to Earth the same brightness for the same object. This was also true for the background light from the interstellar hydrogen that was falling into the solar system. But the Voyager instrument, the one that Donahue had used to search for cometary water, reported glows from the same interstellar hydrogen that were two or three times brighter than these. That is a big difference, especially if you are going to claim you can measure something to within a few percent and you fail to agree by 200 or 300 percent with what everyone else says is going on. Donahue was aware of the calibration problem. "But," he said, "who's right?"

Dessler got angry when Donahue's paper appeared in *Nature*. Donahue had not even submitted it to *Geophysical Research Letters*, where the debate had been taking place. The *Nature* paper got a lot of publicity, and this no doubt made Dessler very jealous. But, in any case, it did not take long for everyone to repeat Donahue's blanket statement that this newly found hydrogen had nothing to do with my small comets. But if there was all this hydrogen gas out there, how could anyone really know what it came from? It could come from automobiles or sick cows. The real significance of Donahue's finding was that there were objects out there that had never been detected before. It did not matter whether they were small comets, cometesimals, automobiles, or sick cows.

CHAPTER 16

The Verdict

I felt obliged to respond to Donahue's discovery of cometesimals. His paper had appeared in the British journal *Nature*, one of the great keepers of current wisdom in science in the Western world. Current wisdom is what a large body of scientists in a given discipline believes and holds sacred. Current wisdom is usually—but not always—correct. Most scientists will jump through hoops to get their papers published in *Nature* and its American counterpart *Science*, as both journals are very popular and widely distributed. But I have had little to do with either one.

I was published in *Nature* as a junior co-author with James Van Allen some three decades ago. And recently I had a paper on a subject not related to small comets rejected by *Science*. It was returned to me without review or comment. For the keepers of current wisdom it seems that the small comets are now the equivalent of bad breath. It matters little to me. I prefer to publish in journals where you can generally count on the competence of reviewers; this has not been my experience when the journal is a keeper of current wisdom. For my work, a competent reviewer means a peer with good skills in basic science.

I am not very good at responding effectively to folklore and faulty current wisdom. I was reminded of that when I tried to respond to Donahue's claim that the excess hydrogen found with the Voyager observations had nothing to do with the small comets that I proposed, even though they are what led him to search for the hydrogen in the first place. What troubled me was his claim that only his model, his number of objects, could account for the hydrogen. The simple fact is that a large number of object sizes and types could account for the Lyman alpha, the dim glow from the thin hydrogen gas in interplanetary space.

So I decided to write a paper saying, in essence, that the small comet hypothesis could also satisfactorily explain Voyager's Lyman alpha emissions. In fact, almost everyone in the United States could have their own model of the number of objects and how their surfaces behaved that would give the same hydrogen light as seen by Voyager. There was nothing miraculous about the contents of my paper. Several colleagues recommended that I submit it to *Nature*, given that Donahue's paper had been published there. My paper was an alternate interpretation of the excess Lyman alpha brightness and *Nature* seemed to be the appropriate place for it. I should have known better.

The referees assigned to my paper made it quite clear that they were not interested in seeing anything else on the subject. The first referee thought that *Nature* should decide "whether the small comet proposal has been so discredited that further discussion of it" was even appropriate. More specifically, he wanted me to explain how the comet mantles would adequately suppress water production and what happens to the water from the comets inside

Earth's orbit. I had tackled the first problem earlier in *Geophysical Research Letters*. The second problem I took care of in my revision. They had me extend the computations far beyond what Donahue had done in his paper and I satisfied all the requirements.

The first referee also raised one of the most ridiculous objections I had seen in some time. He criticized our paper because we had not proven that ice melts when it is heated above its melting point. What world had I entered? How do you answer such a question? If you heat ice above its melting point will it melt? So I answered that in my experience every time I had heated ice above its melting point it had turned to liquid water. But perhaps the referee had another experience. It was quite apparent that his emotions far exceeded his common sense.

The second referee agreed with the first, except on the melting point of ice. He wrote that I had failed to provide "a compelling proof of the mini-comet hypothesis." That, of course, was not the point. He noted that I had referred to many of my arguments from previous replies to criticisms of our work and that these replies had not been reviewed by others before publication. This led him to conclude that we had "built a self-referenced, unreviewed, house of cards that will fall to pieces if even one of their previous replies is incorrect." I had always said myself that only one bit of evidence was needed to prove the small comet hypothesis incorrect. That in fact was what the debate was all about. But so far no one had been able to come up with it.

The comments by the second referee were also personally insulting. This is often the case with less than top-notch referees when they fail to understand the contents of a paper or are emotionally attached to current wisdom. The fact that they are anonymous sometimes brings out the worst in them. They seem to relish this sort of lion-hunting in a cage. The referee said I was "using hand-waving arguments and approximations," and on seeing my reply to the first referee, accused me of Nixon-like behavior. I was stonewalling, he said.

Thinking the British editors of *Nature* so naive as to be unfamiliar with the term, the referee provided an in-depth explanation: "Stonewalling is the process of repeatedly denying an unacceptable statement in the hope that the critic will get tired and give up. It was used most notably by the Nixon administration when they denied any involvement in Watergate, no matter what evidence was presented to the contrary." This was clever, but it was not science.

So the paper went back and forth, back and forth, and then came the final decision from the editor in charge of the *Nature* office in the United States, David Lindley. He rejected the paper. "At this stage," he wrote, "we are satisfied that we have taken a representative poll of current opinion in this field, and the verdict goes against you." I could not believe it. My paper was rejected not on the basis of something being wrong with the arithmetic or something being wrong with the science, but because *the vote went against me.*

I was astonished by the arrogance of Lindley's statement. The verdict had gone against me. You would think that they could have voted the small comets into office had they been popular enough. But on second thought I must say that I became very enthusiastic about this "physics by vote." This kind of voting process would allow us to dispose of the laws of thermodynamics. Perhaps next year the community would revoke them and I could look forward to becoming younger. It would also mean a quick end to the energy crisis.

It was mind boggling. I thought that science was what was correct and verifiable by experiment and observation. Now, suddenly, it was being done by the vote of the members of a given community. This was a very distressing learning experience. To think that Newton's laws might be voted out. Perhaps someone would take Ockham's razor and slash them from our universe.

I had one other brush with *Nature* during the small comets controversy. It occurred just as my own paper was being reviewed at *Nature*. They decided to prepare a story on the small comets for the "News and Views" section of the journal and Paul Feldman was asked to write it. Feldman is a professor of physics and astronomy at Johns Hopkins University in Baltimore and is one of the country's leading experts on comets.

At the time, Feldman and I were competing for an imager on the NASA Polar spacecraft due to be launched in 1993. The camera had nothing to do with small comets. The competition was intense because the opportunity to fly an instrument into space these days is so rare. Feldman is a good physicist. He is bright and energetic. He was not too familiar with the small comets, but he understood what they meant for the solar system and for science. Feldman accepted the assignment.

Feldman told me that he would treat the small comets in a kindly fashion but the story he wrote was downright nasty. Feldman showed a masterful use of the thesaurus, but he put absolutely no science into it. He apparently relished poking fun at me. He compared me to René-Prosper Blondlot, who in 1903, was the only person able to observe the non-existent N-rays.

He also labeled as "Velikovskian" the way small comets were able to account for so many of the solar system's mysteries. He was referring to the late Immanuel Velikovsky and his controversial theories of recent catastrophes in the solar system. Velikovsky's use of the world's myths and legends as proof of his theories has long been ridiculed by science. Feldman also accused me of constantly "reengineering" or adjusting the properties and characteristics of the small comets as I responded to each of my critics. But Feldman could not have read my papers because there was, in fact, no change.

I did not mind. Feldman is a friend. It is difficult for me to be angry with him since he is as child-like as I am when it comes to things of this sort. I got what I expected. We compete nose to nose. He wins some. He loses some. That's life. He got to write that article for *Nature*. I got to do the imager on the NASA Polar spacecraft. Feldman told me privately afterwards that he would never write anything like that again.

I am a member of the American Association for the Advancement of Science and have a profound respect for its journal, *Science*. It claims to represent "the advancement of science." So the news reporters for *Science*, such as Richard Kerr, wield an awesome amount of power. They are given a free hand in their work.

Kerr first approached me to do a story on the small comets at the American Geophysical Union meeting in San Francisco in December of 1987. I told him that I was not particularly interested in having the topic covered in *Science* because the debate was not going on there. Kerr is a very knowledgeable fellow. He has a tremendous task, of course. He has to come up with a couple of articles a week on a myriad of topics.

But write about the small comets, he did. He wrote three stories in less than a year. Kerr thought the idea of small comets absurd. It was, of course, his right to think so. If I were in his shoes I would probably say the same thing. The odds were not in my favor. Why gamble the other way? It did not bother me that he so openly expressed his opinion on the subject. What irritated me was that his articles contained factual inaccuracies. It does not matter when this happens in a little newspaper. But when you get a periodical as prestigious as *Science* publishing something that contains inaccuracies, something has gone seriously wrong.

I let it go the first time. But his second article had even more errors than the first. So for the first time in my life I wrote a letter-to-the-editor trying to set the record straight. It was a futile effort but it did allow me to vent some frustration. There were several factual errors in the Kerr stories. The first error was that he made John Craven a graduate student. John Craven is a senior research scientist. He has not been a graduate student for about twenty-five years. He also stated that the small comets with their protective crusts would release ten million times more hydrogen in interplanetary space than observed, when in fact this was the value for comets with no protective crusts—a big difference.

This information—misinformation rather—was apparently acquired through hearsay, which is not unusual for such a heated subject. Kerr's second article also mentioned that a workshop had been called in Boulder because of Donahue's discovery of the cometesimals. Kerr was not there. I was. That was not the purpose of the meeting. Nor did he mention the very significant outcome of the meeting: that the most important goal for the Galileo spacecraft during the initial part of its journey should be to try to make measurements of the Lyman alpha light from hydrogen in the Earth's vicinity. It was just sloppy reporting. This type of journalism is not a service to readers. I was very disappointed. But I learned that when it comes to heated topics, there was obviously no guarantee of accuracy and fairness, even in the scientific press.

The letter had some effect. Kerr began calling around. He learned that Craven was indeed not a graduate student. *Science* printed a correction in the tiniest possible print in one of its subsequent issues. No other mistake was corrected. But Kerr's next article on the subject showed a marked improvement. I have to compliment him because in the end he tried to correct things. He tried to find out more about the issue. But the whole episode made me question the other news in *Science*. Could I still trust them to tell me what was going on, especially on a dispute I knew little about? Would I get the real picture when I read about a controversy in medicine? It was another loss of innocence on my part.

But my biggest disappointment with the press came not with these big fish but with one of the smaller fish in the scientific pond. I belong to the American Geophysical Union and have been to its meetings since I was very young. I have actually been chosen from its large group of members as a Fellow of the Union.

The Union has a weekly paper called *EOS*. It reports everything from Nobel-Prize winner Hannes Alfvén getting the Union's top honor down to who stubbed their toe last Tuesday. The small comets issue has involved more people and caused more controversy and more discussion than most issues that come up in the Union. But one thing that hurt me—and most of these things do not bother me—is that I was so heavily censored on the small comets issue that not one word has ever been written in that newspaper about the topic. Not a single word. Others have asked about it and they are as amazed by the situation as I am. Meanwhile, reports on my other work have continued to appear in *EOS*. Perhaps the editors thought that the issue of small comets was of no importance to the studies of the Earth.

Sour grapes? Yes, in part. It has been a very frustrating experience for me. People tend to forget that scientists are human, too. Though I cherish the world of science and its complexity, my childhood on the plains of Kansas and in a small rivertown on the Mississippi was spent with a different kind of people, in a different kind of world. In the course of this debate I would sometimes return to this other world to forget my frustrations with the intrigues and the public ridicule surrounding the small comets. I would jump in my truck or my car and head the sixty or so miles east to the Mississippi and its rundown taverns and smoke-filled pool halls.

The real attraction for me there was the people. They are truck drivers, mechanics, businessmen, even Hell's Angels. These people accept you as you are with them and not for what you might be elsewhere. No backdoor scheming, no abrasive confrontations, just a simple and direct interaction. Two very different worlds. There is no basic difference in intelligence, only a matter of education, which is lost here anyway. A short while spent with these people renewed my contact with the basic realities. It was a deep pleasure, a moment away from the paddling line.

CHAPTER 17

The Competition Fizzles

The sides in the comet competition were clearly drawn. It was Donahue versus Frank. The inner solar system was not big enough for the both of us. We knew that a sign in the heavens could declare one of us the winner, but this sign unfortunately was not visible from the Earth. What we needed could only be had from interplanetary space: a portrait of the hydrogen light from these objects in the Earth's vicinity.

If the cometesimals were real, this portrait would resemble a shell of hydrogen left over from the vaporization of these objects by the Sun as they reached the Earth's vicinity. Consequently, this shell would be centered on the Sun and located in Earth's orbit. If the small comets were real, the portrait would instead show up as a torus, or ring, because the small comets, unlike Donahue's cometesimals, are confined near the planes in which the planets travel about the Sun.

Donahue and I knew that this observation of the Lyman alpha light in interplanetary space could only be made with the proper spacecraft. That spacecraft, we realized early on, would be Galileo. It was scheduled to be launched by the space shuttle Atlantis in the fall of 1989. Galileo was a mission to Jupiter. It was designed to drop a probe into the planet's atmosphere while the orbiter itself returned pictures and other data on the planet's atmosphere, the powerful radiation zones that encircle it, as well as its many moons. Because Galileo missed its scheduled flight date due to the Challenger disaster, it would have to swing by Venus and Earth for gravitational boosts to reach Jupiter. It was like an interplanetary pinball game. But instead of two years, it would now take six to get to Jupiter. The spacecraft was never meant to go to Venus, but now that it had to make that flight, no one wanted the side trip to be a waste.

There was much work to do. In June of 1988 some of the scientists on the Galileo mission gathered for a workshop at the University of Colorado in Boulder. I was there. Donahue was there. Altogether more than fifty scientists had been invited. Practically everyone interested in interplanetary hydrogen attended. A few were curiosity seekers looking for a show of tempers. Don Hunten of the University of Arizona chaired the meeting. Its sole purpose was to discuss and establish priorities for Galileo during the Venus to Earth portion of the mission. There were a dozen different experiments on the table but only a limited amount of data storage available on the Galileo tape recorder. During this time we would argue whether the spacecraft could do anything useful on the long-debated topic of just how much hydrogen there was in interplanetary space. The fate of the small comets and the cometesimals hung in the balance.

The best time to look at the hydrogen light is on the way to Venus and back again, and

Galileo was well equipped to make this observation. There would be two ultraviolet instruments on board and each was capable of measuring the ultraviolet light from hydrogen known as Lyman alpha. One of them, the University of Arizona instrument, would be making observations at a very low rate, but these would be transmitted to Earth immediately. Observations from the other ultraviolet instrument, the pride and joy of the University of Colorado, however, would have to be stored on the spacecraft tape recorder. These observations could not be transmitted to Earth until the spacecraft had returned from Venus, about one year after the launch of Galileo.

The Universities of Colorado and Arizona are fierce competitors in this ultraviolet business, but curiously enough Colorado had helped Arizona get its instrument on board. Arizona's ultraviolet imager was actually an unused spare from the earlier Voyager mission and was added to Galileo when the Challenger disaster delayed the mission. I, too, had argued for it, and John Casani, the Galileo project manager, had also been extremely supportive. It did not matter to me that this was one of the groups that had hurled the most severe criticism at me during the small comet debate. I thought that it was in the best interest of science. I felt it would be very worthwhile for imaging the great doughnut of gases and plasma that encircle Jupiter, so I worked hard to get their instrument on board.

It was an easy decision. Science first. This is the tie that binds. And its threads run throughout the Galileo project. Casani, who I consider to be the world's best space project manager, is interested in advancing science as well as implementing a successful spacecraft. We had first worked together when I was twenty years old. He was a younger engineer on the first lunar probes. He would bring the tiny Pioneer 3 and 4 spacecraft to Iowa and we would calibrate them with an x-ray machine. James Van Allen was the principal scientist. It would usually take all night and at the end we would sit on the steps outside the lab and, with beers in our hands, watch the Sun rise.

Just a few days before the Boulder workshop, two scientists from the University of Arizona dropped a bombshell. Donald Shemansky of the Lunar and Planetary Laboratory had been amassing spacecraft observations and began to have doubts regarding Donahue's claims of an excess of Lyman alpha in interplanetary space. So he had asked his student Doyle Hall to compare Donahue's results with the results from their own independent inspection of these measurements. Doyle did not find an excess in Lyman alpha light and his upper limit was only a small fraction of that claimed by Donahue, if indeed there was any extra light at all. Hall and Shemansky concluded that Donahue had erred.

I read a copy of Hall and Shemansky's as-yet-unpublished paper that was circulated at the meeting and I found their claim of how well they could measure the Lyman alpha light a little overconfident. But I think they had carefully and correctly determined that Donahue had indeed made an error. Donahue said that he would go home after the meeting and redo his calculations. But everybody there seemed convinced that he had made a mistake. The news did not bode well.

But in the end, Donahue's error did not affect the group's decision. The group recommended, in a paper known as the "Hunten Report," that the most important thing for Galileo to do between Earth and Venus was to search for the hydrogen from unknown objects. "The primary objective has high priority whatever the actual comet population," wrote Don Hunten. Everyone realized that this was the perfect opportunity to get good measurements of the

Lyman alpha light in interplanetary space. "Controversial or not," Hunten wrote, "the [small comet] hypothesis is certainly interesting and deserves to be tested."

After the Boulder meeting Donahue checked his calculations and found that Hall and Shemansky were correct. "I blew it," Donahue told *Science*. He explained that a graduate student had committed the error. The student had made a slip in a decimal point. It had been a clerical error, in other words. But I looked at the author list on the original Donahue paper. There was Tamas Gombosi, a senior research physicist at Michigan, and Bill Sandel at the University of Arizona. But there was no graduate student. In the acknowledgments Donahue thanked Tom Ahrens, but he is a professor at Caltech, and Harold Masursky, who is well known for his work on the channels of Mars. I was surprised that a graduate student who had so much to do with the result had not even been acknowledged.

But perhaps it was a graduate student's fault. Donahue, at the time, was very busy as chairman of the Space Science Board. This is a tremendous responsibility, so perhaps he had to rely on others and did not have a chance to directly oversee the data analysis and computations. But in matters like this, the advisor is not only responsible for checking the work, but also, ultimately, for the conclusions. It was a very embarrassing situation.

The final version of the paper that Hall and Shemansky submitted to *Nature* in July began with a statement saying that the excess hydrogen light found by Donahue with Voyager was about ten million times less than what would be expected from my small comets. I could not believe it. Donahue had made the error but right away I was being put on the spot. The Lyman alpha light would have been ten million times brighter only if the small comets had no mantle whatsoever. Never did I say that the comets would have no mantles. But I suppose Hall and Shemansky had to come up with something startling—even if it was incorrect—in order to get published in *Nature*.

In their eagerness, Hall and Shemansky went on to cite a nonexistent paper, which I had supposedly written for *Geophysical Research Letters*, on how Donahue's Voyager data fit my small comets proposal. This was the paper I had submitted to *Nature*, not *Geophysical Research Letters*, and it had been rejected. I had written to Shemansky in August informing him of this incorrect citation and asked that it be corrected. But it never happened. The whole shooting match—incorrect citation and misleading figures on Lyman alpha from the small comets—was published in *Nature* in September.

The mistake rapidly snowballed. Richard Kerr, writing for *Science*, picked up the information for his story from the Hall and Shemansky paper without bothering to check with me to see if it was right. This imaginary "off by a factor of ten million" would later come back to haunt me at the next Galileo Project Science Group meeting. It was quite a mess.

I would think the Hall and Shemansky affair caused Donahue the greatest disappointment. His cometesimals had nothing left to stand on. But I have to give him credit. At least he had gone in there and tried to resolve the issue. Few had done anything at all. But he had worked very hard at it. So it did not surprise me that, in the end, Donahue tried to take my small comets down with his cometesimals. When Hall and Shemansky proved him wrong, Donahue said, in effect, *Well, that kills Frank's small comets, too*. Never mind that when he had first proposed his cometesimals, he had insisted that they had nothing to do with the small comets. It was a clever move by Donahue. People tell me that was not fair. But fairness had nothing to do with it. There was nothing dishonest about it either. People fail to understand that science, quite simply, is a savage little game that we play with a great deal of devotion.

The final go-ahead to search for the hydrogen from a sea of small comets had to be made by the entire group of Galileo scientists at a Project Science Group meeting at the Jet Propulsion Laboratory in Pasadena the following January. The membership of the PSG, as this group is known, is truly impressive. It includes James Pollack of the NASA Ames Research Center, who helped develop the notion of "nuclear winter"; James Hansen, head of the NASA Goddard Institute for Space Studies, who believes that global warming is happening now; James Van Allen, who is the recent recipient of the Crafoord Prize, considered the equivalent of the Nobel Prize in space science; Andrew Ingersoll of the California Institute of Technology, who is best known for his insight into the weather of planets other than Earth; and Christopher Russell of UCLA, who has instigated the heated debate on whether or not there is lightning on Venus. Some of the members of the PSG have instruments on Galileo. Others simply intend to use the data from these instruments to develop their own ideas on the nature of Jupiter, its moons, and its immense radiation zones. I am a member of the PSG and have an instrument on Galileo that will look at the ions and electrons in Jupiter's radiation zones.

A PSG meeting lasts a day or two. The stresses and tensions increase as the meeting progresses. Strong personalities are involved and what one group wants to do with the Galileo spacecraft often conflicts with the desires of others. Galileo could not possibly satisfy all of their demands, even though it is probably the most advanced space exploring robot to be flown before the turn of the millennium. But this is where the buck stops; all conflicts must be settled at these meetings. It is not a pretty process. The cheery smiles seen on the first day soon give way to faces etched with grim determination. As the tension builds, there are abrasive exchanges of insight and wit. But compromise and resolve must prevail in the end, as every bit of treasure is squeezed from the two-billion-dollar Galileo mission.

Everyone at this PSG meeting in Pasadena was fighting for resources. I was part of a long parade of scientists, all pitching their own projects for the Galileo mission. The competition for the limited data space on the spacecraft's tape recorder during the Earth-to-Venus-to-Earth leg of the journey was especially intense. Some people were arguing that the instruments should be used to study Venus lightning. Others wanted to use them to study the motions of the Venus atmosphere. Everyone was trying to get a slice of the pie. And, as expected, the idea of using the severely limited time on the tape recorder for observations of interplanetary hydrogen produced by the small comets quickly came under attack.

I got up to defend the taking of Lyman alpha measurements for interplanetary hydrogen. Mike Belton, who is in charge of the imaging on Galileo, pointed out immediately, for anyone not already aware of it, that I was the small comets person. Jim Pollack followed Belton. He asked why we were even discussing the small comets since no one had ever shown they were real. There is no supporting evidence for them, he said. Pollack was biased. We all were. If Galileo did not do the Lyman alpha measurements, Pollack could get his measurements on the Venus atmosphere.

Then Belton whipped out the Hall and Shemansky paper and said we did not need to go look for hydrogen in interplanetary space from small comets. He quoted Hall and Shemansky as saying that if the small comets were real then the Lyman alpha would be about ten million times larger than had been previously detected anyway. I expected that. We used everything we could against one another. But I pointed out that Hall and Shemansky's statement was incorrect. Belton got a little angry at me. I got up and asked him what else he was going to quote. There was no response.

I decided to sit down. There was no reason to stand up and face the wrath of this group in support of science to be done by instruments belonging to the Universities of Colorado and Arizona. It was a no-win proposition. I half expected some words on my behalf from my arch-critics at the University of Arizona. After all, they had gotten their instrument onboard Galileo only recently and, in part, due to my efforts. But to my surprise, it was Ian Stewart, in particular, and Charles Hord, both from the University of Colorado, who strongly supported the search for hydrogen in space. It was their direct support of a search for small comets that would assure time on the Galileo tape recorder. Ian Stewart, who was one of my most vocal critics, would lead the charge. Strange bedfellows, indeed!

CHAPTER 18

Why the Moon Doesn't Ring Like a Bell

Space should hold many clues to the existence of the small comets. Even before the issue of interplanetary hydrogen arose, the Moon had been the focus of the debate regarding the small comets. Everyone realized that the Moon presented the first real test outside of Earth for the existence of the small comets. The Moon is our companion here. It is the first outpost of that fortress known as Earth. Anything that happens to the Earth is likely to happen to the Moon, and vice versa. If small comets are bombarding the Earth, small comets must be pelting the Moon as well. Likewise, if there is no evidence for small comets on the Moon, then small comets would have to go the way of green cheese and other lunar folklore.

Even before our original papers were published we had considered the Moon problem. The primary issue was this: If an infall of comets similar to what we proposed for the Earth hits the Moon, should we not have seen a far greater number of seismic impacts on the Moon than were observed with the detectors, called seismometers, placed on the lunar surface by the Apollo astronauts?

Between 1969 and 1972 seismometers were left on the Moon to measure ground movements due to moonquakes and meteoroid impacts. This network of instruments detected 12,500 moonquakes and 1,743 meteoroid impacts during its eight and a half years of operation. These instruments recorded, on average, a little more than one meteoroid impact every two days. My calculations, on the other hand, which are based on the comet flux at the Earth and on the Moon's smaller size, indicated that some 300 small comets were hitting the lunar surface every day. This apparent disagreement had everyone convinced that the small comets could not possibly be real.

Alex Dessler received a large response regarding the problem these comets posed for the Moon. One came from Ralph Baldwin of the Oliver Machine Company in Grand Rapids, Michigan. He thought that the Moon should quite literally have been sandblasted by these comets. But the small comets are loosely packed porous bodies, not rocks, and there is a world of difference in what happens when such bodies collide with the Moon.

Nearly everyone would make the same mistake, including Paul Davis of the Department of Earth and Space Sciences at the University of California at Los Angeles. Davis had calculated the infall of comets into our celestial companion, based on its area, which is one-thirteenth that of Earth's. He assumed that the Moon's tidal forces would break up some of the small comets at an altitude of seventy-five miles, but that many comets of greater internal strength would reach the Moon intact. Finally, he calculated the seismic response these small comets would produce upon impact from the disturbances known objects had produced on the Moon—the Saturn IV boosters and lunar modules that were sent crashing into the Moon following each Apollo mission. He concluded that every small comet hitting the Moon should have been easily detected by the lunar seismometers. Davis said, given the high impact rates of the small comets, the Moon should be in a constant state of vibration. It should appear, he said, "to ring like a bell."

Davis was essentially correct. If you treat a small comet like a rock, then the very sensitive seismometers that were placed on the Moon by the astronauts should have picked up the tremors created by the impact of these comets on the lunar surface. If you treat them like rocks then every comet that hits the Moon should have shown up on the seismometers, no matter where you placed these instruments. If these objects hit the Moon like hundred-ton rocks and the seismometers detected every one, then there would be more than 100,000 events due to these impacts on the Moon every year. But the lunar seismometers never picked up that many events or signals of any kind. There were less than 2,000 events a year. If you treat the small comets like rocks, then clearly that kind of discrepancy would end the issue of small comets.

Of all the Comments we received, the issue of cometary impacts on the Moon was the most difficult problem to work through. Unfortunately, I do not think many people understood either the Comments or my Reply, because everyone still seems to make the same mistake. The general complaint is that these large snowballs would have the same effect on the Moon as that of large rocks. They would produce the same craters and the same seismic waves. But these comets are not the same as rocks.

The problem can be stated in more familiar terms. Say I throw a one-pound snowball as hard as I can at my car. I then take a one-pound rock and throw it at my car with the same speed. It may come as a surprise, but the results are not the same. And they are not the same on the Moon either. I have a hundred-ton rock and a hundred-ton loose snowball. They are going to hit the Moon at the same speed. But the result will not be the same. All the arguments that were made against the small comets were based on the assumption that it made no difference whether you hit the Moon with a rock, a snowball, or a pillow of goose feathers. But given the choice, I would rather throw a one-pound snowball against my car than a one-pound rock. Davis obviously would just as soon throw the rock.

So would Dessler. Though he served as editor over all the Comments and Replies during the small comets debate, I do not believe he ever really understood the topic. In September of 1986 Dessler wrote me, not on official *Geophysical Research Letters* stationery, but on Rice University letterhead. He did this, he said, to make absolutely clear that the views expressed in the letter were not to be regarded as editorial decisions, but as helpful advice from one colleague to another. He was concerned about the quality of my Replies. He could not understand how the impact of such a body would fail to leave a huge crater on the Moon. He urged me to consider not replying to the Comments or saying simply that I did not agree with them. He thought that if the comets turned out to be fantasy, these Replies would compound the damage to my reputation.

But Dessler was wrong about the small comets. A snowflake shot at the speed of a bullet is not going to crack open a rock. But Dessler has always insisted on treating every impact in terms of a nuclear explosion. So I suppose that given the chance to throw either a snowball or a rock at the windshield of his car, half the time Dessler would choose a rock. Dessler was also wrong about the Replies. Everyone I spoke to thought that a lack of a Reply to a Comment would be regarded by the community as an admission of defeat. I could agree with

Dessler on only one point. If the small comets turned out to be real, he wrote, no one would remember who my critics were, much less what they said.

Few people ever understood the essential difference between a stony meteoroid and a cometary snowball. But Yosio Nakamura and Jürgen Oberst, both of the Institute for Geophysics at the University of Texas at Austin, did. Nakamura was one of the scientists involved in operating the network of seismometers that had been placed on the Moon. So he and Oberst knew what these instruments could and could not do. They had worked with the data and had analyzed them carefully for many years. So, with the help of Stephen Clifford and Bruce Bills from the Lunar and Planetary Institute in Houston, they considered in their Comment several conceivable scenarios to explain why the lunar seismic network might have failed to detect the small comets.

They realized that the problem could be resolved if the comets were objects of sufficiently lightweight material. Such a low-density object would be unlikely to produce a large Moon tremor. They calculated that if tidal stresses dispersed the small comets some sixty miles above the Moon, then all that would remain of each object upon impact is a cloud of snowflakes and ice crystals scattered over four miles. Such an impact would not penetrate the lunar topsoil. And the resulting disturbance would be so slight as to be undetectable by the seismometers. But for this to occur the break-up of the comets would have to be complete and no piece could be larger than the average-sized ones used in their calculations.

One way to get around this problem would be to assume that the objects are nothing more than large, thin clouds of matter long before they even reach the Moon. But this then poses another problem. How would such thinly dispersed objects survive for such a long time, given the heat from the Sun, in interplanetary space?

Nakamura and Oberst also tried to account for the failure of the lunar seismometers by assuming that these objects hit the Moon very softly. But they could think of no mechanism that would put the brakes on these objects sufficiently. Finally, they wondered if I had perhaps not grossly overestimated the size of these objects, for this too could account for the lack of seismometer readings. But I did not.

Nakamura and Oberst really tried to see the problem my way. They realized there is an inherent difference between a soft, flimsy object and a rock. They were aware of that. They realized what the problem was. They would not throw the rock through the windshield of their car. But they could not get from that insight to the response of the lunar seismometers. They stopped just short of understanding what these small comets were and what effect they would have on the Moon.

Even before the small comets came along, there had been some unanswered questions about the number of objects hitting the Moon. The number of large objects impacting a square mile of the lunar surface should be about the same as the number of large objects hitting a square mile of the Earth. The area of the Moon is about thirteen times smaller than the Earth's. Therefore, neglecting the larger gravity pull of the Earth, the Moon should receive about thirteen times fewer objects than we do. The lunar seismometers were calibrated so that scientists could triangulate the readings from the various instruments and determine the location, as well as the size and number, of the impacts. The number of large impacts on the Moon should have been about one-thirteenth the number of objects that had been seen to fall as meteors into our atmosphere. But lunar seismic data indicated that far fewer large objects hit the Moon than that. The discrepancy was substantial. There were about a thousand times fewer Moon shocks.

But the reason for the discrepancy is simple. It can be traced to the insensitivity of the lunar seismic stations for detecting objects that are loosely assembled like the small comets. Frank Press, a celebrated physicist from MIT and the president of the National Academy of Sciences, had worked on the physics of lunar impacts. He, along with Nakamura and a host of others, led the effort to have seismometers placed on the Moon. They calibrated these instruments by looking at the seismic signature of everything from nuclear blasts in the Nevada desert to bullets shot into loose sand and sand bonded with epoxy glue. Every event that would occur on the Moon would be scaled from nuclear blasts or from bullets hitting sand. No one ever worked out what effect a bullet of snow would have on such a surface. A lightly packed snowball cannot hit a surface the way a rock does. It simply does not, even if they are going at the same speed.

My Reply to the Moon problem was quite straightforward. The small comets do impact the Moon. No disruption above the lunar surface is likely since the tidal forces of the Moon are considerably weaker than the binding strength of the small comets themselves. So what happens when these forty-foot, low-density objects smash into the lunar surface at six miles per second? The effect is quite unlike the explosive scenario that accompanies the impact of a high-density stony meteoroid. To understand what happens when a snowball, or a porous body, hits a surface at high speeds I had looked at the brilliant work of a Russian physicist named Yakov Zel'dovich. He had done work on what happens to materials, including porous materials like water-snow, under high temperatures and pressures, though he did not apply his work to the lunar surface.

The evidence of a large rock hitting the Moon is a big crater. This is a permanent record. Most of the energy from a rock just buries itself into the Moon. For a small comet the same energy comes in, but it is spread out over a wider area. It does not end up in the lunar surface even though the same amount of energy is involved. The snowflakes of a small comet are widely separated from one another. So when the comet hits the lunar surface the flakes end up crushing one another. The comet does not instantaneously transmit all of its energy into the lunar surface. Most of the energy produced is crushing energy. So the front half of the comet comes down and gets crushed. It heats up and turns to steam. More and more flakes pile into one another. The crushing energy produces heat. Water temperatures may reach several thousand degrees, while some of the water itself is broken down into hydrogen and oxygen atoms. But very little of the energy is transmitted to the surface. Most of the energy goes into turning the snowflakes into water and the water into a hot steam. Then once the collision is over, the vapor cloud cools and expands rapidly, kicking up dust. But the fast-moving water particles escape because the Moon's gravity is so slight.

The lunar surface suffers very little damage from such an impact, just some wear and tear. A small comet hitting the Moon will not develop as much pressure upon impact as a rock. The pressure at the area of contact between the cometary water and the lunar surface is relatively low, though I would not want to be standing under it when it hits. It would still crush and vaporize me. Where the lunar topsoil is relatively hard the mass of hot water will essentially bounce off the Moon leaving no crater. On weaker lunar surfaces a shallow dent may appear. The resulting crater would most likely be flat and shallow-bottomed and measure about forty feet in diameter.

When a rock comes down and hits the surface, it hits hard enough and with sufficient force that a shock goes out from it, like the shock wave from a supersonic airplane. On the ground we experience this shock as a sonic "boom." The shock wave is produced because the object is moving faster than the particles in front of it can get out of the way. A large wave is produced, much like the bow wave in water from a high-speed boat. A rock that strikes the lunar surface is like the supersonic plane striking the air in front of it. It creates a shock wave in the lunar surface that propagates outward from the impact. Of course, a rocky meteor is destroyed as it tries to plow itself through the Moon. But if a flimsy object like a small comet impacts the Moon then no large shock wave is produced. And if it is flimsy enough and very lightweight then no shock is produced at all.

The extent of the vibrations produced by a ball of water-snow hitting the lunar surface cannot be as great as a rock's. It cannot move the Moon up and down and shake the seismometers the way a rock does. But it can still jiggle them. The instruments should still see a large number of events. But there is another factor involved and that is the frequency of the disturbance moving outwards from the shock wave. The lunar seismometers were tuned to a frequency of one cycle per second or one hertz, which corresponds to one up-and-down motion of the Moon's surface each second. An object hitting the Moon and creating any other frequency will not be as noticeable to the seismometers. Only if two tuning forks are matched does the movement of one fork cause the other to oscillate. In this case one tuning fork is the seismometer and the other is the Moon's surface.

The lunar seismometers were sensitive to the slightest fluctuation. But those fluctuations had better be tuned to about one cycle per second. The frequency caused by an object hitting the Moon depends on the size of the impact area, and this includes the distance that a shock wave moves before it is spent. The larger the impact area, the lower the frequency of the seismic wave. A rocky meteor creates a large impact area from a large shock wave, which moves the lunar surface at about one cycle per second—just what the seismometers were tuned to sense.

The small comets, on the other hand, produce little or no shock waves. Thus, their impact area is small and the waves travelling from the impact area have a higher frequency, about a hundred cycles per second. So the waves produced by the small comets would have passed right by the seismometers. A seismometer tuned to a one cycle-per-second wave will not pick up a hundred cycle-per-second wave. The instrument will be able to detect such an event only if it happens nearby. And the prediction is that a few will hit near the instruments if you have 100,000 comets impacting the Moon each year. But those may well be among the 200 or so impacts, thought to have been rocks, that were recorded each year by the seismometers. So the instruments had not recorded the small comet impacts simply because they had been tuned to hard rock collisions, while the comets, wrote one reporter, "splattered into the Moon at Barry Manilow wavelengths."

The seismometers' lack of sensitivity to the impact of small comets accounts for the discrepancy in the low number of large objects detected on the Moon relative to the number of such objects that are seen falling into Earth's atmosphere. It is not a factor of two or three, or thirty, but a thousand times or more. The difference is enormous. But the existence of the small comets can account for it. The larger number of objects falling into Earth's atmosphere is determined from the number of bright meteor streaks due to Öpik's puzzling dustballs. These dustballs are assumed to be large conglomerations of dust. But if the dustballs are small comets, then the observed large number of meteor trails in the Earth's atmosphere will be in agreement with the infrequent lunar seismic disturbances.

Oberst nearly admitted it himself. In a letter to Dessler, he said that he had found our Reply "most interesting" and went on to commend our "effort to find a formulation for the problem of lunar cometary impacts and their seismic responses." He agreed that "one of the causes of the discrepancy is indeed that cometary meteoroids are underrepresented in seismic data." But he maintained that cometary meteoroids could be identified on seismic data if "their kinetic energy is sufficiently high."

Once I had finished all the computations and completed our Reply I was personally embarrassed to learn that a pair of researchers had already conducted a detailed study of the impact of low-density materials on planetary surfaces. And they had done a much better job of it than I. John O'Keefe and Thomas Ahrens of the Seismological Laboratory at the California Institute of Technology in Pasadena had compared the effects of solid ice, silicate meteoroids, and porous snowball-like objects hitting solid surfaces at high velocities. Their computer simulations had shown that low-density objects would sometimes produce "a flat-floored crater with a central peak," but that "for very low density impactors," the excavation depth was "near zero."

Their work had been published in 1982 and I had missed it. But Dessler had given me just four weeks to come up with a solution to a very difficult physical problem, and I did not have the time to search all the literature before completing my Reply. If I had found their work beforehand, I might not have had to work it out independently. But I was pleased to learn that my schoolboy calculations had yielded the same results as their very thorough simulation with a large computer.

If the small comets are bombarding the Earth's atmosphere, they are likely falling into the Moon as well. For centuries, strange glows and lights have been reported on the Moon. On May 23, 1985, Greek astronomers captured such a transient flash in frame 4 of a sequence of exposures taken about 8 seconds apart. The impact of a small comet on the Moon would produce just such a momentary glow.

I think we have seen the effects of the impact of small comets on the Moon. On the evening of April 21, 1972, Thomas Mattingly, the command module pilot on the Apollo 16 mission, reported seeing a bright flash on the Moon's far side. No seismic signals were detected. So the scientists responsible for operating the lunar seismic network are well aware that their instruments did not pick up every significant lunar event. If a rock had hit the Moon to produce this flash, the seismometers would have detected it.

It was not the first time—nor the last time—that a bright flash or strange glow appeared on the Moon. More than 1,400 such events have been recorded by both amateur and professional astronomers and their reports date back more than 400 years. These events are known as Transient Lunar Phenomena. No one knows exactly what causes them, but many observers interpret what they see in terms of volcanic activity, despite the fact that the Moon is supposed to be a dead body.

I think many of the bright flashes and strange glows seen on the Moon can be attributed to the impact of small comets. As a comet slams into the lunar surface its kinetic energy is transformed into heat. The heat turns the water-snow into liquid water and then into a hot gas of water, hydrogen and oxygen. This is a very strange gas because it is under a very high pressure. Just as it blows apart, the hot gas glows. This is just momentary, but it must glow. The gas is so compressed and has so much energy that it glows just like the filament of a light bulb. To an observer on Earth looking up at the Moon with a telescope, it would appear as a brief, dull glow or red light. If that person happened to be looking at the mountains for volcanic action, they would see this glow and likely exclaim, "Volcanoes!"

Early in 1989 Sigwarth came running into my office with a copy of an article from the latest issue of *Icarus*. It was entitled "Photographic Evidence of a Short Duration: Strong Flash from the Surface of the Moon." It was fabulous. On the night of May 23, 1985, astronomer George Kolovos and his colleagues at the University of Thessaloniki in Greece had taken a series of photographs of the Moon with a small portable refractor telescope. In one frame a striking "bright spot" appears that is absent from both the preceding and following frames. Kodak Laboratories in Athens found no fault in the grain structure of the film. And the brightness of the spot ruled out the possibility of pre-exposure on that part of the negative.

The flash lasted less than sixteen seconds and illuminated the region around the crater named Proclus C. What is most striking about the photograph is that the shape of the bright spot is confined by the irregularities of the surrounding lunar features. This strongly suggests that the event is real. And it seems, from an examination of the illumination on the inner edge of a crater some thirty miles from the event, that the flash took place "from a point slightly above the surface of the Moon."

The astronomers did their best to explain the mysterious flash. They ruled out the possibility of reflections, matter-antimatter annihilations, as well as volcanic eruptions given the absence of any sign of plumes in the subsequent frames. They also ruled out a meteor impact because no great cloud of dust was seen. Finally, they concluded that the flash had occurred after the lunar surface had been stressed, probably due to the quickly rising temperatures caused by the rising Sun. An intense stress, they believed, might trigger a cracking of the lunar surface, releasing gases and electrical discharges, and producing a momentary bright flash on the lunar surface. They calculated that the lunar surface would have to be heated to a temperature of about 3000°F to produce this flash. This, of course, is exactly what we had predicted would happen to the gases from the impact of a small comet.

There is yet a more permanent exhibit of the impact of small comets on the Moon and that is the zodiacal dust cloud. This is a cloud of dust and gas in the plane of the planets that extends from about Venus to the asteroid belt, just past the orbit of Mars. In reflected sunlight this cloud is faintly visible from Earth and is known as the zodiacal light. It appears as a diffuse glow in the western sky after twilight and in the east before dawn. Scientists have long debated the origins as well as the stability of this dust cloud over time. The supply rate has been calculated to be several tons a second. That means you have to pump in several tons of dust per second in order to maintain the known dust stream between the planets. Some scientists believe that the dust is kept in its orbits through shepherding by the Earth and the other planets in the inner solar system.

But there is no need for any fancy shepherding effect to get the dust ring in the inner solar system. Some dust must come from the disruption of small comets that are destroyed by the heat of the Sun as they travel in interplanetary space. And some dust has to come off the Moon when it is hit by small comets. The dust is blown up off the Moon with enough speed to be injected into these orbits. My calculations show that if small comets are hitting the Moon and the dust is being blown off of it, then the orbits should extend from about Venus to Mars in a sort of disk. So it seems likely that the small comets are contributing at least some material to this giant cloud of interplanetary dust.

I am not the only person to think so. A Uruguayan astronomer named Julio Fernández, who has apparently never heard of my small comets, recently proposed in a journal of planetology called *Earth, Moon, and Planets* that disintegrating comets may be supplying the zodiacal dust cloud with several tons of dust per second. He includes among them the known population of short-period large comets, as well as a population of "unobservable small ones." He is aware of the scientific consensus that comets with diameters under about a half mile are either scarce or non-existent. But, says Fernández, rather than attribute this scarcity "to the actual absence of small comets in the inner planetary region, we can argue that they do exist but are very difficult to detect due to their faintness…" Amen.

One last problem remained. If 300 small comets do hit the Moon every day, where, some people wanted to know, was all the water on the Moon? One of the strongest conclusions from the study of the lunar rocks brought back by the astronauts is that the Moon is remarkably dry. But this problem is very simply resolved. The lunar gravity is such that practically all the water vapor from the impact of small comets simply flies off. But some of the water molecules may wander around and eventually condense in the crevices near the poles. A number of scientists, such as James Arnold, a chemist at the University of California in San Diego, have argued that water and ice may be trapped in the permanently shadowed regions near the lunar poles. The small comets may be a source of that ice.

Because most of the water flies off, a huge envelope of water should surround the Moon. We know the Earth has its own tenuous shroud of hydrogen gas lying beyond our atmosphere. The impact of the small comets onto the Moon's surface gives our companion a hydrogen shroud also. The cometary water vapor is hot enough that it escapes the Moon's weak gravity and moves outwards into interplanetary space. After travelling for about a day from the Moon these water molecules are broken apart by solar radiation into hydrogen and oxygen. These molecules will have travelled, on the average, over half a million miles. This means that our planet is sitting in a very tenuous lunar atmosphere, an atmosphere of water, hydrogen, and oxygen. Unlike the known comets, which expend their own water supply, the Moon merely releases the water deposited by the small comets that collide with it. Yet for a space traveler viewing the Earth and its companion from a distant vantage point, say Venus, the Moon should resemble a faint comet.

On March 7, 1971, several bursts of water vapor were observed on the lunar surface by a set of instruments, ion detectors actually, that had been left on the Moon by the Apollo 12 and Apollo 14 astronauts. The detectors had been built and the experiment designed by John Freeman and three colleagues at Rice University. Freeman, by the way, had gotten his Ph.D. at the University of Iowa. He was looking for ionized molecules and atoms in an attempt to measure the Moon's very weak atmosphere. Ionization occurs when an atom or molecule is stripped of one or more electrons. This stripping can occur by exposure to solar ultraviolet radiation or through collisions with other particles.

Shortly after analyzing some of the data, Freeman ran across an impulsive burst of water ions that had been detected by both the Apollo 12 and Apollo 14 instruments. It started a pandemonium. Everyone thought there was water coming up from the Moon. The existence of water in the Moon is a question of great importance. One of the problems with establishing a future manned base on the Moon, for example, is the lack of water. But if the Moon had its own water, the astronauts could simply pump it out.

In order to prove that the water had come from the lunar interior, Freeman had to eliminate all other possible sources for the water burst. So he compared the features of this event to a variety of known events. The only other occurrence that had produced rates as high as that of the 7th of March was the engine exhaust gases released during the takeoff from the Moon of the Apollo 14 lunar module. In that case, the water vapor was present for only two minutes, while on the 7th it had lasted, intermittently, about fourteen hours. Besides, the lunar module had left the Moon a month prior to the mysterious event of the 7th.

Freeman also considered the possibility that his data actually represented a measure of the dumping of nearly ninety-five pounds of wastewater from the Apollo 14 command module in lunar orbit a month prior to the incident. Urine samples were collected and their contents were compared to the water burst of the 7th. But Freeman ruled out this possibility as well. The event could not have occurred so long after the command module had left the Moon. Nor could the quantity of water produced by the waste dump have been as large as that observed on the 7th—anywhere from half a ton to eleven tons, depending on how far the instrument was from the source of water itself.

A number of other possibilities were considered and rejected, including a comet as the source of the water burst. Freeman ruled out the impact of a comet for several reasons. March 7th was not the date of any known meteor shower or the date when the Moon was near any known old comet orbits. Comets are also thought to contain significant quantities of other materials such as carbon dioxide, and these were not detected by the sensors. Finally, the notion of a cometary hit was "firmly ruled out," Freeman wrote, "by the lack of any seismic activity characteristic of such an impact."

Of course, no one at that time knew of the existence of small comets.

The outer planets are also likely to be bombarded by small comets. This image from NASA's Mars Global Surveyor shows the surface of Mars is rippled with channels and tributaries, suggesting that water once flowed there in great quantities. This water may well have come from an influx of small comets.

CHAPTER 19

Mars and the Outer Planets

The Moon and the Earth are not the only targets in this cosmic shooting gallery. Much of the rest of the solar system is likely to feel the impact of the small comets as well, though nowhere can we expect their presence to cause such a bounty—oceans, atmosphere, and possibly life—as on the Earth.

I expect very few small comets to survive inside of Earth's orbit where they would be destroyed by the Sun's heat. No small comets will make it to Mercury, few will make it to Venus. But it is quite a different story for the remainder of the solar system, especially Mars, Jupiter, Saturn, Uranus, and Neptune. The small comets have to be there. Their numbers are large, their effects considerable.

The flux of the small comets in the vicinity of Mars is about the same as it is at the Earth. But Mars is about half the size of the Earth, its gravity is a little more than a third, and its mass is just a tenth that of Earth's. So my estimate of the rate at which the small comets impact the Martian atmosphere is approximately one every twenty-five seconds. That is quite a bit less than the twenty per minute for the Earth, but it is still enough to bring a large amount of water to Mars. The small comets would deposit about three feet of water on Mars every 400,000 years. So, everyone wanted to know, where is all the water on Mars?

Pictures of Mars taken during the United States Viking mission show what looks like a desolate red desert. Nowhere is there any running water. But the Martian surface displays remarkable evidence that water once flowed on it in great quantities. The terrain is rippled with long channels complete with branching tributaries, teardrop-shaped islands, and braided silt deposits. "These networks resemble terrestrial stream networks that are the result of collected rainfall," according to a thorough analysis of these features by a team of U.S. Geological Survey scientists headed by Harold Masursky.

Masursky determined the age of these features using the standard method of crater count. The more heavily cratered an area, the older it is thought to be, because meteor impacts are assumed to happen at more or less the same rate over most of the lifetimes of the planets. Therefore, thought Masursky, since many Martian channels show so few craters, they must be quite young. These stream channels, he concluded, "may have formed during relatively short periods of time at several different times." There must then have been episodes in the history of the solar system when the atmospheric pressure and the temperature of Mars were great enough for liquid water to flow freely.

The atmosphere of Mars today is too thin for liquid water to form on its surface. But the planet does have ice caps at both poles that advance and recede with changes in the Martian seasons. These ice caps were originally thought to be largely dry ice, or frozen carbon dioxide, with very little water. So they certainly could not be a reservoir for an ocean or a fraction of an ocean of water pumped in over time. But scientists have come to believe that there may actually be more and more water in those polar caps than originally thought. The evidence now suggests, according to Hugh Kieffer of the U.S. Geological Survey and his colleagues, that, "the water deposits are fairly thick" in the polar area.

What happens to a planetary atmosphere if it is all water? No one knows really, but such a scenario was worked out years ago by Andrew Ingersoll of Caltech to explain the early water loss on Venus. I think the same scenario is applicable to Mars. If you have a water atmosphere, it will eventually get hot. It will be just like the greenhouse effect on Earth, but it will not be due to carbon dioxide. Water can also produce a greenhouse effect. So as comets come in and deposit more and more water into the atmosphere, it gets hotter and hotter. Eventually the polar caps melt, producing even more water. Mars has little gravity and does not keep its atmosphere as well as the Earth does. So eventually the Martian atmosphere would get so large and so hot that it would finally have to just blow off into space. There is a name for this among planetary scientists. It is called hydrodynamic blow off.

Here then is what may be happening on Mars. The small comets come in and penetrate to a Martian altitude of about twenty-five miles. They vaporize. There is enough air in the weak Martian atmosphere to vaporize the comets. They do not hit the surface—not the ones that are hundred tons—just as they fail to hit the surface here on Earth. So they enter the atmosphere, vaporize and then condense as ice at the polar caps. But eventually there is so much water and ice on Mars that the incoming water vapor cannot be collected out of the atmosphere. Then you have the greenhouse effect. Solar energy is absorbed and trapped in the atmosphere. It gets hotter. Once the surface gets to the melting point of ice, you get more water melting from the polar caps. Water begins to flow in the channels on Mars. This is the Martian springtime. It must be an amazing scene.

But how long does it last? What time passes between springtimes? Will there be violent thunderstorms? Can simple forms of life develop during the springtimes and hibernate during the long droughts that may last for millions of years? We do not know. We have never seen Mars in such a warming stage. But then again we have been keeping tabs on the planet for but a tiny fraction of its four-billion-plus-years' history. Eventually, however, its atmosphere would grow so large and so hot that a great loss of water into space by hydrodynamic blow-off would occur. Within a short period of time all the water would be gone and the process would start all over again.

Such a view is not inconsistent with some models of a cyclical Martian climate. Some planetary scientists believe that Mars has experienced such periodic warmings in the past. Direct evidence of such a warmer period, complete with liquid water, has been reported by a team of scientists headed by Ted Roush at the NASA Ames Research Center. The evidence lies in the planet's dust, which periodically rises into huge dust storms that engulf the Martian surface for months at a time. Roush has found signs of volatiles, substances that evaporate at relatively low temperatures, in spectroscopic observations of the Martian dust. These dust grains, he said, may act as permanent reservoirs for carbonates, sulfates, and water.

Fraser Fanale, a planetary scientist at the University of Hawaii in Manoa, has another idea to dispose of, or hide, a large amount of water on Mars. It did not occur to me because I had read the papers about the polar caps and it seemed clear that there was no ocean of water there. Of course, Fanale was not trying to account for the small comets. He was involved in another fierce argument about whether or not there is more water on Mars than we think. People are unhappy with him because he also is water prone.

Fanale's idea is that the water condenses on top of the polar caps, just like the frost that collects in a refrigerator. Gradually it builds up, putting more and more weight on the polar caps, until the ice slides under the surface dust. He suspects there may be as much as a mile of water under the polar caps. Others have had similar thoughts. Back in 1980, Stan Zisk of MIT's Haystack Observatory reported finding an area of Mars which appeared to have liquid water some twenty to forty inches below the surface. His evidence came from a set of Earth-based radar measurements of the Martian surface carried out a decade earlier by the Jet Propulsion Laboratory's Goldstone Facility.

The notion of subsurface water on Mars is a clever idea. Some future exploration of the Martian surface may one day provide us with the deciding evidence. But until then current wisdom alone is not sufficient to eliminate the possibility of a large influx of small comets on Mars.

Also certain to feel the impact of these objects are the moons of Mars and those of the outer planets. Phobos, a Martian moon, shows extensive alteration by liquids and is thought to conceal ice below its surface. Jupiter's Ganymede, the largest moon in the solar system, resembles a huge, cratered ball of ice; Callisto is also covered with ice. Most of Saturn's moons are relatively small but they too are composed of rock and ice. And many of the natural satellites of Uranus are also ice-clad. Where did all this ice come from? Quite likely from the infall of small comets.

But determining the effects of small comet impacts on planetary moons requires more than fleeting thought. The farther a planet's position from the Sun, the less the impact speed of small comets on its moons. On the other hand, the closer the moon is to a giant planet, the higher the speed of cometary impacts because of the boost in speed due to gravity. For our Moon, the impact speed is such that cometary water is blown off into space. But the middle and outer moons of the giant planets should retain the water from the impact of small comets, particularly if the comets penetrate into a snowy or icy moonscape.

One such candidate is Neptune's Triton, which is now regarded as the coldest body in the solar system. Its exotic surface, an incredible icescape of what appears to be frozen nitrogen oceans, was revealed during the Voyager 2 flyby in 1989. Some think the large shapes on the surface are made of water ice since ices of methane and nitrogen could not be used to build the large vertical structures that were observed. Most intriguing are the volcanic-like geysers that rise out of the icescape and are swept leeward by Triton's atmosphere, looking like chimneys trailing smoke. Some believe the geysers are produced when dark material on the moon's surface absorbs sunlight and heats the nitrogen which then escapes through cracks in an ice layer about eight feet thick. But how can this be? The Sun's heat is about one-thousandth that at Earth and the dark features have been there a long time. The heating of frozen nitrogen, however, could easily be produced by the impact of a small comet. But no one has mentioned this obvious possibility.

The origin of planetary rings poses another puzzle for planetary scientists. While such ring systems are absent from the smaller inner planets, all the giant outer planets have them. There is a single, narrow wispy ring that extends out from Jupiter almost 186,000 miles. Saturn's rings, of course, are broad and magnificent. Uranus has ten dark narrow rings and

Neptune has five faint rings. These rings are all composed of icy particles. Where did all the ice come from? Planetary scientists now reject the notion that these rings are composed of material left over from the time the planets were forming. And some are beginning to realize that these ring systems require a continuous supply of new material. I think the small comets may be one such source. Perhaps some small comets are colliding with rocks and tiny moons orbiting in the rings and leaving some of their water ice and mantle material in orbit to continually rejuvenate the rings.

I believe that the fingerprint of the presence of the small comets is actually visible as dark spokes in Saturn's rings. It seems that dark, radial, wedge-shaped spokes are continuously appearing and disappearing in Saturn's B ring. Fresh spokes have been observed to form within twelve minutes across part of the rings. A small comet is fast enough to do this. So these spokes may well be the tracks of small comets moving across the rings as they fall into Saturn's atmosphere.

The influx of small comets into the giant outer planets—though necessarily just rough estimates based on their population at the Earth—is in all probability absolutely tremendous. Take Jupiter, for example. It is about three hundred times more massive than the Earth. And because it is bigger, its gravitational field is larger. Consequently Jupiter probably sucks in some 500 small comets per second. So over a period of four billion years Jupiter acquires the equivalent of one Earth mass in cometary water. Such an infall of material will produce a lot of heat.

I was shocked when I first calculated the heat produced by these small comets at the outer planets. I thought I had stumbled onto something important and I mentioned it briefly in our original set of papers in *Geophysical Research Letters*. Later Donald Morris at the Lawrence Berkeley Laboratory brought up the subject of planetary heating in an early version of his Comment on the small comets. It was an amusing situation. He first used it as a way of eliminating small comets. Then, when the heat from the small comet impacts was shown to be reasonable, Morris claimed the idea as his own. We ended up having quite an argument about it. Alex Dessler took his usual route and eliminated these tantalizing suggestions for planetary heating from both the published Comment and Reply. It was another blow for scientific inquiry.

The topic is an important one, however. The Sun is twenty-five times dimmer at Jupiter than at Earth. But Jupiter radiates more heat than it receives from the Sun, about ten percent more, in fact. Its neighbor, Saturn, radiates a great deal more heat than it receives from the Sun. And Neptune also radiates more heat than expected. Only Uranus, among the giant outer planets, does not appear to generate excess heat, although earlier reports claimed that it did. The determination of excess heat is difficult.

Current wisdom interprets this excess heat emission in terms of an internal heat source. In the case of Jupiter, the excess heat is thought to be the energy released by slow gravitational shrinkage. This excess heat is one of the major features in the modeling of the history of the large planets and what lies in their deep interiors. But it turns out that the 500 comets that slam into Jupiter every second are enough to account for at least some, if not all, of the excess heat coming from Jupiter without resorting to internal heat sources.

The energy flux from the comets is big enough, within a factor of two or three, to account for the difference in thermal energy production of not only Jupiter, but Saturn and Neptune as well. Uranus remains an anomaly for everyone. So if this excess energy is coming from the small comets rather than internal heat sources, then many of our accepted ideas on the evolution of the large planets and their interiors are wrong. Our current ideas regarding the history and nature of the entire solar system may have to be revised.

But there is no need to worry about the Earth. The impacts of these small comets provide little heat for our planet. The Sun produces by far the largest share of the energy on the Earth. Additional heat sources are radioactive material in Earth's interior and tides produced by the Moon. Besides, the Earth is a small planet. Its gravitational field is much smaller than those of the giant outer planets, so the small comets are not sucked into the atmosphere as efficiently or at such high speeds. The small comets merely supplied us with most of the basic material for our bodies, as well as the benign stewpot for the evolution of life itself.

The Earth is pelted by about 10 million small comets a year. Mars receives a smaller annual share, less than a million and a half. Giant Jupiter is bombarded by an incredible 16 billion small comets each year. Smaller Saturn gets by with about four billion, while Uranus totals 260 million and Neptune with 300 million. Tiny Pluto is blessed with a sprinkle of only 500 thousand small comets a year. That brings the total annual budget of small comets for the solar system to something over 20 billion. That is a lot of comets to store somewhere over the age of the solar system and I quickly began to wonder where all these small comets could possibly come from. That was the next subject we had to tackle. It would not be an easy task.

CHAPTER 20

Where the Small Comets Come From

With so many small comets colliding with the planets, it seemed that any attempt to trace their origin would spell their demise. Many people thought the debate would come to an end because there was just no possible source for ten million small comets plummeting to Earth alone each year for more than four billion years. Where could they all come from? Where could we store so many objects? Unless we could find a likely reservoir, I realized that the small comets could not possibly exist. The basic problem—and one I did not have solved at the time my two original papers were published—is where did these objects come from?

We had to find a believable source for this massive collection of small comets; we had to find a way of mathematically sticking the toothpaste back into the tube. There were several possibilities. The first that came to mind was a reservoir of small comets in the space between the planets. This is a convenient solution, but unlikely. Between pulls caused by the gravity of the planets and destruction from heating by the Sun, it is just impossible to store so many comets in the neighborhood of the planets over the age of the solar system.

The next possibility is that we live in a very special time in the nearly five billion year history of the solar system. What would be special about it is that somewhere in the solar system a very large comet exploded and made a lot of little comets. It would be a unique comet in unique times. The comet would need to be a million times larger than Halley's Comet to get the number of small comets we now see streaming by the Earth. But I worry about all the dust that would come from such a comet. The small comets have very little dust. So while I cannot say that this did not happen, I think the chance is very slim.

For a time we thought the most obvious supply for our large population of small comets would be the Oort cloud. This is the home of more than a trillion large comets and one of the best advertised features of our solar system. It was named the Oort cloud after the distinguished Dutch astronomer Jan Oort who first proposed its existence in 1950. This vast cloud of comets is located at a very great distance from the Sun yet is still loosely bound by the Sun's gravity. Its outer edge is thought to be located at a distance equivalent to 100,000 trips from the Earth to the Sun. Sunlight takes only about eight minutes to reach the Earth but it takes almost one year to travel to the edge of this distant comet cloud.

No one has ever seen the Oort cloud. Its existence has been inferred from the occasional large comets that we see wandering into the inner solar system. Oort showed that if you trace the paths of these comets back before their encounters with the planets, you can find the region from which these comets come. He showed that these comets were uniformly distributed around the Sun in a very distant spherical cloud. The objects that make up the cloud are moving in all directions like a bunch of hovering bees.

The Oort cloud, because it extends out a huge distance, is weakly bound to the Sun and therefore subject to disruption by passing stars and gravity tides from the galaxy. It is believed that the large comets we see have been scattered from this cloud and sent plummeting into the inner solar system. There are new large comets coming into the solar system all the time. Most are not as dramatic as Halley's. Some have orbits that bring them into view only once during our lifetime, during all of written history, or longer. Others will have smaller orbits and will pass by more often. All it takes is a little tug out in the distant Oort cloud and the comets come hurtling in.

If the small comets did come from this Oort cloud, our storage problem would be solved. But such a solution fails to explain the motion of the small comets. We know from the movement of those little fly specks in our images that the small comets circle the Sun in the same direction and in about the same orbital planes as the planets. But an arrival of objects from the Oort cloud would have a greater variety of orbits. Some of the objects would move in directions opposite to the Earth's motion and their orbits would not all be near the orbital planes of the planets. So if the small comets did indeed come from the Oort cloud—or from any passing interstellar cloud from another part of the galaxy, for that matter—you would see them coming in from other directions and at different velocities. But this is not what happens.

So the Oort cloud is not the reservoir that gave birth to the small comets. But if you go back to Oort's paper, published in the *Bulletin of the Astronomical Institutes of the Netherlands*, he points out that, if the cloud was there at the birth of the solar system, by now passing stars and galactic tides would have wiped it clean. It would not have lasted nearly five billion years. All of the comets in the cloud would have been stripped out of it. But we know the cloud is there, so it has to have a supply of large comets. Something must be feeding it.

Most work on the subject just ignores the fact that you have to supply new comets to this cloud. Oort proposed that the cloud was being fed by an inner disk of comets, but most people do not refer to it. When the solar system was formed, the Earth and the planets condensed from dust and rock, and out beyond the planets the comets formed from left-over icy grains and dust in a great disk aligned with the orbital planes of the planets. So a big reservoir of large comets was formed, circling just out beyond the planets. They are closer than the Oort cloud itself, and more tightly bound to the Sun's gravity, so a passing star would not jiggle them as much. This is a more secure storage place. But every once in a while a star might pass very near or through this inner disk and scatter some comets. Some would feed the outlying Oort cloud, some would be lost into interstellar space, and others would plummet into the inner solar system.

In 1981 Michigan State University astrophysicist Jack Hills confirmed that the Oort cloud itself could only be supplied by this inner region of comets, the Oort disk, and then showed, contrary to popular opinion, that many more large comets actually come from this inner region than the Oort cloud itself. He calculated that the total number of comets which have entered the planetary system directly from this Oort disk is about twenty-four times the number that have come from the Oort cloud itself. The total mass of this inner disk, says Hills, may be a hundred times greater than the Oort cloud.

This storage ring beyond the planets is now used by many people to stow away large comets. It is not my idea. I just stuck some small comets in with the large ones. I thought it would be reasonable to place the large comets near the inner edge of this great disk and the small comets in the outer portions of the disk. My reasoning is quite simple. Some scientists believe that the dust in the early solar nebula was concentrated on the inside of the nebula, while the ices or volatiles were located primarily on its outer rim. It would seem, therefore, that the present Oort disk would be populated with large dusty comets at its sunward edge, while small comets made almost exclusively of volatiles such as water would reside further out.

I think that this relatively massive disk of matter beyond the planets left over from the formation of the solar system nearly five billion years ago is the only possible source for the small comets. If somebody could show that there is no disk out there, then there would be no small comets either. There is no other way of supplying the small comets. Of course, we cannot see this disk within our own solar system. But we know that such features exist in the cosmos. Recently, astronomers have been able to obtain images of other stars with vast disks like this.

The Infrared Astronomy Satellite, a joint Anglo-Dutch-American endeavor, found several stars surrounded by such disks in 1984, the most famous one of which is the great disk of Beta Pictoris as seen edge on. Soon afterwards, Bradford Smith of the University of Arizona and Richard Terrile of the Jet Propulsion Laboratory were able to photograph this enormous disk of matter in visible light. They did so by blocking out the light from its hot bright sun so as not to overwhelm the more feeble light reflected off the disk. It is a beautiful picture, a first peek at planet formation.

We know much more about the birth, life, and death of stars than we know about planets. Because we can point our telescopes at the billions of stars of all ages and types, we can determine what happens to a star from the moment of birth onward. But we cannot do this with planets because the only planetary system that we can observe directly is our own. We have no pictures of planetary systems younger or older than ours, so we must guess how the planets were born and how they age. The recent detection of large disks of gas and dust around newborn stars, such as the one around Beta Pictoris, is a giant step toward that elusive goal.

The disk around the star Beta Pictoris extends out a distance equivalent to almost a hundred trips from Jupiter to the Sun. Planets are thought to be forming very near the starward edge of the disk. But while this may be where most of its material lies, there is considerable matter far beyond it. The disk actually thickens as the distance from the central star increases. Smith and Terrile believe the disk may be partly or entirely made of small grains of silicates, carbon-based matter, and ices, primarily water-ice.

Beta Pictoris is in its early formative days, but its features are applicable to our solar system because there is no way to strip off such a large disk of matter over some five billion years. It is too tightly bound to its star to have other stars carry it off. If these observations are applicable to our solar system—and there is no reason to believe that Beta Pictoris is unique—then it is likely that a disk of similar matter lies somewhere out beyond our most distant planets, Neptune and Pluto.

Astronomers are beginning to realize that such a storage bay of comets just beyond the planets is not only likely but necessary. Martin Duncan of the Lick Observatory and Thomas Quinn and Scott Tremaine, both of the University of Toronto, recently found that the most likely origins of large comets moving near the orbital planes of the planets is a disk of such objects near the orbit of Neptune. This is less than a thousandth the distance to the outer edge of the spherical Oort cloud.

Duncan and his colleagues came to this conclusion after extensive calculations using

The large, known comets, as well as the small comets, are thought to come from a large region that lies beyond the orbits of Neptune and Pluto. Evidence that such a belt of matter might lie outside the known planets comes from many sources, one of which is the telescopic observations of other stellar systems such as this one, known as Beta Pictoris. Though planets are thought to be forming at the very center of this disk, a good deal of matter lies far beyond it. Likewise, a large amount of the matter in our solar system probably lies beyond the known planets.

months of computer time. They were looking for a way to explain why the hundred or so large, short-period comets all lie primarily in the planes of the planets and orbit the Sun in the same direction as the planets. The orbits of this special class of large comets are remarkably similar to the small comets. They found that even a long game of "gravitational pinball" with the giant outer planets could not herd large comets into these orbits if they came from the outer Oort cloud. So they went back to the view offered in 1951 by Gerard Kuiper. He had proposed that cometary material left over from the formation of the solar system lies in a disk just beyond Neptune's orbit. But Duncan and his co-workers faced one bothersome point in their proposal: They could not understand what shakes comets out of their storage ring and sends them plummeting into the inner solar system.

But once the existence of an inner cometary disk is established, the solution to this puzzle and to the origin of the small comets is largely a matter of Newton's laws. It has to do with the motion of bodies moving in each other's gravity. But while it is easy to describe a pencil, or a plate of food, falling on the floor, when we are talking about the solar system, we are talking about many bodies moving out there, pulling on each other all at once, and it is no longer so easy to describe. It is, in fact, enough to give anyone a headache.

Just because there is a disk of comets out there beyond the planets does not mean that any of them are ever going to reach the Earth. Left on their own, the comets will remain in the disk until the end of time. But a passing star might bring them in. Such a star would have to run a nearly parallel course through the disk, however, and that is an unlikely occurrence. Otherwise the brief passage of a star through the disk will only produce a drizzle of comets. Not enough of them will reach the Earth to account for the numbers of small comets seen by the Dynamics Explorer satellite. This is a severe problem. If a large object is not there a long time to scatter the small comets, and the small comets do not reach Earth, then we are not here either. All the volatiles needed for the origin of life on Earth are out there.

I think the most likely candidate for the task is a large undetected object the size of a planet. This planet has to somehow scatter ten million small comets into the Earth's atmosphere each year. Many more than this must reach the Earth's orbit without colliding with our planet. This is a huge requirement. What happens if a large planet plows through this rotating disk of comets? It scatters some of them at just the right angle to send them into the inner solar system and the Earth. Those that do not strike the Earth, the Sun, or the other planets, will return to the distant disk and eventually travel back to the planets again. That is how you get the small comets to Earth. But this number of small comets is too low. It will not work. And the reason the number at Earth is too low is that it takes a long time for the small comets to zoom clear out into the disk and back. They do not orbit by the Earth often enough to provide the flux of small comets I need. So the scattering of small comets in the inner Oort disk by a distant unknown planet does not by itself account for the large flux of small comets at Earth.

Something more is needed. What is needed is some way to decrease the orbital times, and thus the orbital distances, of the small comets. If the orbital times decrease then the number of times that each comet crosses the Earth's orbit increases. In this way the same number of comets gives you a far greater flux at Earth than you would have otherwise. This is where the work of physicists Edgar Everhart and Julio Fernández came in. Both used computers to see what would happen to comets in long period orbits when they were disturbed by the giant outer planets. They were talking about large comets, of course, but their work applies just as well to small comets. They found that some comets are destroyed by the Sun, others are lost to interstellar space, and some comets end up in orbits with short orbital times around the Sun. This happens because some of those that reach the inner solar system have to pass close by Jupiter and Saturn. The gravities of these planets scatter the comets into short period orbits that swing out only as far as Jupiter and Saturn. This allows them to go by the Earth several thousand times, rather than once during the same period of time if they had orbits that took them back to where they came from, 300 times further away.

The giant outer planets are responsible for this kind of amplifier effect. It turns out that about one in a hundred comets ends up with a short period orbit. These comets move in the same direction as the planets around the Sun and stay near the orbital plane of the planets, just like the small comets. To give the giant outer planets sufficient time to capture a large enough number of small comets to get the proper infall rate at Earth, the comets would have to have a lifetime greater than 100,000 years. Fortunately, that lifetime is within the range I had estimated for the small comets given the protection of a carbon-based mantle.

This is an incredible series of events. First you need a large storage disk of comets and a large, errant planet. But this planet cannot have a circular orbit like the others, so you need a star as well that long ago passed close enough to the planet, but not so close as to destroy it or pull it away from the solar system. So now you have this large planet wandering through the disk and scattering some comets. Then you need the presence of the giant planets like Jupiter and Saturn to draw the comets into shorter period orbits. Only then do we get the right number of small comets to produce the oceans, a portion of the atmosphere, and, most likely, the organic material for life on Earth. It bothers me that this is such a unique system. But it is plausible.

Yet this set of special conditions makes life on planets around other stars much less likely than previously thought. So perhaps our lack of radio contact with extraterrestrial civilizations is due not to a quarantine for our barbaric social behavior or to the weakness of our radio telescopes, but because we are simply alone.

CHAPTER 21

Death of the Dinosaurs

In looking for the source of the small comets, we ended up explaining far more than anyone ever expected. Any way of herding small comets into the Earth's vicinity was likely to bring in large comets as well. But while the everyday showers of forty-foot comets may have brought the ingredients for life to this planet, a storm of mile-wide comets would most likely spell death and destruction for much of life on Earth. Suddenly many parts of the puzzle began falling into place, beginning with why the dinosaurs disappeared 65 million years ago.

Much of the groundwork for such a scenario was laid in 1980 and the following years by a team of four scientists that included the late Luis Alvarez, a physicist who won the Nobel Prize in 1968 for developing bubble chambers to investigate the structure of the nuclei of atoms; his son Walter Alvarez, a geologist at Berkeley; and Frank Asaro and Helen Michel, both nuclear chemists at the Lawrence Berkeley Laboratory. They had come to the conclusion that a mountain-sized object had hit the Earth 65 million years ago and caused the extinction of much plant and animal life. They had focused mostly on the extinction of microscopic marine animals, the phytoplankton and zooplankton, as well as some larger marine organisms. But the dinosaurs had died rather suddenly and at the same time as well.

Their startling proposal created considerable controversy. The very thought that our destiny on Earth could be affected by an extraterrestrial object conflicted sharply with current wisdom and the emotional issue of our isolation from the rest of the solar system. The intense debate that resulted forced scientists to seriously consider whether or not our destiny might be tied, at least in part, to extraterrestrial events of this kind. Since then science has come to accept this possibility, though reluctantly, and only to explain events of the past. But that such celestial intervention might be ongoing today is still, for science, quite a taboo subject.

Much of the evidence for the Alvarez idea involves an unusual layer of clay that separates two periods in geologic history, the Cretaceous and the Tertiary. This boundary not only contains the last burial ground of the dinosaurs but an unusually high concentration of iridium, a rare element that is, however, much more abundant in meteorites than in the Earth's crust. Walter Alvarez amazed everyone with his discovery that the bottom of this clay layer contained 300 times more iridium than the rest of the Earth's crust. This strongly pointed to an extraterrestrial event.

To explain this mystery the Alvarez team of researchers proposed the following scenario. An object measuring about six miles across struck the Earth 65 million years ago. The material from the explosion enveloped the planet in a blanket of dust. This dust fell onto the ground and into the ocean and produced this clay layer in the geologic record. The settling dust would also have cut out sunlight, halted photosynthesis, and caused much of life on the

Earth to die of starvation. The lack of sunlight might also have chilled the Earth causing yet more deaths. But perhaps the most important killing mechanism was due to the initial shock and heating of the atmosphere produced by the expanding fireball. This fireball would have caused unusual chemical reactions in the atmosphere and produced a deadly acid rain.

Geologists have now found three periods of iridium enhancement in the geologic record. This one 65 million years ago, another 38 million years ago, and the last just 11 million years ago. These dates suggest that something deadly occurred with surprising regularity every 26 million years or so. A similar periodicity has appeared in the work of two University of Chicago paleontologists, David Raup and Jack Sepkoski. Their vast collection of observations of ancient marine life in rocks shows that eight major episodes of extinction have occurred in the past 250 million years, each one about 26 million years apart. Additional evidence for the extraterrestrial connection comes from the work of Michael Rampino and Richard Stothers, both of the NASA Goddard Institute for Space Studies in New York City. They show that large terrestrial craters have been produced at nearly the same intervals. Evidently something from space has been killing off life on the Earth with striking regularity.

The Alvarez idea that something crashed into the Earth 65 million years ago was a good one, but it did not fit all the facts. The recurrence of extinctions at other periods in time suggests the long-term effect of showers of objects, such as comets, rather than a single object. It is difficult to believe that the Earth would get struck by one large rock on a regular time schedule. Besides, paleontologists examining the fossil record have noted that the periods of extinction did not occur suddenly but gradually; they took place over several million years. And within these several-million-year periods they have identified separate periods of extinction, a phenomenon they call stepwise extinctions. Only a shower of large objects over time could produce such an effect.

A shower of comets can occur only when a reservoir of comets is disturbed. The culprit can be either a star or a planet. The chances of an errant star crashing through the solar system and sending comets careening toward Earth on such a regular basis are slimmer than your chances of winning the Iowa lottery. But people have tried to prove it. Richard Muller, a physicist at the University of California at Berkeley, tried to account for 26-million-year extinctions with a solar companion called Nemesis.

Muller places Nemesis in an orbit around the Sun that takes 26 million years to complete. Every 26 million years it zooms through the Oort Cloud and produces a shower of comets that causes periodic extinctions on Earth. The problem is that Nemesis could not possibly survive in such an orbit. It barely would be within reach of the Sun's gravity. Most likely, it would get thrown out of the solar system by passing stars or gravity tides that roam through the galaxy. It would not give you many 26-million-year cycles. Its orbit is just too big.

An unseen planet nearer the Sun is a better bet. Daniel Whitmire, a physicist at the University of Southern Louisiana, realized this and created Planet X. I think he had the right idea, but the wrong orbit. He did his homework correctly without a lot of hocus-pocus. He thinks that just beyond the known planets is an as-yet-undetected planet about the size of the Earth. This is not so far away, about a hundred times the distance from the Sun to the Earth. That means it is wandering out there about where Pluto is. In about the same location or a little further out Whitmire has this belt of large comets. This is his view of the Oort disk.

Whitmire places Planet X at this distance so that it will not be easily stripped away. The planet will not wander away as a star passes by. He is safe on that score. But he has another

To explain the constant bombardment of the Earth by small comets, I have proposed the existence of a large, dark, as-yet-undiscovered planet that regularly passes through the outer part of the comet belt where the small comets are thought to be located. These transits of the belt by the Dark Planet send swarms of small comets streaming into the inner solar system and the Earth itself.

problem and he deals with it very cleverly. When you shrink the orbit of the planet, the period around its orbit becomes smaller, so you can have no 26-million-year period for an object this close to the Sun. The best you can do is only a thousand years. So what Whitmire does is tilt the orbit of Planet X up a little bit out of the orbital planes of the planets. This orbit will very slowly change its position relative to the comet belt because of small tugs caused by the gravity of the planets. Most of the time Planet X will not be in the comet belt and hence there is no storm of comets at Earth. The only time it does so is when the most distant part of the orbit is located within the comet storage disk and that happens every 26 million years.

I like Planet X but it will not do for the small comets. Whitmire's scenario does not allow me to shovel in enough comets to the Earth. First of all, his comet disk is not big enough. I need to extend it out a hundred times farther. I have to have a disk that is at least comparable to the one seen at Beta Pictoris. Actually, I need one that is larger. I think the pictures we have of the disk at Beta Pictoris are just not sensitive enough to reveal its full extent.

The other major drawback to Planet X is that some of our spacecraft have now reached these distances and have discovered no such planet out there. Pioneer 10, for instance, has

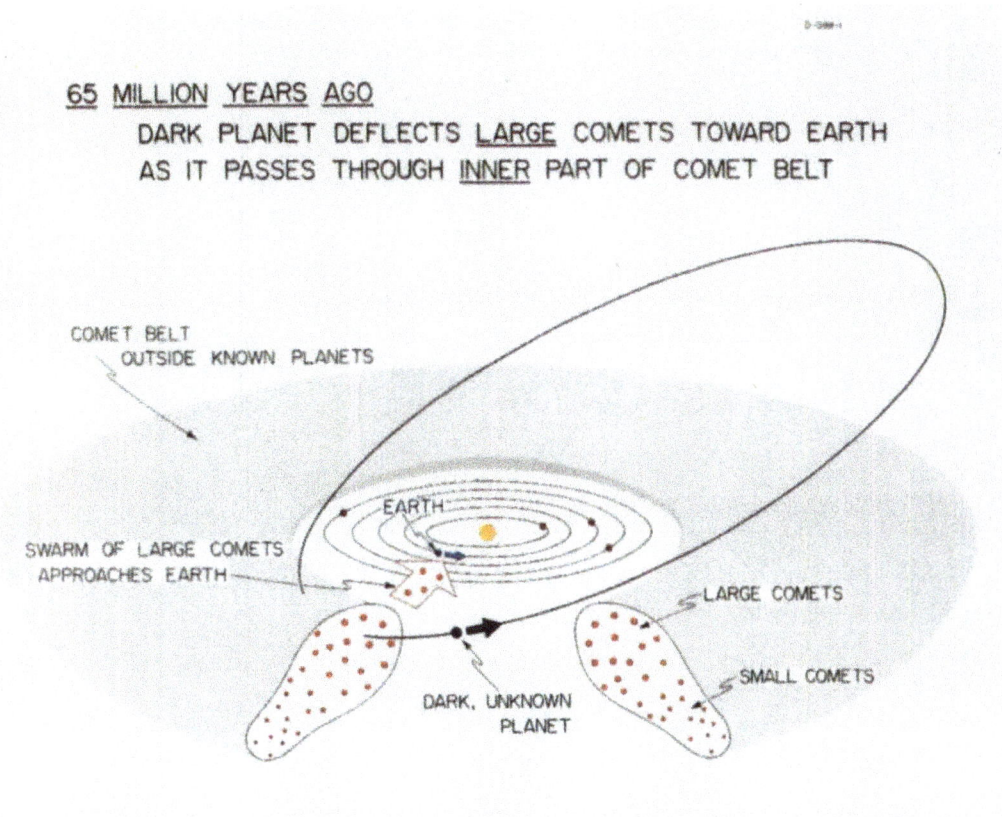

I believe that the eccentric orbit of this Dark Planet causes it to transit the inner portions of the comet belt—where the large, known comets are thought to come from—once every 26 million years or so. When it happened 65 million years ago, the Earth was bombarded by a swarm of large comets that created a "nuclear winter" scenario on Earth, killing off many living species, including the dinosaurs. The next swarm of large comets, caused by such a transit, is not expected to reach the Earth for another 12 million years or so.

now moved to destinations outside the positions of the planets. If an object the size of Earth was located out there, we would have seen the gravitational pull of that object on this spacecraft by now. But we have not. So the object is not there.

But Whitmire's idea is magnificent. All I did was revise it. I first made the planet larger to create a greater disturbance in the Oort disk. Then I extended the orbit of this planet, but not too far, so that at its greatest distance the planet is still within the firm grasp of the Sun. This larger orbit allows me to reach out into that zone where the small comets are. This also means that the planet need not have been detected yet. Mine is not a new idea. It is just Planet X with more mass and a larger orbit. But these differences require that the planet have a new name. For obvious reasons, I have called it the Dark Planet.

I think there exists a Dark Planet that moves around the Sun in a great looping orbit that shakes billions of comets loose from the Oort disk. It takes about 500,000 years for the planet to complete its orbit. At its closest, this orbit brings the Dark Planet near the orbits of Neptune and Pluto or just beyond; at its farthest, it is perhaps another 150 times farther out. Most of the time the Dark Planet plows the outer regions of the Oort disk where the small comets are stored. If it is large enough, about the size of Jupiter or Saturn say, then the planet will disturb enough small comets to explain their steady rain on Earth. But the great number of small comets in the disk also creates a slight tug on the Dark Planet that moves its orbit slowly so that every 26 million years the planet enters the inner edge of the Oort disk, a region rich in large comets. When this happens tens or hundreds of large, ten-billion-ton comets plummet into the Earth over a period of about 10,000 years. The consequences for the Earth are enormous.

The environmental havoc resulting from such an intense, repeated bombardment of Earth should cause the extinction of various species and leave its mark just about everywhere. Alvarez believed that the event 65 million years ago was the result of the impact of a single great asteroid. But the distribution of iridium from a single asteroid should be heaviest in the vicinity of its impact. A large number of impacts caused by a comet shower, on the other hand, should produce a fairly even salting of iridium over the Earth's surface, and this is what is observed. It also happens that the duration of such a cometary shower is about the same as the interval that geologists have estimated it took to deposit the iridium in the Cretaceous-Tertiary boundary.

Also expected at this geological boundary is carbon from the incomplete burning of the comets as they crashed through the atmosphere. This has been found by geologist Wendy Wolbach and her colleagues at the University of Chicago, though they interpret its presence in terms of global wildfires ignited by the impact of one enormous meteor. Shocked quartz and large craters are also expected from the impact of large dusty comets. This shocked quartz has also been found and a survey of the ages of large craters on Earth shows a periodicity of about 28 million years.

The case for a single impact is further weakened by the evidence for the stepwise extinction of species, a series of catastrophic events, during this period of time. But these stepwise extinctions are easily explained by the motion of the Dark Planet's orbit. Several successive orbits of the Dark Planet will cross the inner region of the Oort disk during each 26-million-year interval, since the planet's orbital period is short compared to the slow turning of its orbit. Each of these crossings will, in turn, send a brief shower of large comets towards the inner solar system. This means that all together several brief showers should occur within the entire event. This not only helps explain why paleontologists find extinctions to be gradual, occurring over several million years, but it also tells us why there are several episodes of species loss, or stepwise extinctions, occurring within each of these longer periods of extinction.

Many people thought the gradual rise and fall of extinctions, and the stepwise extinctions, were inconsistent with comet showers. They believed comet showers could only come from the Oort cloud and that these showers would begin quickly then decline over millions of years. But stepwise extinctions do not last this long. They did not realize that you can get brief showers if you scatter comets from a disk rather than from a cloud, and this is just what you would need to produce stepwise extinctions as well as the overall gradual rise and decline of extinctions every 26 million years.

I did this work in the summer of 1987. I did not do the fundamentals. The idea came from Oort. It was refined by Hills, Everhart, Fernández, and Whitmire. All I did was insert a population of small comets in the scenario. I took their population of large comets just beyond Neptune and extended it with a large belt of small comets. I simply assumed that the size of the comets decreases from the inside of the disk outwards. So the smaller bodies are found toward the outer rim of the disc, and the larger ones on the inside. And I deduced that there has to be a Dark Planet out there that scatters the comets toward the known planets. Without this Dark Planet, there would be no small comets here today and no storm of large comets every 26 million years.

I am convinced that comet showers are the answer to many scientific mysteries. With a storage disk of comets and a Dark Planet I can produce comet showers and with that I can account for the death of the dinosaurs, the 26-million-year cycle of mass extinctions, the shorter periods of life loss during each of these cycles, and the present shower of small comets into the Earth's atmosphere. These findings, of course, irritated everybody.

I wrote a paper on the topic called "A Hypothesis Concerning the Inner Oort Disc and Its Relationship to Comet Showers and Extinction of Species" and sent it to the journal *Icarus* in November of 1987. It was not well received. One reviewer commented he could do better sitting at a bar drinking his beer. I was too candid and transparent. That was the problem with my paper. I wanted to get the point across, which so many had missed, that a single undiscovered planet could produce brief comet showers as well as long ones. I should have known better. A paper in which everyone can understand the mathematics looks trivial and gets rejected. This one did.

My first public presentation of the scenario did not fare any better. It was at a meeting of the Division of Planetary Sciences of the American Astronomical Association in Pasadena, California. Many people in the audience responded with ridicule and disbelief. I was being disrespectful of current wisdom in astronomy. Another part of the problem concerned language. I am a physicist. Most of the people there were astronomers. Most laymen would not think so, but there is something of a language barrier between us. I spoke the wrong language. Fortunately, I could still understand the signs on the restroom doors.

And so it was that my efforts to find a source for the small comets led to a mechanism that also explained how a storm of large comets may have been responsible for the death of the dinosaurs. Beyond that, I am sure the most frightening aspect of this new hypothesis escaped no one. If these periods of extinction occur like clockwork every 26 million years, they will, no doubt, happen again. A storm involving hundreds of large comets will once more hit the Earth and devastate life upon it. The good news is that we are presently between two such periods of extinction. It will be another 12 million years or so before another storm of large comets precipitates yet another major wave of extinctions here on Earth.

CHAPTER 22

The Debate Comes to an End

Sometimes while working in the early hours of the morning I amused myself by thinking of the nature of this seemingly deadlocked struggle over the small comets. During my more cynical moods, I thought of several brontosauri that made the effort to enter a quicksand swamp to overwhelm a comrade who was foolish enough to wander there in the first place. Their comrade's name might have been Frank, and those entering the swamp could have been Donahue, Hanson, and the others. One named Dessler could have pointed the way to the conflict. Along the edge of the swamp would stand a multitude of other creatures with lesser capabilities and courage but all secure in the final outcome since the struggle was so one-sided. They would throw a few rocks at the hapless comrade without knowing why. The brontosauri would thrash and struggle and finally sink slowly into the swamp. So it was with the small comet debate that took place in *Geophysical Research Letters*.

Alex Dessler was caught in this debate in a very peculiar way. He was the editor of the journal at the center of the conflict and the small comets had made him famous. He decided to publish our two papers against the advice of his referees and sometimes people held him more responsible than they held me for all the fuss the small comets had created. He probably thought that the debate on the small comets would end quickly, one way or another. But it lasted well over a year.

Dessler received so much heat for publishing our papers that at the end of his first year as editor he felt obliged to explain why he was publishing controversial material that sought to overturn conventional scientific wisdom. "The importance for scientific progress of the occasional new idea that proves correct is out of all proportion to their number," Dessler wrote in an editorial. "Everyone would agree that not all controversial ideas are wrong. Because it is not possible to tell in advance which new idea is correct, it is best to get new ideas into the open literature where they can be discussed, attacked, tested, or supported as the will of the community and the soundness of the idea dictate. This is the way science has advanced, and no better way has been demonstrated."

Dessler ran this editorial at a time when he thought I was at least partially wrong about the small comets. On the one hand, he believed that the atmospheric holes in our images were nothing more than instrumental noise, yet, incredibly, he thought that there could still be such things as small comets. No doubt Donahue's paper on cometesimals had, at one point, helped convince him of that.

"Scientists would like to think that they are steely-eyed, going after a goal in a nice straight line," Dessler said at the time. "Forget it. There's serendipity and good fortune involved, too. A good man makes his accidents turn out well. And here, I think, is a good example. I think Frank is going to turn out to be not quite right. Nevertheless, it will have done a lot of good for science, advancing it in a left-handed way." At the same time he wrote to me saying that if he lacked objectivity, his beliefs were skewed in my favor. "Only if the small-comet hypothesis is verified will my editorial decision to proceed with the publication of your original papers be vindicated," he said. "All editors are biased toward vindication." He realized that his future hung on the fate of the small comets.

There has always been some tension in my relationship with Dessler. The crucial episode goes back some twenty-five years. Dessler was just starting the space science department at Rice University. Everyone at the time was trying to duplicate the success of the Iowa program. The first satellite, Explorer 1, was worked on at Iowa and so was the first successful space probe sent past the lunar orbit, all under the guidance of James Van Allen. We became a center of excellence in space science and good students were applying. Other universities soon wanted their own space science programs.

To quick-start the program at Rice, Dessler began hiring people out of Iowa. He decided, in fact, to gamble his space science program on a person he had persuaded to come to Rice from Iowa. I felt that Dessler was making a mistake. I told Dessler that. So did Van Allen. But Dessler to this day says we did not tell him that this person, who shall remain nameless because he has suffered enough as it is, might be a problem. This individual suffered a nervous breakdown and within a few years the Rice satellite development program collapsed. The episode was something of an embarrassment to Rice University. I think they expected too much, too quickly. But just a couple of years ago, at a dinner in Arizona, Dessler was still swearing at me. He insisted that the debacle had been our fault.

During the debate on the small comets Dessler and I corresponded and talked on the phone constantly. When he accepted those papers, it was like saying, "I do." We had something akin to a close marriage for three years but not a pleasant one at that. He often gave me advice and I think he meant well by it. There were times very early in the debate when I thought he was really worried about my health.

I began to feel the pressure after the first six months. I was spending hundreds of hours on the Replies. It was my obligation. I had made a terrible mess and I had no choice but to defend my actions. That was the penalty I had to pay. But preparing those Replies very nearly did me in. I would work until the wee hours of the morning and then Sigwarth would come in later and check my work. He would redo all the calculations to see whether I had made any errors. Sometimes I did. Sigwarth is very good.

Towards the end of the debate Dessler had us rewrite the Replies over and over. That did not help matters. He was putting on the pressure because he did not like what we said. He also had people rewrite their Comments. He was trying to gain control over the debate that had long slipped from his grasp. This was particularly true with the last published Comment, the one by Wasson and Kyte concerning the composition of the small comets.

I had a little trouble with Wasson and Kyte at the beginning because they made fun of small comets. They likened them to the "living comets" proposed by the well-known British astronomer Fred Hoyle and suggested that the small comets might be coated with live viruses. Hoyle and his colleague, the Sri Lankan astrophysicist Chandra Wickramasinghe, have proposed that living cells shed by comets marked the beginnings of life on Earth and that comet-borne viruses and bacteria continue to arrive here from space and may be the cause of sudden, widespread epidemics on Earth. Most scientists have not taken these ideas seriously.

But I think that Wasson and Kyte later realized that perhaps their analogy was not so funny and removed the remark from their final draft.

My troubles with Dessler were not so easily resolved. He wrote to me complaining so much about the quality of my Reply that I wondered if he were jockeying for a spot as co-author with Wasson and Kyte. But his real objection, apparently, had to do with the length of the Reply itself. He pointed out what he thought would be "a nice, appropriate response," crossing out all but the first five sentences of my entire four-page Reply! I was flabbergasted. Dessler also seemed to be co-authoring the Reply.

So I rewrote my Reply and kept it down to the required two pages. But Dessler was still not happy. He seemed fixated on associating the effects of small comet impacts with nuclear explosions, accused me of making changes in my model, and he, too, began comparing me to Hoyle. "As I see it," he wrote, "you are painting yourself into a corner, much as Fred Hoyle has. You ought to think about the parallel between you and Hoyle. If the comets exist, you will get full credit if you answer Wasson and Kyte simply by saying that your comets are different from their comets in that your comets have virtually no dust and they assume 50 percent dust content. If the small comets exist, it doesn't matter what you say, you are going to win a Nobel Prize."

I again shortened my Reply, told Dessler why small comet impacts were not like nuclear explosions, and explained that my model did not change. But I did not book my ticket to Stockholm. Nobel Prizes are won for true advances on the frontiers of science and not for generating turmoil in our perceptions of Earth and the solar system, as I have done. I consider such talk about the Nobel Prize nonsense.

I went on to tell Dessler that there was only one similarity between Hoyle and myself. The phenomenon we each proposed profoundly affects the Earth in the present. The reception that such proposals receive among scientists is significantly more intense than those that concern some event in the remote past or the far future. They threaten our beliefs that Earth is hermetically sealed, isolated from the rest of the cosmos. Otherwise his comparison to Hoyle was like comparing apples to oranges. My attempt to estimate properties of the small comets was clearly not the same as "painting myself into a corner."

It became clear to me that Dessler was not familiar with work being done in many of the important topics that were being brought up in the course of the debate. I do not think that Dessler ever understood the small comets. He was convinced that the comets had to stop at high altitudes and leave too much water up there. He was certain that the break-up of the comets at high altitudes would form ice clouds in sunlight that would be easily visible to observers at night. And he was deeply obsessed with the subject of lunar impacts. He was convinced that each small comet collision with the Moon carried the force of a small nuclear bomb. If the small comets were real, then according to Dessler we would see mushroom clouds on the Moon.

But Dessler was wrong on all of these points. I think he was simply unwilling to spend the time doing the reading and calculations that would elevate his conclusions to something beyond personal beliefs. And his apparent fixation with nuclear explosions was understandable as well. I finally went to the library to see how old he was in 1945. It turns out he was a very impressionable seventeen.

Several times during the course of the debate Dessler called and said: "Did you change the numbers? Did you change the size?" But I never changed anything. I made no adjustments. People said that was cruel. Some people made adjustments to allow ten times more comets to fall into Earth than they previously did. Dessler hoped things would change so he could say, "Well, that's the end of it, a mistake was made and Frank is making adjustments." But all along the number and size of the small comets have been the same, their movement in the solar system has been the same, and their composition has been the same. That is all I ever stated about them. That is all there is.

Most people did not work through the series of Comments and Replies that constituted much of the "official" debate on the small comets issue. Most people were only interested in knowing who won. It was a natural curiosity. But no one won or lost. My critics did not win because they did not have any solid facts to show that I was wrong. On the other hand, I did not win either. I did not have any piece of evidence to show that I was absolutely right. What I had done, however, was present evidence which, when taken all together, favored the existence of these small comets.

It was a long, tough year and a half. Dessler knew what he had put me through. At the end of the debate he sent me the "Alexander J. Dessler Award" in recognition of my "truly amazing array of responses to criticisms of the small-comet hypothesis." But the humor only hid the bitterness that would soon surface.

In November of 1988, Rice University had a party to celebrate the twenty-fifth anniversary of their space science department and Dessler's sixtieth birthday. Several people were asked to make contributions. One person ended up giving Dessler a oneway ticket to Venus for his editorial role in a vicious debate about whether or not there is lightning on Venus.

I, too, was asked to contribute and looked for something amusing yet pointed to send Dessler. So I had my secretary, Barbara Hermeier, pick up a copy of Clement C. Moore's "A Visit from St. Nicholas" in the children's section of the Iowa City Public Library. I then sat down with a can of Diet Pepsi and reworded the poem. My satire of Dessler's misunderstandings on the small comets issue was read at the party.

'Twas the night before Christmas and all through the Texan bog
Not a creature was stirring, except Alex and his dog.

His dog murmured something about small comets falling this night
And Alex responded with hopes that this could not be right.

Alex was nestled all snug in his bed
While visions of giant snowballs danced in his head.

And with visions of mushroom clouds hovering the moon
Alex settled down into a feverish swoon.

When out on the lawn there arose such a boom
Alex sprang from his bed and out of his room.

Away to the door he flew like a flash
And out into his yard he made with a dash.

The moon on the giant pile of new-fallen snow
Gave a luster of midday to objects below.

When, what to his wondering eyes, should happen,
As the surface of the heap of snow seemed quickly to blacken.

With blinding flashes in the air so lively and quick
Alex knew in a moment that he must be sick.

More rapid than eagles the small comets came
And Alex eagerly offered each comet a name.

Now, Soter! Now, Donahue! Now, Rubincam and Hanson!
On, Davis! On, Nakamura! On, Chubb and Wasson!

To the pile of snow Alex quickly went
To see how large a crater the small comet had sent.

Now, dash away! Dash away! Dash away, all!
There was no crater there even so small.

So up into the sky Alex anxiously peered
To see from where these small comets were steered.

The legions of small comets at large distances could not be found
And Alex responded with comments not so profound.

The red point of Mars grinned down from the sky
Why aren't you blue, Alex said with a sigh.

His clothes were all tarnished with ashes and soot
As he kicked the comet debris with his foot.

With a wink of his eye and a twist of his head
Alex dashed into his house to get into bed.

And laying his finger aside of his nose, with a grin on his face
He dreamt of all the current wisdom that he would erase.

***

The small comets story would outlast Dessler. His term as editor came to an end on the last day of 1988. He applied for another term but did not get it. He probably thinks the small comets controversy did him in.

The small comets are now all that Dessler has left. He has been giving talks on the small comet hypothesis. The same so-called snags in the hypothesis still trouble him. But he no longer discusses these issues with me directly. For a while he sent me letters, signed "Tex." Tex is Dessler's dog. "I'm still convinced that the small comets exist, but I am having trouble convincing my master," Tex wrote early in April 1989. Tex obviously represents that part of Dessler that still believes in the small comets. To convince his master, Tex wanted "any good physical arguments supported by calculations." Dessler had enclosed a photo of Tex looking up in the sky, ostensibly for small comets.

"Now I know what poor Galileo went through, and you too, poor guy," Tex added. "What should I do? I have considered the unthinkable (i.e., biting him), but I am his best friend, so that is out of the question." I found Tex to be a very perceptive dog and thought his master should reward him.

Two weeks later there was another letter. It began "Dear Mr. Excitement." There was "a disturbing development," Tex wrote. "We have another dog—just a six-month-old puppy—a female named Spot. She is a real nothing of a dog." But the real problem, Tex explained, is that Spot "has been convinced by my master that small comets do not exist."

I did not reply.

CHAPTER 23

Astronomers Enter the Fray

To prove to anyone that the small comets really did exist still required the final test. The small comets would have to be seen with a telescope. It was the only decisive way to resolve the issue. It was the most important thing anyone could do. We had to look with a telescope and get a picture of one. This search would constitute the next and final phase in the discovery of the small comets.

Steven Soter planted the seed for this search back in February of 1987. It was not his intention, I am sure, but it was thanks to Soter that someone eventually went out there and looked for the small comets with a telescope. At the very least, he made it happen a little sooner, he gave the idea a little fertilizer.

Soter was an astronomer with the Center for Radiophysics and Space Research at Cornell University in Ithaca, New York. Thomas Gold, the noted astronomer, was also there. Gold, you will recall, had been very upset by my small comets proposal. Perhaps he felt I was treading across his personal territory. So it did not surprise me to see Gold acknowledged at the end of Soter's Comment that ran in the February 1987 issue of *Geophysical Research Letters*. But Soter did bring up a good point: We now had the technology, specifically new sensing equipment for telescopes, capable of spotting these objects.

Soter believed that many small comets should have been observed by telescopes used to search for man-made satellites. One such system is an observatory southeast of Socorro, New Mexico, which served as the prototype for the U.S. Air Force's Ground-Based Electro-Optical Deep Space Surveillance (GEODSS) network. The observatory consists of two telescopes mounted side-by-side and equipped with a special television camera that permits high sensitivity readings onto videotape. This sensitive television camera, unlike photographic film, allows a telescope to spot rapidly moving objects at low-light levels.

The Air Force has now deployed four such state-of-the-art systems around the world and uses them primarily to track foreign satellites. The Space Surveillance Group of MIT's Lincoln Laboratory developed the system for the Air Force during the late '70s and early '80s. Laurence Taft, an astronomer, was in charge of testing the first of these telescopes in New Mexico. He made a lot of measurements of artificial satellites, satellite debris, and meteors, but he never conducted any organized search program. All he had were notes of his observations, such as "I saw an object." Taft himself admits that "precise records were not always kept."

Taft spelled out the capabilities of the telescope, as well as his results, in a handful of papers. In one test, Taft made a count of man-made satellites and their debris. In the telescope's "deep space" search mode, at altitudes above some 900 miles in other words, Taft spotted about one piece of debris, as small as eight inches, per hour. In a search for near-Earth debris, at altitudes below about 900 miles, conducted in January and February of 1984, Taft found eleven times more objects out there than were cataloged by the North American Defense Command. These objects were as small as four-tenths of an inch.

Looking back over Taft's results, Soter came to the conclusion that if the small comets did exist, then "this system should have detected hundreds of them, even if they are as dark as the least reflecting known objects in the solar system." In the deep space mode, Soter said, the observatory, operating at its limit of detection, should have spotted these objects as far away as 62,500 miles from the Earth. He calculated that, at best, the system should have been able to pick up some 250 small comets per hour or, at worst, about forty per hour.

But when Taft used the telescope in this mode for more than 200 hours during 1977 to 1981, mostly in September through October and in February through March, the detection rate was only about one object per hour. And all of these objects, said Soter, could be identified as belonging to distant man-made satellites, not comets. He thought that even if the comets were very small and reflected very little light, the system should still have picked up several per hour. So, Soter concluded, it "seems unlikely" that natural objects as large, close, and numerous as the small comets, could have been overlooked.

But Taft's actual work left many questions unanswered. If you look at his papers, it seems he saw objects he could not really identify. These objects had no light curves. A satellite turning and reflecting back sunlight would appear to blink on and off—that is known as a light curve. Objects without a light curve are either not spinning or not artificial. In any case, Taft never identifies these objects. Many of the objects he saw at large distances from the Earth did not exhibit the periodic variations in brightness of a rotating spacecraft. He simply assumed they were debris from man-made satellites.

I had questions about his work so I called Taft myself. I asked him how many of these objects he had seen at large distances and could not identify. He said several per hour. I had him repeat it. He said three to five per hour. I think he felt awkward. Later, when I began citing his figures, Taft got mobbed by scientists and the press. He told other people he did not see any small comets. To one reporter, he said: "Frank is wrong. There aren't any small comets out there. If they were there we would see them. We have never seen anything that even remotely resembles them in hundreds of hours of observation. So if they are there, then they are really tough to find."

I think Taft felt the pressure. Suddenly he was right at the center of a very emotionally charged debate. It was an awkward position to be in. Anyway, life is always more pleasant in a cheering crowd than as the lonely supporter of an unpopular renegade.

Though Soter's paper got people thinking about telescopic observations of the small comets, his arithmetic was incorrect and so were his conclusions. He greatly overestimated the detection rate of small comets by Taft's telescope. His rates for the detection of the small comets were a hundred to a thousand times too high. He failed to understand the importance of their dark mantles and their motions past the Earth. But even more important he did not realize that some of Taft's observations took place during a period when the number of small comets was at its lowest.

I had known since the middle of 1986 that there were variations in the numbers of small comets over time. We found that there were ten times as many small comets in early November as there were in mid-January. This overall decline also showed variations on shorter time scales, such as a decrease on about December 14th. No one was aware that there were large time variations in the occurrence of atmospheric holes in the Dynamics Explorer images. No one knew because my paper on the subject, which I had submitted to *Geophysical Research Letters* in the summer of 1986, had been rejected by Dessler.

These time variations quite clearly were a crucial element in any discussion regarding the possibilities for detection of the small comets by a telescope. I realized that this was a splendid opportunity to publish a large portion of my previously rejected work on the topic. This upset Dessler tremendously. He kept wanting to restrict my answer to the single point Soter had raised. He insisted that eighty percent of my Reply had to be a direct response to Soter and that only twenty percent could be new material on time variations. Nor would he let my Reply suggest a new research program. It took Dessler a while to realize that the time variations, when tied into detection rates, were very relevant to my Reply. And so it was that my paper on time variations came to be published after all.

I redid Soter's work taking these time variations into account. It turns out that the rate of detection by Taft's telescope for small comets far from the Earth could be as low as one every ten hours to as many as one per hour. These rates are not inconsistent with the rates reported by Taft. His papers even stated that he saw objects with an increasing frequency as he looked out farther from the Earth. And they did not have the light curves that are typical of spinning man-made satellites. The brightness of a small comet will not vary as it spins because its surface is even; it does not have the solar panels and flat metal surfaces characteristic of artificial satellites.

I think Taft spotted the small comets. He just assumed these objects were spacecraft and spacecraft debris. He said these must be old, retired spacecraft that were not spinning because they were not being used. But old or not, once a spacecraft is spinning it will remain spinning for the rest of its life.

I also had to determine the rates for the detection of small comets when the telescope looked for near-Earth objects. The small comets fragment at altitudes somewhere between about 600 and 1,900 miles. For at least a brief moment after break-up, these small comets probably brighten. Their speeds are large enough to produce a streak during an exposure time of one-thirtieth of a second on Taft's telescope. My estimate of the detection rates of small comets at low altitudes with this telescope range from about one event every forty hours at 1,900 miles to one event every 360 hours at 600 miles. According to Taft, the hourly rates for streaks due to satellites and meteors in the telescope frames are about twenty events per hour. That meant that one out of every 800 to 7,200 of these streaks would be due to the passage of a small comet. But these streaks were like needles in a haystack and to identify which ones belonged to a small comet would have required a tedious examination of the image frames. Taft had done no such thing.

The work that Soter put me through was not a waste of time. It would be necessary for anyone who later decided to go out and look for these objects with a telescope. Soter's numbers were not right but they were close enough to establish that it might well be possible to detect the small comets with a specially-equipped telescope. They would be difficult but not impossible to find. The search would have to be carefully done. The observer would have to be dedicated to this task. You could not do it as by-product of some other random search. You had to have a detailed plan.

This public confrontation between Soter and myself also made it clear that even with this new technology, finding these objects was not guaranteed. You could easily waste one or more years of your life doing such a search. You had a chance of spotting the small comets only if it was done the right way.

Within a year the search would be done the right way.

CHAPTER 24

Where Are You Now, Galileo?

Astronomers never really bothered to search for the small comets, though it was, quite clearly, their responsibility. Even more troublesome, astronomers, by their denials, actually hampered this effort. They just did not care to look.

I should have expected it. Astronomers have traditionally shown little interest in near-Earth objects. They have always sought to see as far out into the universe as they could, not as close to Earth as possible. The only searches in the vicinity of the Earth are for satellites and rocket debris. Telescopes are simply not available for any other kind of search. But even if they were available there is a psychological block. There is a presumption that the Earth floats in a void. Other than a few flecks of dust that cause the bright meteors we see when we look up at the sky, the Earth, everyone believes, is isolated.

For a long time even the dust and rocks were hard to accept. For years such eminent bodies as the Royal Society in Britain were skeptical of stones that rural folks claimed fell from the heavens. Scholars were appalled at the ignorance and credulity of people who reported such things. It was not until 1803, with the spectacular meteorite shower that fell at L'Aigle, seventy miles east of Paris, that such a notion actually became scientifically respectable. But even afterwards there were doubters. Thomas Jefferson seems to have been among them. He supposedly said in 1807, "I could more easily believe that two Yankee professors would lie than that stones would fall from heaven."

I was raised believing that the Earth sits here in relative isolation. There would be a little dust floating around, certainly, and there was always the fear that a large rock might come by and hit us every ten million years or so. But, otherwise, we are isolated. The universe stopped revolving around the Earth long ago. There is nothing coming in from out there. There is no need to look. It did not seem to matter to astronomers that I, a physicist, had evidence of a stream of objects arriving here from the outer rim of the solar system. And never mind the profound effects such a situation would have on our existence, our history, our oceans, our atmosphere, and the outer planets.

Though the will was lacking, the technology to conduct the search was not. There was a new technology available, as Steven Soter had pointed out, that represented a remarkable leap beyond the photographic film used by most telescopes to record their observations. Film is very sensitive but very slow. If a telescope remains pointed at a star for a long time, a half-hour or an hour, the photographic plate collects the faint light from the star and produces a very good picture. You can afford to spend that much time with a star because stars do not streak through the sky. You can sit there and integrate, or gather, the light from the star for an extended period of time to get your picture. But a rapidly moving object passes over the photographic plate in seconds, so the integration time is short, and unless the object is extremely bright, it will not show up in your picture.

My comets, being so small and dark, are only capable of being seen by reflected sunlight. Such faint objects are difficult to find amid the many bright points of light in the cosmos. To find them required a relatively new technology called array detectors, which are attached to the eyepiece of the telescope. Array detectors are made up of thousands of little photoelectric eyes. Being much faster than photographic film, they are able to record fast objects streaking through the telescope's field of view. These array detectors, which are also called CCDs, or Charged Coupled Devices, have revolutionized astronomy.

Still, few telescopes in the country are presently equipped with this rather expensive technology and, among those that are, not all are suited to conduct a search for near-Earth objects. In 1987, one of the few facilities that could seek the small comets was the Spacewatch Telescope at Kitt Peak in Arizona. The telescope is in the hands of Tom Gehrels, an astronomer at the University of Arizona. Gehrels had been drawn into the small comet debate by Dessler, who had asked Gehrels to review a paper I had submitted to *Geophysical Research Letters* in the summer of 1986. In his review Gehrels said that he could not take the small comets seriously because their numbers were a million times greater than what he thought was reasonable.

Gehrels is interested in the minor bodies of the solar system, especially those such as asteroids, that cross the Earth's orbit. These are potentially catastrophic if the Earth is in the right place (the wrong place!) to get hit as they come by. This is the kind of object that Luis Alvarez, Gehrels, and others believe caused the death of the dinosaurs. They are interested in these objects because they think that one of them once hit the Earth with catastrophic results and might do so again someday.

There is something of a three-way race to discover asteroids approaching the Earth. That is what Tom Gehrels does. That is what Laurence Taft does. That is what Eugene Shoemaker, an astrogeologist at the United States Geological Survey in Arizona, does. Shoemaker and his associate Eleanor Helin, at the Jet Propulsion Laboratory, use the Schmidt telescope on Mt. Palomar. Their telescope has the largest view of the sky. They use photographic plates, which are perfectly acceptable for asteroids because these objects do not move very quickly against the field of stars. With the Schmidt telescope, Shoemaker and Helin discover more asteroids than anyone else.

Steven Soter was convinced that the small comets should have been detected by people such as Gehrels, Taft, and Shoemaker in the searches for Earth-approaching asteroids. But this is unlikely. These searches are carried out by taking at least two pictures sufficiently separated in time to detect the slow motion of a distant asteroid. The apparent motion of the small comets, on the other hand, is sufficiently large (because they must be much closer to the Earth in order to be seen) that these objects are beyond view for at least one of the pictures. So the small comets would most likely not be seen during searches for asteroids, even if photographic film were sensitive enough.

I worried that no one was going to look. Taft had found nothing using the MIT prototype telescope in New Mexico and was out of the picture. Shoemaker could not look, as he lacked the proper equipment, though he insisted that the small comets were not there. That left only Tom Gehrels and his Spacewatch Telescope. Dessler suggested to Gehrels back in 1986 that he look for these objects with his telescope. But Gehrels would not do so. He was certain that the small comets could not be there. He knew the answer without looking.

Gehrels' posture has been the same all along. In a paper he wrote for *Physics Today* in 1985 he prepared a table of the number of near-Earth asteroids and comets. He listed objects weighing from about a hundred thousand tons to a hundred billion tons and their numbers in the vicinity of the Earth. He then extended his table to small-sized objects, for which he had no observations, to find the highest possible number of hundred-ton objects hitting the Earth. He concluded that one would impact the Earth every 100 years. Our estimate was one every three seconds.

Like Dessler, Gehrels also argued that the impact of each of the small comets on Earth would be the equivalent "of several times the energy expended at Hiroshima in August 1945." He could not believe that twenty Hiroshimas per minute were going off some fifty miles above our heads. Of course, he was talking about rocks, not comets. On the basis of what Gehrels knew, objects of this size fell on Earth so infrequently that it was just pointless to use his telescope to look for them. He just did not want to do it. But it is actually quite possible that Gehrels, like Taft, had already detected the small comets during his time on the telescope. He had just not identified them or had assumed they were satellites. The observation of small comets with a telescope would be an infrequent event.

So the rumor spread quickly. It had started out as "Gehrels is not going to look because he knows the small comets are not there." But within two months Gehrels' stance had been transformed into reality. It was rapidly understood by astronomers that Gehrels had looked for the small comets but had not found them. So naturally there was a lack of interest by astronomers. Gehrels, the world expert with one of the few telescopes properly equipped to spot the small comets, had looked and could not find them. Besides, who would believe that astronomers could miss ten million house-sized objects falling into our atmosphere every year, when they can look at Jupiter and newborn stars, and see to the very edge of the universe some 15 billion light years away? They could not possibly miss so many objects just beyond the Earth's atmosphere.

Astronomers should have learned their lesson long ago. Galileo, in his day, could not get people to look through a telescope to see the moons of Jupiter because people just knew that Jupiter could not possibly have any moons. Everyone just knew that all heavenly bodies circled the Earth. If Gehrels had been alive back then, I think he might have refused to look. But now, three centuries after Galileo, Gehrels, who as an astronomer should know better, was saying, in effect: I don't want to waste my time using my telescope to look for small comets because they don't fit in with my thoughts on this matter.

Why had astronomers not learned the lesson of their great mentor? The answer is simple. Like the moons of Jupiter in the 17th century, the small comets today are an emotional issue. Everyone chokes on the subject. It does not follow current wisdom. I think that skepticism is healthy in science because it acts as a filter to quickly sort out the few true findings and hypotheses from the great number that prove to be incorrect. But skepticism carried to the point at which further observations are not made to decide the issue and opinion is substituted for fact is clearly a detriment to the progress of science.

Soon after I had published the first two papers in *Geophysical Research Letters* many people, including Dessler and Donald Hunten, told me to relax. People would go out and look, they said. People would do the experiments and use their telescopes to search for the small comets. All I had to do was sit still for a year or two and it would all be straightened out. I was naive enough to believe that this would happen. Long before George H.W. Bush made the term popular in his presidential campaign, I adopted a "kinder and gentler" period of my own. But two years passed and no one—other than Donahue, Olivero and Mendillo—did anything. No one looked. No one did the critical thing. Of course, there was a problem. Imagine being an astronomer. What if you should find them? How then do you answer the question that everyone had been asking: How could you ever miss these things? You would be embarrassed.

Someone eventually would search for the small comets. But it would not be an astronomer. And a disaster would have to intervene. It was January of 1986. We were temporarily distracted from the small comets. After nearly a decade of effort, the Galileo spacecraft, the most complex and most sophisticated robot spacecraft ever built by man, was being prepared for its launch to Jupiter in May. I was readying the instrument I have on Galileo that will measure the ions and electrons in the vast radiation zones of Jupiter. Then, suddenly, the space shuttle Challenger went down in flames. There was massive confusion in the Galileo project. The spacecraft had been specifically designed to go up on a shuttle flight.

A month later we pretty well knew the impact this disaster would have on the exploration of space. The shuttle would not fly for a long time, and Galileo would not be launched until the fall of 1989. But now, instead of about two years to get to Jupiter, it would take six. Having missed its best launch opportunity, the spacecraft would now require a gravitational boost provided by a Venus flyby and two such boosts by the Earth in order to reach Jupiter. It was still worth it. As long as it happened within my lifetime, I could wait.

So could Torrence Johnson, the project scientist on the Galileo project for the Jet Propulsion Laboratory (JPL). He could work on Voyager, which was making its magnificent tour of the solar system. But some scientists were stuck in the project with little else to do. There were good scientists like Clayne Yeates who depended on the launch. He was science manager for Galileo and had been since 1977, when we first met. Yeates, in fact, was one of the key people overseeing the science investigations on the Galileo mission, including mine. There was some worry at JPL whether anyone on the project would stick around. Would they wait three more years for launch and then another six years for the arrival at Jupiter, essentially doing nothing with their lives until then? It was a long time not to do any science.

Yeates found himself with little to do after Challenger blew up. I sat in his office right after the shuttle disaster listening to him wonder whether he was going to stay on the Galileo project for all the years it would take to recover from this tragedy. It was questionable. He wanted something interesting to do. Of course, I had just the thing. Yeates was curious about the small comets. He had already seen the Comment by Steven Soter and had become interested in using a telescope to search for them. Yeates is an excellent scientist and is willing to look at new problems. He is able to work with equations and is capable of gaining an insight into their meaning. He is also bright and strong-willed and unlikely to break under pressure.

John Casani, the project manager for Galileo, was really interested in having Yeates get into something so they would not lose him. He was too valuable to the Galileo project. So in the summer of 1986 Casani asked me to talk to Yeates and encourage him to work on the telescope problem so that he would not notice all these years going by in his life.

Yeates agreed. But he got more than he bargained for.

CHAPTER 25

The Great Search Begins

"I thought it was pretty wild." Clayne Yeates' first impression of the small comets was no different than anyone else's. "There were a lot of reasons that made it seem unlikely."

But unlike everyone else, Yeates decided to do something about it. "It looked like a real challenge to make the telescope observations because I could either rule out the small comets with a negative result or I could show that there was a population of these objects. It seemed worthwhile to do and nobody had really ever done it before."

Yeates realized that the one telescope capable of detecting very small, near-Earth bodies was actually geared up and available at the time. This was the Spacewatch Telescope located at Kitt Peak and run by the University of Arizona's Tom Gehrels and Jim Scotti. So in March of 1987, after familiarizing himself with the properties of this telescope and spending many months trying to figure out how best to use it in a search for small comets, Yeates approached Gehrels about such an investigation. Gehrels first responded that these objects could not possibly be real. He had long ago made it clear that he did not wish to spend his time looking for them.

Yeates recalls: "We were sitting in a little refreshment room at the University of Arizona at the time, having a Coke or something, and there were a bunch of people in the room. So I said, 'Hey, Tom, we're scientists aren't we? Maybe we ought to look. We should be objective about this and just take a look. It would be worthwhile if we could show that these things aren't there.' Then Gehrels changed his tune and said, 'Well, maybe we ought to look at this.'"

Some time later, during a telephone conversation, Gehrels asked Yeates how he planned to fund the search: Can you get some money? Yeates asked how much. Forty-five thousand dollars, Gehrels replied. Yeates felt the cost was a bit steep for a couple of nights of observation but told Gehrels he would secure the necessary funding.

Yeates then spoke to the influential Donald Hunten at the University of Arizona about the idea, and together with Hunten and Gehrels, Yeates wrote a proposal to obtain the money from the Jet Propulsion Laboratory's (JPL) Discretionary Fund. This was a kind of cash machine available to the JPL director for interesting projects that need quick funding. As it turned out, JPL director Lew Allen and his chief scientist Moustafa Chahine were both quite interested in the search, so Yeates got the money to rent Gehrels' telescope.

Meanwhile, Yeates completed the details on his scheme to detect the comets with the telescope. "A lot of astronomers were making offhand comments like how easy it would be to see these objects if they were there," says Yeates. "But the more they said how easy it was, the more suspicious I got. So I started to do a bunch of calculations to see just how easy this was to do. And one of the first things that became clear to me was that this was not very easy at all. It was very difficult in fact."

Yeates knew there was only one way to find these small, very dark, fast-moving objects. He had to bank on our predictions from the atmospheric holes, concerning their number, their speed, and their direction of motion past the Earth. Otherwise there were just too many possible ways to aim the telescope. But to get a snapshot of these objects in the Earth's vicinity, Yeates concluded that the telescope would have to be operated in a novel fashion. He would have to move the telescope at a rate that matched the predicted motion of these objects as they passed by the Earth. He would have to aim the telescope as he would a rifle in shooting skeet. Only in this case the clay targets were small comets.

This was the only way, he realized, to hold the objects in view long enough to see them. The objects would be very dim, but by skeet shooting with the telescope, the small comets would be tracked long enough for the telescope's array of photoelectric eyes to have a chance to accumulate enough light from these objects so they could be seen. But without the basic information on the objects' speed and direction, Yeates would have been hard pressed to find these comets. If you are at a skeet shooting range and you have no idea where the clay targets are coming from, or how fast, chances are you are going to miss them.

Yeates counted on our predictions and did all the calculations and worked very hard at it. He did it all himself. I did talk to him from time to time and I checked his calculations. But this was his work. He was the one who had the perseverance to look up all the numbers, find all the necessary information, design the aiming of the telescope, and conduct the actual search. He spent a year doing this. And everyone thought it was a trivial thing to do.

But when Yeates explained to Gehrels how he planned to use the telescope to search for the small, Earth-approaching comets, Gehrels was aghast. He insisted that this was not the way to look for these objects with his telescope. Gehrels wanted to scan the sky for them just as he did for asteroids. He would aim the telescope in a fixed direction to catch the objects as they headed toward the Earth, down the length of the telescope, so to speak. Yeates explained why this scheme would not work. The chances of spotting a small comet this way were almost nothing. Yeates thought they were far more likely to spot these objects by moving the telescope across the sky in such a way as to catch the small comets as they sped by the Earth at a distance almost halfway to the Moon. Gehrels did not seem to understand what Yeates was trying to do and a heated discussion ensued. In the end, Yeates refused to fund any search using a scheme other than his own. Gehrels conceded.

Meanwhile, Yeates submitted a paper to *Geophysical Research Letters* on the method he would use to search for the small comets. Gehrels' 36-inch Newtonian telescope would be pointed away from the Sun, eastward near the horizon at sunset and westward near the horizon at dawn. And the telescope would have to be used at its maximum range, or threshold, which is about 85,000 miles, a distance almost halfway to the Moon. This would allow the observation of the largest possible area of the sky and increase the likelihood of spotting the small comets. The shutter on the camera would remain open for twelve seconds during which time the telescope drive would be shut off. This drive normally allows a telescope to follow the motion of the stars across the night sky. But by shutting off this drive, the Earth's rotation would provide just the right motion to "skeet shoot" with the telescope.

Yeates' skills at this kind of cosmic skeet shoot were unknown, but if he should chance to almost lock onto a passing small comet, the object should show up as a dim, short streak in his images. It would only be as bright as a star of the 18th magnitude, which is about one hundred thousand times dimmer than what the naked eye can see in the sky at night. But by using the telescope in this way he thought he should spot as many as one or two small comets an hour.

The reviewers did not like Yeates's paper. One referee criticized the fact that "very restrictive assumptions are made by the author" regarding the characteristics of these objects. What the referee failed to understand is that this search was conceived as a test of the small comet proposal. If you are going to search for objects, it is good to know what they look like and where to find them. If the small comets existed, their character and orbits should be as I had described. Any other assumptions regarding these objects would not be a test of my proposal. Incredibly, the referee even objected to Yeates' use of the words "small comets."

The other referees also seemed to misunderstand the paper, including Eugene Shoemaker. He found the detection rates Yeates had calculated too high. "I went through my own calculation of the detection rates with the Spacewatch system ... about eight years ago," Shoemaker wrote, "and concluded that this much more powerful system should yield a few objects per year with diameters near [forty feet], assuming about 500 hours of scanning per year." Yeates estimated something on the order of one or two per hour. Shoemaker, like the other referee, seemed ignorant of the fact that the search was a test of my idea, not Shoemaker's.

Shoemaker's review also included the stock answer. The Spacewatch Telescope, he noted, had been operated for a great many hours without finding any near-Earth objects. "The failure to find Frank's microcomets with the present accumulated observing time on the Spacewatch Telescope," he said, "is a pretty good indication that Frank is wrong." There was no need for a search, Shoemaker seemed to say, because he already knew what the outcome of such a search would be.

The referees saw no need to publish a paper on how to look for the small comets with a telescope. Dessler agreed with them and rejected it. But he indicated to Yeates his interest in publishing such a paper along with the results of a search. It was a reasonable decision.

In late October 1987 the search began. But the first attempt never got off the ground. Gehrels and Yeates tried a test run together, but the array detector did not work and nothing was accomplished. The next available date on the calendar was the 20th and 21st of November. Gehrels was in India at the time, so Yeates and Jim Scotti took two full nights of exposures. "It was a pretty crude system," Yeates recalls. "We had to set the telescope in motion by pushing a button with your hand and then push another button by hand that would open the shutter on the camera. It was not all computerized. We had to sit there all night."

Another run was made on January 23, 1988, by Scotti alone this time, though Yeates was in contact with him at the telescope to insure proper conduct of the experiment. Following Yeates' instructions, Scotti also took calibration pictures of stars with known brightness. These would serve as points of comparison that would allow Yeates to estimate the brightness of any small comets that appeared in the images.

The runs had now been completed. The exposures were on computer tape and needed to be analyzed. But Gehrels did not seem interested. Later, Yeates would have to make copies of all the observation tapes; Gehrels wanted to erase his, apparently to save a few dollars with their reuse. The analysis could not be done at Arizona anyway, as Gehrels had no data displays that could be used to look for extremely dim objects. We did it at Iowa. So did the

Jet Propulsion Laboratory. But it would have been extremely expensive for Yeates to hire the help necessary to do the analysis at JPL. So I made our facilities at Iowa available to Yeates, and Sigwarth also offered to help. He was working on his master's thesis at the time and I thought that after Yeates was finished perhaps we could get some of his data and extend the analysis of these images as part of Sigwarth's doctoral thesis.

Yeates came to Iowa a number of times in the winter of 1987-88. He searched the images himself, using the computer programs that Sigwarth had designed to display them on the screen. He spent hours looking at the images, trying to identify every single item in them. It was an exhausting job. Gehrels had known that it would take an immense amount of time to process images of this kind. Yeates did not think he was going to find anything. But he thought the search was worth gambling yet another year of his life.

Yeates studied each picture in painful detail. Consider that each picture is made up of over a hundred thousand pixels or photoelectric eye readings in the array detector. It took a while simply to get used to looking at all the information they contained. The most obvious features were the horizontal streaks of the same length that were due to the motion of stars across the image. These were about a hundred pixels in length. There were also dead spots in the array detector. These were pixels that had failed to operate and were always dark.

Other spots in the array detector were essentially overworked. They were hot and not functioning, but they were bright and unmistakable. There were also bright vertical lines and features in the lower right-hand corner of the images due to other flaws in the array detector. These were always there. And then there were dark patches in the images due to ordinary dust specks on the telescope mirrors and array detector. It took a while for Yeates to learn what and where everything was. It was a matter of endurance. He nearly went blind doing it.

Sigwarth did all he could to help Yeates. Occasionally, I would go up to the image laboratory in the early hours of the morning and see how they were doing. We would look at the images projected up on the screen. After eliminating all the known objects and false signals, sometimes a short, faint, unidentifiable streak remained. The first one I saw was really a thrill. It was actually up there on the screen. The streak had the motion of a small comet and the right brightness. Yeates had only to find out how many of them were out there. Their number had to agree with the number of the atmospheric holes. Of the 1,200 images he took, Yeates analyzed 171 images that way. Thirty-six streaks of unidentified objects appeared in these images.

Yeates considered, then eliminated, anything that could possibly masquerade itself as a small comet. Satellites caused long streaks, much longer than a star. And most often they would be crossing from north to south across the field of view. Besides, the telescope had deliberately been pointed well above the high geostationary orbits, those satellite positions that remain fixed in the sky above you throughout the day and night. And anything in low orbit would cross the sky very quickly, so that also was not a problem.

Earth-orbiting debris from exploded rockets and old satellites were also taken into account. Yeates had been in touch with Don Kessler, an orbital debris specialist at NASA's Johnson Space Center, who had furnished him all the relevant orbits and positions but none had been in view with the telescope at the time of the observations. Nor could the unidentified streaks be accounted for by meteors. If these streaks were meteors, then the meteor rate in the atmosphere would have to be a hundred million times more than that presently known to account for them.

Yeates also identified the effects of cosmic rays, those charged particles that travel at nearly the speed of light and come from remote places in the galaxy. And he had to make certain that the unidentified streaks in the images were not being caused by the array detector itself. In order to be able to recognize such false streaks, Yeates had Scotti do some runs of the array detector without the telescope. He put a black hood over the array detector, oriented it in the same direction as they had the telescope during the actual observational runs, and opened and closed the shutter on it. Cosmic rays would still penetrate the black hood and hit the array detector and any flaws in the detector itself would still show up.

And, indeed, cosmic rays produced large signals in a few pixels and gave a short little streak. These were easily identified in the images. Also unmistakable were the short bright streaks produced by high-speed electrons emitted by the slightly radioactive material in the array detector itself. All the standard tests were done, in other words, and then some. In the end Yeates showed that the possibility that he had made an error in finding these objects was less than one in ten million or so.

But Yeates did not limit himself to a mere visual inspection of the images. Once he had identified possible comet streaks by eye, he conducted a rigorous analysis to establish that these streaks were due to the actual presence of an object. Features that repeatedly showed up in each of ten pictures—five taken before the picture in question, five afterwards—were subtracted in order to identify the temporary streak of an unidentified object in the image. In this way the exact positions of all the candidate comet streaks were located. The brightness

In November of 1987 and January of 1988 Jet Propulsion Laboratory physicist Clayne Yeates searched for the small comets using the Spacewatch Telescope at the Stewart Observatory at Kitt Peak, Arizona. By using the predicted motion of the small comets from observations of atmospheric holes, Yeates managed to get the first optical pictures of the small comets with this specially equipped telescope. The large streaks in the image are star trails, but a faint, small comet trail appears as well (the slanted line in the outlined box). These small, dark, fast-moving objects have so far only appeared in images taken at the very limits of detection.

of all the pixels surrounding each candidate streak were then averaged and compared to the average brightness of all the pixels in the candidate streak itself. The pixels in the streak turned out to be considerably brighter than the background around the streak. The streaks, in other words, were real.

Yeates had come here, looked at the images, and found the streaks from the small comets. He had sorted out everything else. He knew where the blemishes were in the images, he knew where the cosmic rays were, he knew where the stars were, and he knew where the satellites were. If there were no small comets out there, there should have been nothing left on those images. But the streaks were there. "The whole thing worked beautifully," Yeates said.

The streaks left in the pictures by the small comets were usually about ten pixels long. Granted, they were fairly dim streaks, but that is exactly what we had expected. After all, these objects were black balls measuring just tens of feet in diameter and travelling at about 20,000 miles per hour at a distance almost halfway to the Moon. Yeates now had his streaks and could calculate how many of these things were out there. And the mind-boggling thing was that the number he calculated was within a factor of two or so of what we had predicted for the atmospheric holes. His accuracy may have been only a factor of two, or maybe ours was in estimating the number from atmospheric holes. Anyway, what does a factor of two mean when we are talking about a million times more than the accepted number of objects of that size?

Yeates was stunned by what he had found. He had not expected to see streaks. All he had to do was look at a hundred images, find no streaks, and the small comets would have ceased to exist. But he found them. And the objects had the right motion in orbit, the right speed, and they were about the right size and darkness. It was fantastic.

CHAPTER 26

Double Trouble

Even before the pictures taken with the Spacewatch Telescope had been analyzed, Gehrels had drafted a paper on the small comet search. According to Yeates, the paper concluded that these objects had not been found. There were a few blanks left in the paper for the numbers, such as the number of exposures, which Yeates was supposed to fill in. And the paper listed Gehrels as the first and main author. Yeates was relegated to second author, even though the search and its design had been his idea from the very beginning.

This did not go over well with Yeates. He was very angry. It was one thing to say you were not going to look because you knew the objects were not there. But writing up a negative paper even before you did the analysis and putting yourself as first author, when someone else had done all the work and rented your telescope like a Hertz car, was too much.

Gehrels had sent Yeates a copy of the paper in January of 1988 and asked if he was ready to announce that they had not seen anything. "I was shocked that he would do such a thing without analyzing the data," says Yeates. "You could not just sit at a terminal and look over the images. There was a lot of analysis to be done."

The scientific community had learned of the search for the small comets by Yeates and Gehrels the previous month, at the fall meeting of the American Geophysical Union (AGU). So for months that followed reporters pestered them for the results of their search. "Gehrels was telling the press he didn't see anything," says Yeates. "I think he led people to believe that he had analyzed the data." The press would then call Yeates, who was reluctant to release his findings until his work was complete.

But the pressure finally led to the release, at the end of March 1988, of his preliminary results. A press release issued by the Jet Propulsion Laboratory on the next to the last day of the month said that Clayne Yeates had "apparently confirmed one of the decade's most controversial astronomical theories—that millions of small water-bearing comets strike the Earth's atmosphere every year."

Two months later Yeates went on to present his findings in public, without Gehrels, at the spring meeting of the AGU in Baltimore. Yeates had by now ruled out all other possible explanations for the unidentified streaks in the images. "What I was left with," says Yeates, "was a class of objects that behaved like Frank's. It was the same number of objects. They had low inclination, prograde orbits. And they were about the right size. I am convinced that these are real objects." The objects Yeates found were actually somewhat smaller than what I had proposed for the small comets. "But," Yeates says, "if these objects were actually slightly darker than we assumed they were, then they could be as big as Frank's small comets."

It was clear to many who saw the photographs at the AGU meeting that Yeates had found something. There was something there that had not previously been accounted for, something truly astounding. But some were equally certain that the streaks did not represent real objects and were mere noise-fluctuations in the data that are due to chance. What no one knew at the time is that we had other undeniable evidence to show that the streaks did represent real objects. These pictures, however, had not yet been fully inspected at the time of this meeting. What we had managed to do was take two different pictures of the same object. And we had done it several times. We had consecutive images of the small comets. They showed the same object in two pictures.

The idea of consecutive images had first been suggested by Dennis Matson of the Jet Propulsion Laboratory in a conversation that took place in early 1988 with Yeates and Torrence Johnson, also of JPL. As it so happens, all the work Yeates had so carefully done to show that the objects were real could have been avoided by simply getting two consecutive images—one taken right after the other—on the same object as it sped by the Earth. This was the standard of proof in astronomy. Astronomers generally agree that two images of the same object is sufficient proof of its existence.

Yeates told Gehrels that he wanted to take some consecutive images with the telescope and Gehrels said they could do it in April. They were trying to set up a time when Yeates could get down to Arizona, but on the clear moonless night of April 19, 1988, Gehrels, who had some time available on the telescope, decided to make the runs himself. In one run, he operated the telescope the way he thought you should do it to see the objects if they were there, which just happens to be the wrong way. But in deference to Yeates he took two runs the proper way. Unfortunately, there were problems with the shutter, which seemed to stick occasionally, and as a result only forty-eight good pairs of images were obtained. Gehrels sent the tapes to Yeates with a note saying, "If you see them here, I'll be convinced."

To obtain these two pictures, the telescope was trained on a particular spot in the sky, the drive was turned off, a twelve-second exposure was taken that was then transferred onto computer tape, and thirty-six seconds later a second image was taken. After each pair of images was recorded the telescope was returned to the original pointing direction to attempt to intercept another small comet streaming past the Earth. This method was designed to insure that if a small comet was in view of the first image then it should be in the second image, unless the object was near the edge of the first image and it moved off the edge of the second one.

Yeates gave us the computer tapes of what we called the double images, as he had agreed to let Sigwarth and me write up the results. These images had to pass a set of very strict requirements to qualify as two different pictures of the same object. The streaks on each pair of consecutive images not only had to be of the same length and brightness, but they also had to be separated by three streak lengths, which represents the object's motion between frames. In addition, the two streaks had to pass through the same straight line with reference to the stars. So the requirements were highly constrained. The two streaks had to pass a total of four tough tests: direction, separation, brightness, and length.

And they did just that. Each second streak looks like the first in the pair and each time it was found in the expected location based on the time delay between exposures. The streaks also appeared at about the same rate as they had in the single frame exposures. Yeates had found one streak in every five or six photographs. This meant that in forty-eight pairs of double images we should find about seven or eight small comets. We found six. And each time a comet appeared in the first exposure, it also appeared in the second one taken thirty-six

The telescope images of the small comets contain a lot of "noise," some of which is identified in this frame. A small comet appears as a faint trail in the box at the top of the image. This is the first in a series of consecutive exposures taken by University of Arizona astronomer Tom Gehrels in April of 1988, and subsequently analyzed for the presence of small comets by Clayne Yeates, John Sigwarth, and myself.

The same comet seen in the previous exposure is seen in the box in this exposure, which began 36 seconds later. Taking consecutive frames of the same object is the standard of proof in astronomy.

seconds later. The chances that the pairs of streaks were due to something other than a swarm of objects near the Earth were one in several million for some images and one in several billion for other images. We calculated that in order to get one pair of streaks that was due to noise and not a real object would require the continuous operation, twenty-four hours each day, of the Spacewatch Telescope for at least one year for some of the sightings and more than a thousand years for others. But only a few hours had been needed to take pictures of the small comets. Such is the power of the two-picture technique.

I wanted to present these findings at the next AGU meeting that was coming up in December 1988. As usual, I had to write a short summary of the work long before the meeting itself. Gehrels had taken these images, so as a professional courtesy I put his name on the paper. I had written to him about this but he was in India at the time. His secretary apparently failed to forward his mail. When Gehrels returned, he was upset that his name would appear on the summary in the announcement book of the upcoming meeting. Gehrels wrote an angry letter to the AGU. He wanted his name off the paper. He did not know the results, but this did not seem to matter. I told Gehrels that I would take his name off the paper at the meeting. I explained that I had put it there simply as a courtesy. After all, it was his telescope. And he had taken the pictures.

During this time, Yeates wrote up his findings on the small comet single exposures. Once his papers were completed, he sent a copy to Gehrels, along with a set of slides. Gehrels was upset. "If only you could have shown me your pictures before talking to science writers!" he wrote. Gehrels looked at the slides and so had, independently, Jim Scotti, Uwe Fink, Wieslaw Wiśniewski, and Bob McMillan, all at the University of Arizona. These were "all experienced CCD observers," according to Gehrels, and together they had concluded that the features Yeates indicated "could be explained by ... noise." Gehrels then begged Yeates to "let up" until he had seen the same streak twice on consecutive exposures. "If it were true," he said, "it would be a major discovery, an upset in our understanding of the solar system." But Gehrels had yet to see anything convincing.

Shortly afterwards, in a telephone conversation, Gehrels told Yeates that he did not want to be a co-author on these papers. Yeates agreed happily. After all, Gehrels had been certain that these objects did not exist, had asked Yeates to pay for telescope time, and did not participate in the analysis of the pictures. He still did not seem to completely understand the search technique that had been employed and he confided to Yeates that he really did not understand standard methods of identifying noise, which means he was not really qualified to judge whether the streaks in Yeates' images were real or not. Yeates later suggested that he read an elementary textbook on the subject. I also spoke to Gehrels and offered to come to Arizona and explain the pictures to him in detail, but he refused.

Later I sent Gehrels one of our pairs of images of small comets. These consecutive images dramatically confirmed the methods used by Yeates to identify the events in his earlier set of single exposures. They also showed that Gehrels never really understood how to search for the small comets. Gehrels wasted half of his observation time that evening back in April looking in a direction which Yeates had explained would yield no small comet sightings. But ironically, Gehrels' mistake had provided further support for the work done by Yeates. No comet streaks, either single or consecutive, were expected from Gehrels' choice of pointing the telescope, and, indeed, none were found.

But the double images failed to convince Gehrels. "We have been in this business a long time," he wrote on October 17th, "making similar observations and discerning radioactive decay images." He then asked why I had only sent him one pair of double images. "Is that all you have?" And why had Yeates released the pictures to the public without showing them to him first?

He also denied saying that he did not understand Yeates's work. "What I did say was: If you have a major discovery and if you show pictures at all, they have got to display convincing images." If we were seeing something, Gehrels suggested, perhaps they were particles from Halley's Comet.

Nonetheless, Gehrels hoped that the small comets "were here to stay, as a new population of the solar system. We will then honor you and Yeates for having pushed them." But he could see no evidence for them and he wanted his name taken off the paper for the AGU meeting.

The pair of pictures of the same object I sent Gehrels was not the only one, I replied. There were, in fact, six such pairs out of forty-eight pairs of images. Opinions as to whether or not the objects are real were simply not relevant. Where is your analysis, I asked Gehrels? He had, of course, never done any work on the images. His casual inspection of the photographs was meaningless. I asked him how "radioactive decays" could produce consecutive streaks in the pairs of images? It was simply impossible.

I then went on to inform Gehrels that it was Yeates' right to release the images. The Jet Propulsion Laboratory purchased the use of the Spacewatch Telescope and Gehrels had not participated in the analysis of the images. I also asked how he could possibly propose an alternative explanation for the objects in the photographs, if he did not even believe the objects were real in the first place. And if he had looked at the size and number of objects involved, he would have realized that his explanation regarding Halley particles was absolute nonsense. I promised to remove his name from the paper.

Gehrels also sent a copy of his letter to Yeates. Yeates sent him a stinging reply. "The complete absence of any quantitative analysis of the data on your part," he wrote, "even though you had the images at your facility, plus your insistence that visual inspection of the images is sufficient to determine if the images represent real objects, leads me to conclude that you are not competent to 'check their validity.'" Yeates reminded Gehrels that he too had long ago offered to come to Arizona and discuss his analysis of the pictures with Gehrels personally, but that Gehrels had refused on the grounds that he was too busy. And Yeates objected to Gehrels' statement that he was "pushing anything." Said Yeates: "I am just presenting experimental results." It was a very trying time for Yeates.

Gehrels replied a week later. "Your reasonings are impressive," he told Yeates. "On the other hand, it would be reassuring if only a few of the pairs of images would be brighter than near the noise level. If there is indeed a major new population in the solar system, it needs a strong confirmation. Astronomers need to know about this situation and they will quickly confirm or otherwise." Gehrels then sent me a brief postcard. "Relax," he wrote. "Life is too short for animosities. The future has the answer."

Gehrels later insisted that the following note be posted with my paper at the December AGU meeting. "If there indeed were as many cometesimal events in the noise of our CCD frames as L.A. Frank and C.M. Yeates believe, I would expect a noticeable distribution in their size and distance such that occasionally we should see a convincingly bright event. The images that have been shown to us are unconvincing, and this includes the observations of 19 April 1988 for repeated events. Anyone can look at the pictures released by Yeates and Frank, and decide for oneself. Before believing the discovery of cometesimals, a new type of population in the solar system, one would want to see a few good images. Independent observations are needed."

There were several problems with Gehrels' statement. He was talking about cometesimals. These were Donahue's objects and they had long ago ceased to exist. What Yeates and I claimed to see in the images were small comets. And contrary to his statement, the sizes of the small comets had been determined from the brightness of the streaks in the images. It was an exciting result and showed that the numbers of larger objects were much less than the smaller ones. And with regard to independent observations, this had been discouraged already by the folklore that Gehrels had looked for, but not found, the small comets.

So instead we posted the following: "We express our appreciation to T. Gehrels of the University of Arizona for obtaining the pairs of consecutive images. Dr. Gehrels declined our offer of coauthorship because he did not want to participate in the design of the telescopic search or the analysis of the images. The fluxes of the small comets are larger by a factor of about 100 million than the fluxes he believes are there but has not searched for."

Meanwhile, Yeates submitted a pair of papers on his search for the small comets to Dessler at *Geophysical Research Letters*. One was a paper on how he had conducted the search with the Spacewatch Telescope, the other was the results of his analysis of the single exposures he had obtained. The reviews that came back were amazing. The first referee was clearly against publication. "What is missing from this paper is a scientific mind-set," wrote the referee. "I see many streaks in the figures. So what?" He went on to say that anyone familiar with near-Earth and deep space artificial satellites could "easily invent explanations" for these streaks. Of course, not enough satellites have been launched to account for the high numbers of streaks in the images. This referee stopped just short of "inventing" a fleet of UFOs to explain the streaks. Yeates was dismayed by the quality of the comments. He agreed that anyone could "invent explanations," but he explained that the streaks were so restricted in their character that no other known population of objects could have produced the observed streaks than the small comets.

The second referee, Eugene Shoemaker, who prefers not to hide behind a cloak of anonymity during the review process, recommended the publication of the paper describing the way to search for the small comets, but felt that the results reported in the second paper did not pass "the normal criteria for acceptance of observations of small bodies." He insisted that the streaks were just noise and went into a discussion of his findings with photographic plates which had nothing to do with the present search. Yeates used Gehrels' CCD, this array of photoelectric eyes, because film was worthless in such a search. Shoemaker then went on to underscore the necessity of "multiple images taken in succession," and ended his review with a bold wager: "I'm prepared to give odds that he will not find any verifiable images, although it is possible that he could get lucky." Little did he know that he had already lost his wager.

Some time later Yeates received the comments of a third referee who Dessler said was a "CCD expert." "This is the only person who really reviewed my paper on technical grounds," says Yeates. In his review, the referee had recognized that Yeates had done "a considerable amount of work on both papers" and felt he should be commended for it. But he questioned the performance of the array detector and had "very strong doubts" about Yeates' conclusions. He could not recommend publication, he said, but admitted that "there is one argument that would convince me of the reality of the streaks. If they could be found on successive images." Yeates prepared a six-page response to the referee's comments and sent it on to Dessler.

At the beginning of September Dessler wrote to Yeates, clearly stating that "for your paper to be accepted for publication, the referees must be convinced that you have seen the same object in two consecutive exposures." This was of "absolute importance," he said. He also asked that Yeates delete all his work regarding the reality of the streaks, as the pairs of images would make this "superfluous." And, quite unbelievably, he also wanted Yeates to eliminate most of the discussion of alternative explanations that might account for the streaks. But it was Dessler's final comment that made Yeates laugh: "Until one of these objects is tracked for several days so that an orbit can be determined, your paper fails the primary test of demonstrating that the streaks are caused by the same objects proposed by Frank…"

Even a high school amateur astronomer would know that a small, dim object going past the Earth at 20,000 miles per hour cannot be tracked for several days. Yeates informed Dessler that his demand was "utterly impossible." At best, Yeates wrote, "an object such as the one I'm observing could be viewed for only several hours during one night if it could be located and tracked." He also told Dessler that he would provide a pair of successive exposures that show the same object, but that the contents of his paper had to remain intact: "The results of the flux, size, [orbital] inclinations, and ranges are all required in order to show the correspondence between the observations and Frank's hypothesis." The double images simply confirmed the single images and showed they were real, he explained to Dessler, but "nothing more about the physical nature or orbital parameters can be learned from the double images which is not already known from the single images."

When the referees saw the double images, they must have been taken aback, for the rules of astronomy were about to change. Yeates returned from his honeymoon at the beginning of December to find a letter from Dessler informing him that his papers had been rejected. Enclosed was a second review by Shoemaker saying that three consecutive images of the same object were needed for him to believe the streaks were not noise. Yeates, quite understandably, was angry. *It seems as if astronomers had decided to change the standard rule of confirmation just for us. It was an honor, I suppose. The existence of these objects could be very embarrassing for Shoemaker. Imagine spending all your time looking for objects that are going to hit us in the future and missing all these objects that are hitting us right now.*

So rather than having two images of the same object, astronomers now randomly decided that three were necessary. But I think that if Yeates had three they would have wanted four. And if he had four they would have wanted five. This was not science. This was not a review process. He felt like a donkey being led on with carrots. Yeates was very unhappy about the poor treatment he received at the hands of Dessler and *Geophysical Research Letters*. The papers on the search were supposed to be reviewed within a month and either accepted or rejected. But Dessler strung out the process for half a year. Yeates realized then that it was a mistake to have submitted the papers to Dessler.

Yeates had answered every criticism raised by the reviewers. But with the rejection letter from Dessler and the second review by Shoemaker was a review from a fourth referee, which Yeates had never seen, as well as a short note from the third referee, asking why Yeates had ignored all his points. Yeates had prepared a six-page response! So Yeates called Dessler and asked him about this. As near as Yeates can remember, Dessler replied: "Well, it wouldn't have mattered anyway. I think I forgot to send it to him." Dessler had simply ignored Yeates' defense of his work. This kind of treatment can crush an individual. But Yeates held up. There was nothing wrong with his papers. People were simply frightened of the result.

Shortly after Dessler's rejection letter Yeates received a call from John Horgan at *Scientific American*. Horgan had heard the news from Dessler and wanted to get Yeates' response. Yeates was furious. Dessler had done what no editor is supposed to do. He had revealed what the referees had said about Yeates' papers. The story, entitled "A Snowball's Chance," told of the evaporating support for the small comet theory and explained why Dessler had decided to reject the Yeates papers. Three referees, Horgan wrote, thought that the streaks produced in the images were noise and one believed that the streaks showed the passage of orbiting debris. It was a free-for-all. Donald Hunten later told Yeates that he had the distinction of being the only scientist he ever knew whose rejection of a paper was announced in the press.

The story had gone on to say that Yeates was considering protesting Dessler's decision. And I, supposedly, was considering setting up an observatory in Iowa to look for the small comets. This was not quite true. Actually what I was going to do was look for the brief flashes produced by their impact on the Moon. But it did not matter; Dessler had already sold Horgan the Brooklyn Bridge. Just before publication, Horgan had called to ask me if consecutive images of the same small comet were available to show that the search was successful. I said they were and sent them to him Federal Express. But to my dismay, the pictures were returned immediately with the comment that there was no space in the magazine for the pictures. Instead, the article contained more silliness from Dessler suggesting that "these efforts" would allow the small comets to persist in the public imagination—just like the Loch Ness monster.

Not content to ridicule the small comets in public, Dessler also did his best to keep any mention of them out of the scientific literature. Barclay Clemesha, a scientist at the Instituto de Pesquisas Espaciais in Sao Paulo, Brazil, provided a recent example. Clemesha apparently had a run-in with Dessler over a paper Clemesha had submitted to *Geophysical Research Letters* on a phenomenon called the "Sudden Sodium Layer." The SSL, as it is known, is a formation of thin layers of enhanced sodium concentration that sometimes appears in the upper atmosphere and is not well understood. Clemesha thought the small comets, which he called cometesimals, could contain sodium dust particles and be the source of this curious phenomenon. The usual mechanisms cited to explain this sodium layer, said Clemesha in a letter to me, "are highly improbable, and I still believe that something related to cometesimals is the best of a bad lot. Unfortunately, Dessler only accepted the paper on condition that I remove all reference to cometesimals!"

There was a war going on out there. We had to fight to get our papers published. We had to fight to let astronomers know that these objects existed. If Yeates had gotten negative results his papers would have been accepted for publication within forty eight hours and he would have been proclaimed a hero. That is the way the system works. But these findings were embarrassing. How could astronomers miss these things? So when the comets were seen through the telescope no one believed it. Gehrels was not convinced. Shoemaker was not impressed. Torrence Johnson, an astronomer at JPL, thought they could be real, but insisted that the rate at which these objects passed the Earth had yet to be determined. Others simply said that, regardless of the evidence, the objects could not possibly be there. Astronomers had forgotten the lesson of Galileo.

No one is going to open their arms to you if you go against the mainstream. My own paper on the double images, which I prepared for the December AGU meeting, was relegated to the poster session, where papers are literally posted in a room rather than presented orally. Some reporters came up to me and said, "Well, this must not be very important because it's down here in the poster session." But I made no fuss about it. I knew they would do that. That is the way the game is played. It did not seem to matter that the appearance of these objects in two consecutive pictures was probably one of the most exciting things to happen in the field of astronomy in the past decade.

CHAPTER 27

The Turning Point

The search was over. The existence of the small comets had been confirmed. But few believed it. We had won after nine innings, but the others insisted that the game go on. There would be no closing time. We would just continue to run up the score.

The 23rd of March 1989 was like any other day. Except that the cosmos threw a curve ball at Earth and astronomers struck out once again. An asteroid measuring about half a mile in diameter had sped by the Earth unnoticed at about 46,000 miles an hour. This boulder missed the Earth by a mere half million miles, which is about twice the distance to the Moon. In cosmic terms, this was a very close call.

Had this huge rock hit us, it would have left a mile-deep crater in the ground the size of the District of Columbia. Had it impacted the ocean, waves hundreds of feet high would have swept over coastal areas and caused widespread destruction. And one day it just might. Scientists are convinced that the asteroid would likely return sometime to hit either the Earth, the Moon, or Mars.

The event itself is alarming. But more worrisome is the fact that this huge rock crossed the Earth's orbit undetected. No one knew about it until almost two weeks after it had passed. The object was found by Henry Holt, a retired U.S. Geological Survey scientist now working on a NASA-funded project to detect and track asteroids. Holt discovered the object in a set of photographs which he and Norman Thomas, a retired astronomer from the Lowell Observatory in Flagstaff, had taken on March 31 using the eighteen-inch Schmidt telescope at the Mount Palomar Observatory in California. When the photographs were processed several days later Holt noticed a streak of light against the background of fixed stars in two pictures taken an hour apart. This streak indicated that the asteroid was moving across the Earth's path and not on a collision course. A NASA scientist told the press that an asteroid heading right for the Earth would look just like a small fixed star and would be even more difficult to detect.

This is how little we know: A mountain-sized object narrowly misses the Earth and essentially escapes the attention of astronomers. It was more than a million times more massive than a small comet, but no one saw it approach the Earth. No one knew it had come so close until it had already passed. If astronomers cannot even see a mountain-sized object as it approaches the Earth how could they expect to see the small comets? Whatever gave them the impression that they would be able to detect something the size of a small house, and as dark as a black sheet of paper, in the Earth's vicinity in the first place?

The close call seemed to mark a turning point in the small comets controversy, though the first faint glimmerings of acceptance began to appear several months before. In January of 1989 NASA had awarded me a $30,000 grant to work on small comets. This is not a lot of money. I get millions of dollars to do scientific research, but until I received this grant I did not have a legitimate license to work on small comets. Of course, the amount of money itself was far less important to me than what it signified. This was official recognition for the legitimacy of small comet research. Though my proposal asked for financial support on several investigations, only one of which concerned small comets, all but the small comet funding was rejected.

The new social situation took me by surprise. A few months earlier I learned that my career apparently had not been sacrificed by my pursuit of the small comets. The University of Iowa, in recognition of my work, and perhaps in spite of the worldwide controversy over the small comets, gave me the Carver/James A. Van Allen Chair in Physics. The Carver is an endowed chair and is considered to be the highest-ranking academic position at the university. There are only twelve such chairs. The chair is also named after James A. Van Allen, discoverer of the radiation belts around the Earth and my mentor as an undergraduate and graduate student. I could go on doing what I had been doing. And I would do just that.

Yet it was quite clear to me that the scientific community in the United States was still in such a deep emotional state over the small comets that publishing the results of Yeates' recent telescope search was just not possible in this country. So, instead, Yeates and I decided to submit our papers on the search to refereed European journals with excellent reputations. The result was like a breath of fresh air. The comments of the reviewers were helpful and fair. Yeates submitted his paper on the search and the single images of the small comets to *Planetary and Space Science*. The reviews were favorable and the paper was published, after some minor changes, in October of 1989.

Our paper on the consecutive images was submitted to *Astronomy and Astrophysics*. The referees made a few suggestions and wanted a statement in the paper saying that the observations were tentative evidence for the small comets and that confirmation would require a more capable telescope. They also wanted the inclusion of an alternative suggestion that the tracks might be due to a previously unknown population of tiny moons orbiting around the Earth. I agreed to these changes and the paper was published in February of 1990.

We had successfully bypassed Dessler. But he was still up to his old tricks. His three-year term as editor of *Geophysical Research Letters* ended on December 31, 1988, and his application to continue as editor had been turned down. But Dessler spent much of the following year writing a long paper about why the small comets did not exist. It seemed to me not to be a serious statement but a spoof of science written for the entertainment of the reader. Impacts of objects with the Moon were described colorfully in terms of "tinks," "pops," "booms," and "woomps." And as proof that small comets could not exist, Dessler cited magazine and newspaper reports in which Dessler himself had been quoted as saying that Yeates' telescope observations were rejected as noise. Who was he trying to fool?

James Burch followed Dessler at the helm of *Geophysical Research Letters*. Burch is a scientist and vice president of the Southwest Research Institute in San Antonio, Texas. He is known to be objective and mild-mannered. So in June, when the scientists at the University of Calgary decided to withdraw their names from the joint paper we had been writing on the atmospheric holes seen by the Viking spacecraft, we decided to submit the paper to Burch on our own. Sigwarth and I had put a lot of work into it; we had shown that the chance that the Viking holes were not real was about one in ten thousand. This was good enough for us.

Burch sent the paper to two referees, one who had good suggestions for strengthening our arguments and the second who was more interested in venting his frustrations. He amused himself by comparing the small comets to flying pigs. "As credible evidence for the theory that pigs can fly..." this referee wrote, "one would require considerably more than a selection of blurry photographs ... purportedly showing examples of the shadows that pigs have cast on the ground as they flew rapidly by overhead." Burch did not allow us to see these comments until the paper was accepted because they had nothing to do with the proper conduct of scientific inquiry. When we finally did see them, we found them quite amusing. In December of 1989 Burch published the paper showing that the atmospheric holes had appeared in the pictures taken by the Viking satellite. It was a very good month.

Much to his credit Burch was not closed-minded on the subject of small comets. He also published the work of Scott Bolton, a young, capable scientist at JPL who helps control various instruments on the Jupiter-bound Galileo spacecraft with radio commands sent from the ground. While researching his doctoral thesis for the University of California at Berkeley, Bolton found an unusual effect in the data he had examined. The wind of positively charged ions that flows continuously from the Sun into interplanetary space appears to slow down along a portion of the Earth's orbit. This happens almost every year during the months of November to January. Bolton claims that this slowdown of the solar wind is not due to the Sun itself, unless there is some unknown yearly activity, but to some undetected material in interplanetary space. The only source of this material that he can identify is the water vapor that leaks from the small comets.

Our search for the small comets was delayed by the astronomers. We were delayed because they would not look. That is something I never counted on. I never thought that we would have to take the extra time to go in and be astronomers ourselves. That we did so certainly means that our findings are less believable than if astronomers had done it themselves. But we had no choice. It was a very bad social situation. Now, perhaps, astronomers will be forced to go in and take a look. If not, the big breakthrough could come from their far more numerous kin, the amateur astronomers.

In the summer of 1989 I was invited to give a talk on small comets to a large gathering of amateur astronomers at the Texas Star Party at Prude Ranch near Fort Davis, Texas. I was immediately struck by two things when I arrived. I realized that my home in the cornfields of Iowa was not the world's end; this was. I was also very much impressed by the vast array of telescopes on the fields of the ranch. Though I am not an amateur astronomer, I had been asked many times about the possibilities of looking for the small comets with smaller telescopes but had been too busy to do anything about it. Now, I had done my homework.

Amateur astronomers whose telescopes have mirrors or lenses measuring twelve inches or larger should be able to sight the small comets. But they would have to know where to look and what to look for. During the course of a day there are two times for observation, each about one or two hours long. One ends about forty-five minutes before sunrise; the other begins about forty-five minutes after sunset. The small comets will be seen at a distance about 2,500 to 4,500 miles from the observer, so the telescope should be pointed in such a way that it is looking for them at these distances, just outside of the Earth's shadow. Inside the Earth's shadow the objects are not illuminated by the Sun and are invisible.

What would you see? Every two hours or so a small, quite dim object will slowly move across your view, as long as your field of view is about four times the size of the Moon. The object will move a distance equal to the Moon's diameter every five seconds or so. Of course, if there are ten telescopes viewing different regions along the Earth's shadow then the number of sightings by all ten are one small comet every fifteen or twenty minutes.

One amateur astronomer came up to me after the talk and said: "We've seen objects like that, but we just never counted them. We thought they were satellites." But the number of man-made satellites that move in this way is very small and if the number of events seen by the amateur astronomers is high, as predicted, then they must be due to small comets. There are thousands upon thousands of suitable telescopes in the world and their combined observing power is awesome relative to what professional astronomers can achieve with their larger, though far fewer, telescopes. But best of all, amateur astronomers are not afraid to look.

Part 2: Vindication (1990-2001)

CHAPTER 28

The Rest of the Story

With the arrival of the nineties, the small comet debate entered its second decade, but most people were still singing the same eighties tune. That first summer had *Scientific American* declaring that the small comets were "still dead." This was no doubt writer John Horgan's idea of comedy. Just as *Saturday Night Live* had noted "Franco is still dead" several weeks after the death of the Spanish ruler, mocking the way in which the media had obsessively noted every downtick in the Generalissimo's health, so Horgan would do the same for "four dubious but enduring theories." In the story, Horgan put small comets in the same bag as B. Stanley Pons and Martin Fleischmann's cold fusion, Thomas Gold's non-biological petroleum, and Ephraim Fischbach's fifth force.

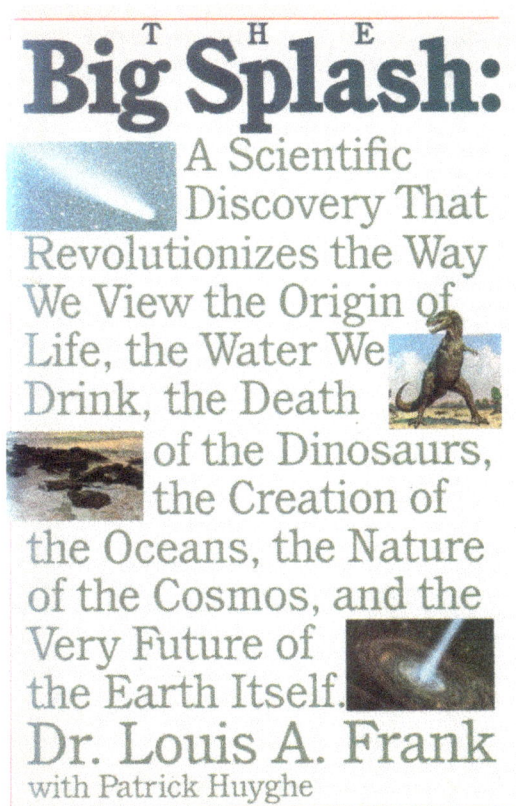

I've been watching this kind of treatment of science for the past 20 years, and I suspect that, as a result, fewer and fewer scientists will be willing to venture out of the mainstream. Anyone caught straying from the science herd is an easy target, of course. But it should be noted that this attack came from a writer who would go on to write a book called *The End of Science*, which essentially argued that science has already made all its big discoveries and that all that is left for science to do is a matter of relatively minor refinements. Having someone like Horgan covering advances in sciences is like hiring a wolf to tend to the sheep.

Just a few months later, I would finally have my say in such matters. In October of 1990 the first edition of this book, called *The Big Splash*, was published by Birch Lane Press, an imprint of the Carol Publishing Group. The subtitle that our editor, Hillel Black, wrote for the book, got the attention of *Publisher's Weekly* as the fall season's longest subtitle: "A Scientific Discovery that Revolutionizes the Way We View the Origin of Life, the Water We Drink, the Death of Dinosaurs, the Creation of the Oceans, the Nature of the Cosmos, and the Very Future of the Earth Itself." It is true, of course, that we touched on all of these topics in the book, as the existence of small comets has widespread implications, but I think the all-compassing and rather grandiose sounding subtitle made some people who were already unfavorably predisposed toward the small comet discovery even more critical. "Here's the answer to *everything*," the book cover seemed to say. It was not, of course, and I never meant it to be so.

Despite this minor annoyance, I realize that it took guts to publish this book. The Carol Publishing Group took on a book that no one else in the industry was willing to touch, for the most amazing of reasons, it turns out. When the book was shopped around, the general reaction from publishers was "Interesting idea, but is it true? Unless Frank is right, we don't want to publish the book." Clearly this is a Catch-22, and while I cannot blame publishers for their caution and endorse their interest in the truth, I have never known publishers to be so concerned about such matters in the past—or since for that matter. It seems that they habitually pass off as "truth" often rather questionable material. Apparently my story deserved special treatment.

When the book appeared, the reviews were, on the whole, quite favorable. The reviewer for *The New York Times* called it "a savage tale of how science works." The *Discover* magazine editor who wrote a review for *New Scientist* said the book was "a highly readable, rather bitter, and entertaining account" of my side of the controversy, and argued at the end that "it seems only fair to keep on open mind." The reviewer for *Sky & Telescope* found himself "violently disagreeing many times" with what I had written but admitted that he had "enjoyed this book immensely." The book also apparently "failed to overcome… [the] skepticism" of the reviewer for the *San Francisco Chronicle*, who nonetheless thought the book was "well worth reading." *Kirkus Reviews* noted that my "wide ranging theory crunches on many scientific toes"—it does—and the *Library Journal* called it "a provocative work in a hotly contested area"—which is also quite true.

Even the kindliest reviewers found it necessary to hedge their bets, however, pointing out that the existence of the small comets was by no means a done deal. I had no problem with that. I did not expect any enthusiastic endorsements. At least the small comet controversy was out there in all its sordid detail and that is what counted. Now everyone could read "the rest of the story," as the veteran radio newsman Paul Harvey was so fond of saying. Of course, most of the established scientific press simply ignored the book. Neither *Science* nor *Nature* ever saw fit to even mention its existence. That was not a surprise, but the reaction of some of my critics certainly was. I heard that Alex Dessler was proud of having played the role of villain in the book, and Tom Gerhels told a number people he thought it was a great book.

A month after the book appeared, Dynamics Explorer 1, the satellite that provided the data for the discovery of the small comets, died. The spacecraft was only expected to last a year before being destroyed by radiation, but it lasted nearly a decade, from the fall 1981 to the fall of 1990, and certainly qualified as a "grand old lady of space." Near the end, the data it produced became increasingly sporadic. There was no degradation of the images themselves, however, just in their number. Finally, the batteries that powered the camera died out. But the black spots kept showing up in its images—all the way to the end.

A few small comet experiments on tap around this time all failed to materialize for one reason or another. The first was our own attempt to look for evidence of small comet collisions with the Moon. During the course of building the next generation of cameras to be lofted into space, we purchased an array of photoelectric eyes that was sensitive enough to look for brief flashes on the Moon with the University of Iowa's own 24-inch telescope. The array would have allowed us to take rapid exposures of the Moon with enough sensitivity to confirm whether the flashes that have been reported on the Moon are there or not, and if so, how often they occur. This would have to be done when the Moon is not too bright, when a large portion of its Earth-side face is in the dark. When the Moon is nearly full and bright, its light would simply swamp any flash of light measuring tens of feet on its surface. But by looking at the darkened part of the less-than-full Moon, we thought we should be able to see the impacts of these small comets on its surface, bright flashes lasting just a second or less. The problem we ran into had to do with the electronics, which would not allow us to read out the camera quickly enough. And we certainly did not want to sit there and look through a million images for lunar flashes; the task would extend beyond my lifetime.

But this matter of transient lunar phenomenon cannot be shoved aside by astronomers forever. A lot of people are still seeing these mysterious flashes and glows on the lunar surface. On April 23, 1994, for example, about a hundred amateur astronomers reported seeing an odd darkening near the edge of the bright lunar crater Aristarchus. Their observation would later be confirmed by data from the Clementine spacecraft, which happened to be mapping the lunar surface at the time.

Others were also looking for novel ways to use new technology to search for the small comets at the end of the 1980s. Among them was Peter Banks, a professor of physics in the Department of Electrical Engineering at Stanford University. He had shown in the June 1989 issue of *Geophysical Research Letters* that the radiant heat from the cometary water clouds should be detectable with modern-day cameras on low-altitude satellites. This heat radiation, also known as infrared radiation, is produced by the collision of the cometary water as it plunges through the very tenuous atmospheric gases at high altitudes. Banks is correct. Hot water turns out to be quite bright in the infrared, which is the same type of radiation that you feel from an electric stove or keeps hamburgers warm at fast food restaurants. Such measurements have probably already been made, but public access is restricted. Missiles and rockets also eject hot water vapor into the atmosphere and the search for such hot water molecules may have already been accomplished in the early stages of the Star Wars Defense program. If so, the vapors from the small comets also may have been detected.

Unfortunately, Banks' infrared search never happened either, and for rather obvious reasons, all of them tied in with what the military is doing. The military is interested in detecting ICBMs launched from such places as China and Russia and can do so using cameras that can pick up the infrared heat coming from the hot water in rocket exhausts. By telling us whether or not the military can detect the water from the small comets, they would reveal the sensitivity of their sensors and that would be a security problem. But whether Banks' calculations are right or wrong, there should indeed be infrared signals from the small comets.

Another exciting experiment was suggested by a question that came up again and again during the debate on the small comets: How often would the space shuttle and other spacecraft run into these objects? It was a natural question, so I did some calculations. The results were fascinating. At high altitudes, an impact of a spacecraft with a small comet that has not vaporized would be disastrous. But these comets are small and the occurrence rate is low, so an average-size spacecraft would only be struck once every 50,000 years or so. This means that one spacecraft in every thousand will be struck in high Earth orbit every 50 years. Has it happened yet? No one knows. But some satellites have been lost and no one knows why.

The situation in low Earth orbit, where the space shuttle astronauts fly, is quite different. They will run into one of these things once in every 200 orbits. But what they collide with are the cometary water clouds, not the comets themselves. The collision of a large cometary water cloud with a spacecraft is benign. The shuttle would survive such a collision very easily. Its surface temperature might go up a few degrees and there would be a slight drag on it, but these effects would be nearly undetectable. So the astronauts have probably flown through these things and not known it. But there is a way for the astronauts to see these cometary water clouds.

Enter Stephen Mende, a physicist at the Lockheed Palo Alto Research Laboratory in California. Mende proposed a very simple, low-cost experiment that would allow the astronauts to see the cometary water clouds as they move across the sky. It turns out that these things can become visible in one of the ways that Halley's Comet has been observed. Halley's has an immense hydrogen envelope around it, produced by the break-up of water into its constituents, a hydrogen atom and a hydrogen-oxygen pair, from exposure to ultraviolet light from the Sun. This hydrogen-oxygen pair glows brightly as it scatters solar radiation invisible to our eyes in the ultraviolet. Halley's has been photographed in this light.

The small comets should produce the same cometary bright glow as they break up above the Earth's atmosphere. If the astronauts had a camera outfitted with a device that allowed them to look at this wavelength, they should be able to see the clouds of water vapor from these comets. This camera would operate like the night-scope of a rifle but for ultraviolet light instead of infrared. The clouds would look like great spheres plunging through the atmosphere. They would be a spectacular sight.

The problem is that this special hand-held camera for ultraviolet light burned up in the Challenger accident. Lockheed did build another one, however, and the subject of what could be done with it came up at a meeting with the astronauts back in the fall of 1988. There were some two dozen astronauts at the briefing at the Johnson Space Center in Houston, and I was there discussing how they might help take better pictures of the auroral lights. Someone then mentioned the small comets, and I said there was a way to see them. There was some snickering, at first, but I think they would love to do it.

Mende and I worked out how such an instrument could be used to search for the small comets. We think that the astronauts would need several hours of observing time and they would have to point the camera in a particular direction to be able to see these large cometary clouds. With luck they might be able to spot one every seven minutes, but it might take as long as forty minutes between sightings. And sightings would only be possible when the camera and shuttle window are not illuminated by the Sun or by the dayside of the Earth because this light would blind the camera.

Eventually astronauts using such a camera are bound to experience a close approach with a cometary water vapor cloud. It is likely to frighten them. It would start as a bright little dot in the star field and expand quickly into quite a large object. Imagine seeing such a thing coming at you at 40,000 miles per hour. You would see it for about a minute. But if you looked away from the camera momentarily to the point in space where the object was hurtling towards you, your eyes would see absolutely nothing.

Mende's instrument has yet to fly on the shuttle, however. NASA could still do it, I suppose, but it just has not been manifested. It takes a long time to schedule these things. Putting something on a shuttle is a big deal. It requires a lot of follow through, and I guess that after being denied over and over again, Mende just lost interest and the project eventually got lost on somebody's desk at NASA. They needed a little bit of money to do this, and NASA apparently decided they did not have that little bit of money. That is how things work out sometimes.

The next year, 1991, brought great tragedy. On April 18th Clayne Yeates, the brave physicist from the Jet Propulsion Laboratory, was on his way to an after-work drink with some friends when he suffered a brain aneurysm and drove off into a ditch. He never regained consciousness. Over the years, he had suffered some very severe headaches that his doctors could never track down. His death was a great loss. He was a good friend and a great scientist. Other than Sigwarth and I, Yeates was the only other major promoter of the small comets; he had gone out there as a skeptic, done the telescope measurements, and had come back a believer. Now, again, there were two.

Then, several months later, the University of Iowa experienced a bloody nightmare. On the first of November, a Chinese graduate student in physics named Gang Lu stood up during a meeting in his department, pulled out a .38 caliber revolver, and began "methodically extinguishing some of the brightest minds the campus has known," wrote a reporter for *The New York Times*. Apparently distraught over his failure to receive an academic honor for his dissertation, Lu gunned down several people in my department, including Lu's dissertation advisor, Christoph K. Goertz, a 47-year-old professor of physics and astronomy; Robert Alan Smith, an associate professor of physics and astronomy; and Linhua Shan, a research investigator and graduate student in the department whose dissertation was nominated for the award.

The mad student then went down to the second floor of Van Allen Hall where I have my office and fatally shot the chairman of the department, 44-year-old Dwight Nicholson. Then, sensing that Smith was not yet dead, Lu went back to the third-floor meeting room and finished his killing of Smith. He then continued his premeditated killing spree, dashing over two blocks to Jessup Hall, the administration building, where he asked for Anne Cleary, the associate vice president of academic affairs with whom he had filed a complaint about the nomination process. When Cleary entered the reception area, Lu shot her in the head. At the same time, he also wounded Miya Rodolfo-Sioson, an academic affairs staff member who was at her desk. A few minutes later, Lu shot himself in a second-floor room above the office of University of Iowa president Hunter Rawlings.

When I heard the gunfire, I began shoving people back into their offices because all the secretaries were popping their heads out when they should have been diving under their desks. I was not a target, but the event did affect me. When senseless killings like this occur, you become more wary of people. The thought that one of my critics might go off the loop like this does bother me. Some scientists feel personally threatened by this small comet discovery. Not physically threatened, of course, but their jobs or their position or their reputations are threatened. Some feel that I am destroying what they have worked for all their lives, and I am very sensitive to people who feel that way. But there is nothing much you can do about it. The University is an open institution with free access to people in it. My home, however, is surrounded by a nine-foot-tall, partially-electrified, wooden fence.

Meanwhile, a shootout of a more civilized kind was taking place between myself and Alex Dessler. The editor who had published my original small comet papers had been working on the mother of all small comet critiques and in August of 1991 it finally appeared in *Reviews of Geophysics*. The editor of *Reviews of Geophysics* at the time, Marcia Neugebauer, sent me Dessler's paper for my comments. She may have hoped that I would make some small comment, that Dessler would make his revisions, and that would be the end of it. But I told Neugebauer that Dessler's paper was so full of errors that I should be allowed to write an equal length reply, and Sigwarth and I ended up doing just that.

Dessler's paper essentially replayed much of the criticism of the small comet hypothesis that had run in *Geophysical Research Letters* for two years, but it was full of false statements and errors of various kinds. Dessler, for instance, neglected such basic things as gravity in his model of how the small comets deposit water vapor in the atmosphere. On the subject of the forces that cause the breakup of the small comets in the atmosphere Dessler miscited the value of tidal stresses at an altitude of about 1,000 miles, and he miscited one of our 1986 papers in stating that "one sixth of the small comets are in retrograde orbits," though we made no such statement. Dessler also miscited us on the source of the small comets, saying they come from the Oort cloud rather than from the inner Oort disk, and he mixed up forward-scatter radar meteor rates with backscatter radar measurements. Perhaps he did not understand these simple but important matters. When calculating the lifetimes of ice particles in space, Dessler's equation has an error in it. And like everyone else, Dessler had problems with small comet impacts on the lunar surface. He asserted that they should mimic the impact of one stage of the Apollo Saturn launch vehicle onto the lunar surface, when in fact this rocket stage does not in any way resemble a low-density porous object like a small comet.

I do not think Dessler cared about any of this. His primary point was that the black spots in the Dynamics Explorer images were due to some unknown instrumental artifact. Everyone said it was an instrumental artifact, and I can see how one would like to accept that. I think that most people did not even read the relevant literature, they just took the word of the next fellow down the line, that an artifact was there, and that it had been found. But no one has identified an instrument artifact responsible for the dark spots in our images. We could not and neither could Dessler, though he certainly tried. His logic was flawed because it was based on numbers that were incorrect. Dessler simply ignored anything that did not support his criticism. He ignored the results of the brilliant telescopic search by Yeates because Gerhels had said it was noise, for example, and he ignored the geophysical correlations that the black spots increase in size as they approach the Earth, and that their frequency matches that of radar meteors in appearance.

Our paper, entitled "Atmospheric Holes and Small Comets," which appeared in *Reviews of Geophysics* in February of 1993, responded to each and every point brought up by the Dessler paper and, as such, is a complete review of the phenomenon in itself. We also used the opportunity to update the probable mass of the small comets. The possibility had been suggested to us, by Arthur L. Lane of the Jet Propulsion Laboratory, that the cometary water was not absorbing the ultraviolet light on a molecule by molecule basis, but that clusters of molecules could also be producing the dark "atmospheric holes" we were seeing. This realization brought everything into agreement with the measurements Clayne Yeates had acquired using the Spacewatch Telescope. Because his data gave us a better idea of the size of these objects and their visual magnitude, we were able to come up with a more accurate picture of their mass. Basically, we concluded that the small comets were probably half as massive as we had originally estimated on the basis of the Dynamics Explorer images alone.

On the heels of Dessler's paper in *Reviews of Geophysics,* but long before our response appeared in the same journal, we had to endure the humiliation of Dessler's dog receiving an award from the American Geophysical Union. The award was presented to Tex, Dessler's dog, on December 10th at the Solar-Planetary Relationships section dinner at the fall meeting of the American Geophysical Union in San Francisco. Tex received the award from section president Bruce Tsurutani for "keen astronomical observations and loyalty to authority." The news appeared in *EOS*, the official newsletter of the American Geophysical Union, in March of 1992. Over a six-year period, since the small comet debate began, this was the first and only mention of the small comets to appear in *EOS*. I realize that the awards are supposed to be given "in fun and are intended to be humorous," but given the American Geophysical Union's total disregard of the small comets in the past, I felt this first mention of the subject in the American Geophysical Union newsletter was tasteless.

While Tex was winning awards, those with my name on them were being canceled. Many of my colleagues labeled me a crank for my unwavering defense of the small comets, and I was blackballed from the community. It is public knowledge, for instance, that I was not elected to the prestigious National Academy of Sciences for this very reason. All in all, it was yet another shameful period in an increasingly sordid tale.

CHAPTER 29

Polar Rising

Such is the nature of space exploration that scientists working in this field often must think about their *next* project years before their current project has gotten off the ground. And so it was for us with Dynamics Explorer. Even before its launch in 1981, NASA had begun selecting participants for a proposed spacecraft called Polar. The purpose of the Polar mission was to study the flow of energy and particles from the sun into the Earth's atmosphere. Initially we were not selected to do the auroral imaging on Polar because NASA headquarters was convinced that our auroral cameras on Dynamics Explorer would be a failure. That was the rumor anyway. So NASA selected a team from Johns Hopkins Applied Physics Laboratory (APL) to do the auroral imaging on Polar instead.

Shortly after NASA's decision, Dynamics Explorer achieved orbit and started sending back these beautiful auroral images of the Earth. NASA was shocked. So I decided to challenge NASA's decision on Polar. We had to convince NASA why they should go with the proposed Iowa imager rather than the one from APL. NASA debated the issue for years, but finally we were successful; it was the only reversal of its kind, as far as I know. I am sure the wonderful Dynamics Explorer images really made the case for us. But the reversal no doubt caused some bad feelings with the people at APL such as Paul Feldman, who was to be principal investigator, and Ching Meng and the others at Johns Hopkins who would have built the instrument.

So it was that we began working on our instrument for the Polar spacecraft several years before the small comets were even a gleam in the eyes of Dynamics Explorer. By the mid-1980s, however, we had the data from Dynamics Explorer, so we knew the small comets were real, but we were still a couple of years away from making our findings public. We began, however, to think about how we could modify our instruments on Polar to include the capability of looking for water clouds directly in the visible wavelengths. Opportunities like that do not come around often in space physics, but we made the changes to our instrument without any risk to the primary objective, which was studying the Earth's auroral lights.

Basically, we made sure that our ultraviolet camera on Polar would have a very large field of view and very low noise so that there would be no question of instrumentation being responsible for the black spots in the images. And then for the visible-light cameras, it was a matter of choosing the right filters. We only had a limited number of slots available for filters. We were going to use the auroral filters obviously, but we were also able to put in a few others that had not been used on an Earth-orbiting space platform before. One, the hydroxyl filter, looked for light from the fragment of a broken water molecule, which has a hydrogen atom paired with the oxygen atom. The other was a sodium filter. These filters would eventually tell us about the dust, water, and sodium content of the small comets. Are they just like the known large comets, only smaller, or are they an entirely different class of objects?

Of course, we needed an ultraviolet imager to do our small comet work and there was already another ultraviolet imager on Polar. So we had to have a very good reason for having our own. But that was easy: it was a safety measure. Our ultraviolet camera would give us a view of the Earth that would enable us to protect our visible imagers. If the visible imagers were, in fact, pointed at the bright sunlight, they might be destroyed from such intense light. Of course, we used a filter on the ultraviolet camera that matched the Dynamics Explorer filter (1,304 angstroms), which we knew from experience was ideal for looking for atmospheric holes.

The other UV imager that would go up on Polar was originally designed by Doug and Marcia Torr, who at the time were stationed at the Marshall Space Flight Center in Huntsville, Alabama. They were concentrating on the ultraviolet, whereas we were concentrating on the visible emissions of the aurora. They designed and built their ultraviolet camera. They understood it. Then one day they lost control of their instrument because of "financial irregularities" and suddenly NASA headquarters had to find someone to take over the instrument. George Parks, one of the Torr's co-investigators, begged for it, and even though the instrument was clearly outside his field, NASA handed him the instrument. Parks was the Torr's co-investigator for plasmas, and he had had a plasma instrument on ISEE, the International Sun-Earth Explorer, which we had been on as well. Space is a big place, but space science is just as certainly a very small world.

The Polar project had more than its share of problems. The spacecraft was constantly plagued by a lack of funds. At one point, NASA even tried to cancel the project and as a result there was a hiatus of more than a year when absolutely nothing was done. Getting Polar going was one long, hard fight, but after almost being cancelled several times for budgetary reasons, the mission was finally approved in the late 1980s.

The delays were so sufficiently long, however, that there were considerable changes to the NASA personnel in charge. Initially, Jack Scudder was in charge, but when the University of Iowa hired Scudder away from NASA, Mario Acuña took over on Polar. Six months before the scheduled launch, however, the project was in shambles. The principal investigators had no funds to develop their computer programs, and there was no staff at NASA Goddard to plan the operations. At that point Acuña convinced a very hesitant Bob Hoffman, who had been project scientist on Dynamics Explorer, to take over as project scientist on Polar. Hoffman immediately discovered there were funds available at NASA headquarters for the project that had never been properly distributed and was able to get things off the ground—literally.

We worked overtime to get our instrument done. We began putting it together in January of 1993, and we met our first delivery of the instrument for assembly on the spacecraft in the fall of that year. NASA had scheduled Polar for a shuttle launch in December of 1995, but that would get delayed until the following year because it interfered with other launches. Polar ended up on a Delta rocket that went up on the 24th of February 1996. It achieved an eccentric orbit with a perigee of about 3,200 miles and an apogee ten times that distance, positioned over the northern polar regions. In this eccentric orbit the Polar spacecraft circled the Earth every 17.6 hours.

We were worried when we first turned on the Earth Camera of our Visible Imaging System (VIS) on Polar. With all the high voltages, all the sensitive electronics, start-up is a very dangerous time for an instrument. In fact, before launch we had a bad arc, much like a small lightning bolt, in the vacuum chamber that destroyed our Earth camera, and that was a heart-stopper. Anyway, the first month in space was largely taken up with engineering tests. The trick is to bring the instrument up slowly, gradually increasing the high voltage to the values at which the instrument will operate. Jeff Dolan of the University of Iowa was at NASA Goddard with John Sigwarth sending commands to the instrument in real time so that they could tell right away what was going on.

By the end of March 1996, we were getting our first images, but they were fairly dim. When we began to get long exposures, however, everything was smeared. We realized then that we had a major problem on our hands. Soon after, launch engineers discovered that a

An atmospheric hole appears as a cluster of pixels above Poland in this global view of the Earth's far-ultraviolet dayglow, or sunlit atmosphere, taken by the Visible Imaging system (VIS) on NASA's Polar spacecraft on April 6, 1996. Images of atmospheric holes taken by the Dynamics Explorer 1 spacecraft 15 years previously showed the atmospheric holes in single pixels, leading critics to dismiss the phenomenon as an "instrumental artifact." These black spots in the image, or atmospheric holes, are caused by water vapor clouds from small comets disintegrating anywhere from 600 to 1,200 miles above the Earth. The water vapor clouds momentarily and locally absorb the Earth's dayglow as viewed from the Polar spacecraft. This dayglow, which is emitted by oxygen in our upper atmosphere, is light at the ultraviolet wavelengths beyond the range of human eyes. A coastline map has been superimposed on the image to better understand the position of the event.

magnetometer boom did not extend out all the way, so the spacecraft engineers had to use the weights inside the spacecraft to try to balance it out. But the weights just did not have the dynamic range the engineers needed. They managed to bring the wobble down a bit, but they could not get rid of it entirely.

There was a lot of wobble. It smeared everything in the images. For example, with the spacecraft at apogee, we could not tell whether a feature in the images was located at Boston or just south of Washington, D.C. What were we going to do with that? It not only jeopardized the auroral measurements we wanted to make but the atmospheric hole work as well. We were very depressed for a while. To have worked all those years and then to have the images come out worse than those from Dynamics Explorer was a real letdown.

But we were lucky. We managed to compensate for the wobble. The computers on our instruments can track certain signals from the spacecraft, sense what direction the camera is pointed, and can lock in to one of the fixed points in this wobble motion and retrieve the resolution. We did have to sacrifice the sensitivity because we had to close the shutter during part of the image exposure. Our duty cycle is therefore less, but we had enough sensitivity to overcome that. Without this ingenious fix, Polar would have been a disaster for us.

So it was not until the first week of April that we were able to get decent images out of Polar. By that time, I was back in Iowa. One day shortly afterward I was on the phone with Sigwarth, and he was reporting on all the engineering details he was monitoring at Goddard. And then, at the end of this rather long conversation, Sigwarth casually said, "Oh, yes, the atmospheric holes are there." We knew they would be. Jeff Dolan saw them, too, but no one else in the control room at Goddard knew what they were looking at. The holes were there and just the size we expected them to be. I was overwhelmed with joy by their report.

Of course, it did not take long for people to start asking us if Polar could see the atmospheric holes. Astronomer Robert Mutel at Iowa wanted to know right away. So did Van Allen. But we did not tell anyone. We just said: "We're working on it." We wanted to do the complete analysis first. We wanted to wait until the first paper was accepted before saying anything at all to anyone. But things did not work out that way.

We began working on our first Polar small comet paper almost immediately. It was easier doing the analysis the second time around because this time we knew what the holes looked like. This first paper essentially stated that the new Polar spacecraft had confirmed the existence of the stream of small comets in the vicinity of Earth that was first discovered with the Dynamics Explorer spacecraft a decade earlier. We submitted this paper, entitled "Evidence for the Presence of Small Comets in the Vicinity of Earth," to *Nature* in early June of 1996.

Three weeks later we received *Nature*'s rejection letter. "I regret that we are unable to conclude that the paper provides the sort of advance in understanding that would excite the immediate interest of a wide, general audience," wrote Leslie Sage. Nor does it address the issues raised by Alex Dessler in his 1991 *Reviews of Geophysics* paper, the *Nature* assistant editor continued. Of course, it did not, we had addressed those issues in our own *Reviews of Geophysics* paper in 1993. There was no reason to rehash those arguments in *Nature*. A year later, the reason for the *Nature* rejection—that there is not sufficient interest in the subject—would come back to haunt them.

Then on September 23, 1996, our camera on Polar captured something totally unexpected—the actual disruption of a small comet in the vicinity of Earth. The image from the ultraviolet camera showed a bright trail that was the record of the final moment of the

This image of sunlit Earth, as seen at wavelengths in the far-ultraviolet on September 23, 1996, was taken with a camera on an Earth-orbiting spacecraft. The image has a mottled appearance because intensities in this spectral window are factors of millions dimmer than those for visible wavelengths. The Earth's limb is seen on the left-hand side and the terminator on the right. The trail in the upper right-hand side of the image is the record of the final moments of the disintegration of a small comet.

disintegration of a small comet. The small comet was 1,300 miles away from the spacecraft at the time of disruption. The cometary gas cloud missed impact with the Earth's upper atmosphere by a distance of about 15,500 miles.

We wrote a paper on this startling image called "The Smallest Comet" and sent it to the journal *Science* early in December of 1996. That paper was also rejected. On the 13th of January, R. Brooks Hanson, a senior editor, wrote: "... relative to other manuscripts under consideration, your paper was viewed as a lower priority rating for our limited space. While your discovery and analysis of a small comet from the POLAR spacecraft is quite interesting, because at this point there is little analysis of the comet and the larger implications for an understanding of the solar system are not clear, we felt that your paper, which is focused on announcing the discovery, is more appropriate for a more specialized journal."

Though a decade had passed, the reaction to the subject had not changed one bit.

CHAPTER 30

Vindication

When the time came to make our Polar findings public, we decided that the best forum for the announcement would be the Spring 1997 meeting of the American Geophysical Union, the same one in which we had announced the discovery of the small comets 13 years previously. So in the opening months of that year we began planning just how we would explain the significance of our Polar results to the public. We realized that we would need the support of NASA headquarters, as well as the assistance of the Office of Public Affairs at the Goddard Space Flight Center in Greenbelt, Maryland. Unfortunately, Goddard had done nothing to support our discovery of the atmospheric holes with the Dynamics Explorer spacecraft back in 1986. It was not just us, however. The entire scientific community at Goddard, and those of us on the Dynamics Explorer project in particular, never received the type of support from the Public Affairs office that we should have gotten.

But this time things would be different. With the help of Bob Hoffman, the Polar project scientist who is based at Goddard, we enlisted the help of several people in Goddard's Office of Public Affairs. One was astronomer Stephen Maran, an associate director, who at the right moment notified his press contacts that an important announcement was forthcoming. William Steigerwald, the Goddard science writer, prepared the official NASA News press release, and Wade Sisler, the director of their audio/visual department, prepared the news video tape, including animations, that would illustrate our findings for the television audience. At the same time, Gary Galluzzo, the science writer for University News Services at the University of Iowa, prepared his version of the press release. We also produced a website on the small comets so that people hearing the story could go to the internet for a complete background on the subject. This left us the time we needed to prepare our presentation and write up the Polar results in a series of four papers we would submit to *Geophysical Research Letters*.

The big day came on Wednesday the 28th of May 1997. This American Geophysical Union meeting in Baltimore was not unlike any of the others I had attended during the past twenty or so years. But there was a distinct sense of anticipation in the air as I prepared to address the TV crews and dozens of reporters at the 10 AM press conference reserved for our Polar findings. Up on the podium with me was one of my critics, Robert Meier, a situation which may have served as a clue to what was coming. Then, immediately after the press conference, I presented the findings to my colleagues and the public in one of the convention center's largest auditoriums. Kathy Sawyer of the *Washington Post* would describe the crowd of colleagues there as "riveted." There were so many lights from photographers that the audience could not see my slides. This was clearly not business as usual.

I began both the press conference and the public session with a little background for those who were not familiar with what we had found a decade earlier with Dynamics Explorer. I probably need not have bothered. But the point was that the Polar results confirmed our previous observations of atmospheric holes by Dynamics Explorer and did so in a very powerful manner. It involved a different instrument with better cameras that provided a better spatial resolution of the objects. Our new data confirmed the dimensions of the holes, their frequency of appearance, and their east-to-west motion across the sunlit atmosphere. Polar detects approximately several thousand atmospheric holes per day, which gives a global rate in the range of 5 to 20 per minute—the same sort of numbers we obtained from Dynamics Explorer. In some cases, the Polar UV camera caught the same atmospheric holes in consecutive exposures as it moved across the field of view. In other images, the spots were doubled, as they should be—because the spacecraft's wobble is such that the same object was recorded twice in the same exposure. None of this is characteristic of noise, of course.

An unintended swaying motion in the spacecraft's platform, which we were later able to compensate for, caused any physical object in Polar's line-of-sight to appear as "cat's eyes" or double in the images. These frames, taken on April 6, 1996, demonstrate the "double vision" effect on both a single star, Beta Chamaeleon (a), and a single atmospheric hole (b) produced by the disintegration of a small comet. Such images provide further evidence that atmospheric holes are not due to instrumental artifacts. The swaying motion of Polar's platform was caused by a misalignment of the center-of-gravity and spin axis of the spacecraft.

What surprised everyone the most, however, was Polar's observations of an entirely new phenomenon—a large population of objects that had never been detected before. In some of our images we found bright trails left behind by small comets exploding at altitudes ranging anywhere from about 3,000 miles to 30,000 miles. The oxygen trails produced as sunlight dislodged the oxygen atoms from these water-bearing objects made for some very striking images. Dynamics Explorer could not see these trails as it was a different kind of imager and simply did not take a picture of its entire field of view simultaneously the way Polar does. The imager on Dynamics Explorer took 12 minutes to scan its entire field of view dot by dot. But with the picture taking capability of Polar we were able to look at the sky and see them burst and produce these beautiful oxygen trails. This very powerful evidence for the existence of small comets has gone totally unchallenged. This discovery surprised not a few astronomers.

The confirmation was one thing. These new observations of small comet trails were another. But I was not finished. Still more impressive were Polar's visible light images showing

An oxygen trail from a small comet, superimposed on a map of the Earth, showing the "snowball" over Germany on September 26, 1996. Some small comets will be disrupted at much higher altitudes than for the vast majority of those impacting the atmosphere. For those bursting at higher altitudes, trails of bright glows from atomic oxygen, which are due to the disruption of some small comets as they approached the Earth, were clearly observed by the Polar spacecraft's high-resolution Earth Camera for ultraviolet wavelengths. These trails are distinguished from artifacts due to the penetration of energetic charged particles into the camera's sensors by a sequence of electronic shutter closings during the exposure period for the image. Thus a real oxygen trail is identified by alternate brightenings and dimmings along the trail. An energetic particle track will not have this distinctive fingerprint. Such images provide undeniable evidence that a new class of objects has been discovered in the vicinity of Earth.

Right: The unprocessed image is actually a "stack" of five consecutive "raw" images with some cosmetic processing to remove cosmic ray hits, nightglow backgrounds, flat-field optical normalizing, distortion removal, etc.

bursts of light from water fragments (hydroxyl). Polar was able to confirm that the atmospheric holes were actually clouds of water vapor. The visible wavelength camera on Polar spotted emissions of the water fragments that form when hydrogen atoms are torn away from water molecules as the water snow is exposed to sunlight when the small comets disrupt. These bursts appear at altitudes of about 1,250 to 2,000 miles, just above where small comets are supposed to disrupt and produce the huge clouds of water vapor that show up as atmospheric holes in our images. And they appear at the same frequency as the atmospheric holes.

This is the "smoking gun" for water, as Robert Meier aptly put it. To be able to look at these small comet water clouds, just as one would look at Comet Hale-Bopp and detect its emissions from the water fragments, which is a standard proxy for water if you are on the ground looking at these large comets, was very powerful, independent evidence of the existence of *these* comets. These bright emissions allowed us to determine the amount of water in the small comet blow-ups and we found that the amount of water was the same as we had deduced for the atmospheric holes—an entirely different phenomenon, entirely different measurements, but the numbers were about the same. The bottom line is that we are indeed looking at small comets coming into the atmosphere in large numbers.

Before Polar, we had no proof that it was water that was producing the black spots in the images. The Dynamics Explorer results merely allowed us to interpret the black spots as being produced by water vapor in the atmosphere. But the visible imaging camera on Polar showed that the water is there. This was a major step because it could have been something exotic, some form of propane, or dust balls, or something else. Until we were able to use the same proxy as for large comets, there was always that uncertainty. But our reasoned guess that they are primarily water-snow is now confirmed. I think that is Polar's most important finding.

The other filters on Polar told us there is no sodium in these small comets and no dust, or so little dust that you cannot see it. While we know that large comets are quite bright in dust and sodium, the small comets did not have any strong sodium signal or dust signal, which means they don't have the dust and the sodium and the metals that the large comets have. If they had, they would have been detected long ago because a dust trail that close to the Earth would have been seen by people just before dawn and just after dusk. It is nice to know there are differences between the small comets and the large comets. This indicates that their history is different.

So we learned a lot from Polar. The most important thing is simply that all the features of atmospheric holes seen with Dynamics Explorer were confirmed by Polar. And two, the measurements were extended by showing that here was water, tons and tons of water, rather than some other exotic compound. There were no big surprises. We did not find out they are made of exotic elements or that their numbers were a factor of 10 or 100 less, or that the trajectories were different, or they are not moving from dusk to dawn across the sunlit face of the Earth, or that the holes in the atmosphere were much larger or smaller than we had anticipated.

This cosmic rain was very real and even the critics were impressed. At the press conference, critic Robert Meier's words were the equivalent of a white flag. "I had pretty much given up on the idea that these are real," he admitted to reporters. "But after looking at the data, I'm convinced the phenomena are real."

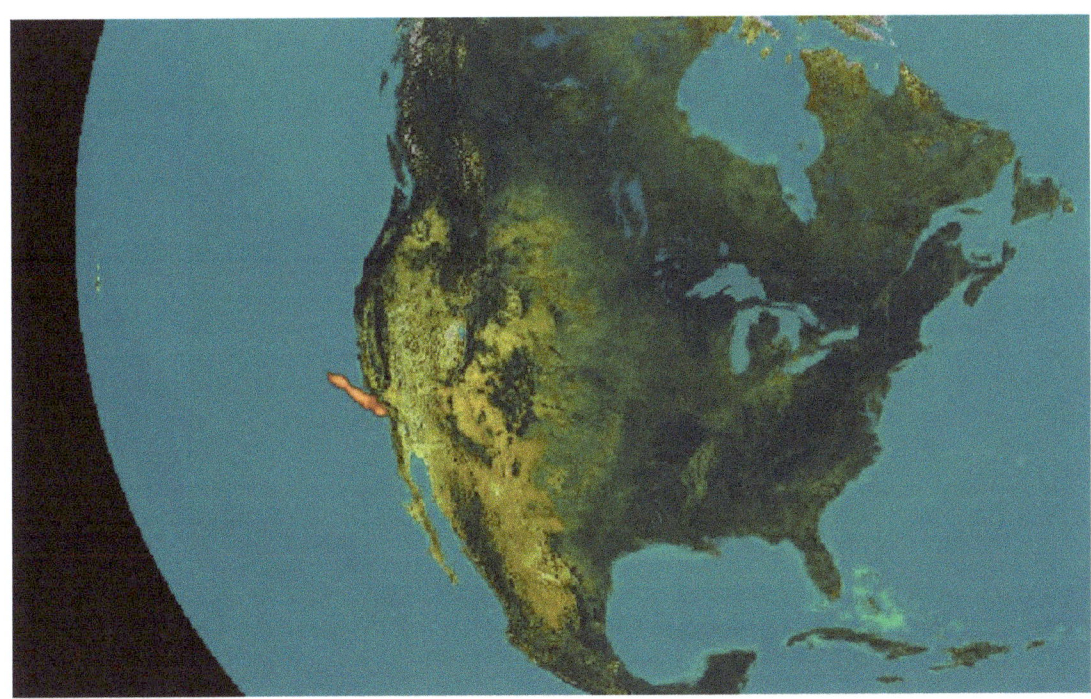

Top: An oxygen trail over California (Sept. 28, 1996). Bottom: a long oxygen trail from the disruption of a small comet passing by the Earth at a large distance (Aug. 28, 1996). The oxygen trails are superposed upon "Face of the Earth"™ maps in order to provide perspective of the view from the Polar spacecraft.

It was a story the media could not resist. I was amazed by the way the newspapers played the story the next day and in the week that followed. It was front page news in several large metropolitan newspapers, including *The New York Times, The Philadelphia Inquirer,* and the *Baltimore Sun*. The story even appeared "above the fold" on the front page of *The Washington Post*, a placement which in the newspaper business signifies an important story.

The press was quick to describe the situation as a "vindication," probably as a result of my old critic Tom Donahue using the word in his comments to the press. "It's exciting that a guy who was pretty badly pilloried is going to be vindicated," he told Richard Monastersky of *Science News*. But Robert Meier was doing it too: "They've certainly vindicated the old observations," he told Monastersky.

U.S. News & World Report was perhaps the most gung-ho of all. "A scientist—and his theory—vindicated" was the title of their story. Donahue had used the "v" word for their reporter, too: "These results certainly vindicate Lou Frank's earlier observations," he had told William Cook. When another researcher told Cook, "If this was correct, we would have to burn half the contents of the libraries in the physical sciences," Cook wrote in reply: "Get out the torches."

Even the science magazines managed to utter a few words of praise. *Scientific American* titled their story "Return of the Snowballs," and noted that "Frank is back" with "improved data" and "a big publicity buildup." Richard Kerr, the reporter for *Science,* called the results "a stunning turnabout." And Robert Matthews of *New Scientist* saw it as "a wonderful tale of scientific David beating reactionary Goliaths."

Some people were honest enough to see the writing on the wall. Ian Stewart, one of my first critics, sent me a nice email message saying, "I guess I just have to swallow crow." Even *The New York Times* chimed in with an editorial a couple of days later that noted: "Just when you thought scientists had pretty well explored the far reaches of this planet and found everything significant worth knowing, along comes a discovery so startling and so potentially important that it seems hard to believe it was overlooked."

John Sigwarth thought we would hear from our detractors right away. But it took them a couple of days to recover from the shock. The messenger was the first to take the heat. Some people tried to pin the blame on NASA for endorsing such a controversial story, but NASA gets flack for everything they do, so they were rather impervious to this sort of attack. They handled the Polar findings just like they handled the stories about life being found in a Martian meteorite. If NASA only paid attention to topics that were free of naysayers, they would not have much to talk about.

Our Polar announcement appeared as an official NASA press release and it deserved to, as this involved Polar, a NASA spacecraft, and I am funded in my work by NASA. A few people even criticized the quote that appeared in the press release from NASA scientist Robert Hoffman: "I'm pleased that Polar's instruments were able to actually detect these objects streaking toward the Earth and disintegrating into clouds of water vapor." As the Polar science manager, Hoffman had every right to be proud of this achievement. The press release also quoted George Withbroe, science director for the Sun-Earth Connection program in NASA's Office of Space Science, as saying: "This is an intriguing result that requires further scientific investigation. We need to look closely at measurements from other sensors to find out if they see related signatures in the atmosphere, now that we have learned more about what to look for." Does Withbroe's statement sound like a ringing endorsement of the small comets? It does not. In a later interview with *New Scientist* correspondent Robert Matthews, Hoffman bristled at the accusation that NASA's news releases were merely "paid advertisements." "NASA has not endorsed the interpretation of the images in any way whatsoever," he said.

We were, of course, well aware that our findings would gain added weight by appearing as an official NASA press release. For that reason, we had to be particularly cautious about what we said and how we illustrated it. We certainly did not want to scare the public by NASA's announcement that the Earth was indeed being pelted by thousands of house-sized snowballs weighing tens of tons each day. So to illustrate our results we picked an image of a small comet coming down over Germany, rather than one coming down on San Diego, believing that such an image might be less frightening to US citizens. We also took care to note that these Earth-bound comets represented no threat to either people on the ground or in orbit, as these objects disintegrate far above the Earth's atmosphere. In fact, we also prepared a diagram showing the relative heights of an incoming small comet and compared it to things that people might be familiar with. At an altitude of about 330 miles the comet vaporizes; this is about the altitude you will find some of our satellites. At an altitude of five miles, where commercial flights are common, the small comet is nothing but a cloud of water vapor being dispersed by the wind.

But that is not to say that one of these small comets could not—on rare occasion—strike a satellite, or the shuttle, as we have previously noted. And indeed, on the 1st of July, 1997, I suspect that a disintegrating small comet may have struck the Mir space station. According to Bryan Burrough's entertaining chronicle of life aboard Mir, entitled *Dragonfly*, on July 1, 1997, the Mir commander Vasily Tsibliyev heard what sounded like two successive "muffled explosions." When he looked outside he was startled to see a "floating blizzard of snowflakes" around the Spekr module. "His first thought is that the flakes are frozen bits of fuel," writes Burroughs. "They look just like the little ice crystals that form when the station's thrusters spit out drops of fuel. But no: there were too many." He tried taking pictures of them with a camcorder, but the flakes were too small. Nothing showed up on the videotape. Months later Tsibliyev remarked: "Everyone was just scratching their heads and could not come up with an explanation."

I had done what I had set out to do. It had taken me a tenth of a century to do it. I had proven that the atmospheric holes are there. I had shown that these objects have water in them. And I had shown that there are 10 million of these things coming in a year. But we still do not know everything there is to know about these objects.

What we have to do now is go up there and meet the small comets at a distance of 600 miles out from the Earth. Polar sees these objects with great resolution but from a great distance. How could we now get up close and see these objects in detail? That is the question that a group of us began trying to answer back in December of 1996. The group included a number of critics who wanted in on this project when I showed them the Polar results. They included Tom Donahue and Michael Combi at the University of Michigan; Paul Feldman at Johns Hopkins University; Robert Meier, George Carruthers, and Charles Brown at the Naval Research Laboratory; and Ralph Bohlin at the Space Telescope Science Institute.

How a small comet deposits its water on the Earth. The incoming comet is very dark because it is coated with a carbon mantle. It contains about 100 tons of water snow and is moving at about 10 miles per second. At very high altitudes, it breaks up into fragments. Exposure to sunlight then rapidly vaporizes the water snow and the resulting cloud of water vapor expands to 30 miles in diameter. The water vapor absorbs sunlight reflected from oxygen in the upper atmosphere, causing the appearance of black spots in the images of the sunlit Earth taken in ultraviolet light. The cometary water vapor cloud slows as it descends and stretches out into a thin, wispy pancake. Upper atmospheric winds then disperse the cloud so that all traces of it essentially disappear. The cometary water vapor gets mixed into the normal meteorological system of condensing water that precipitates onto the Earth. The small comets are responsible for a tiny fraction of the annual rainfall, depositing only about one ten-thousandth of an inch of water on the Earth's surface per year.

The proposal members were not of one mind, however. We all agreed that there is a remarkable influx of objects occurring, but we all disagreed on what they are. Sure, there will be water in them, but what is their structure, and what is in them? That is what our proposal, which we worked on for six months, stated quite frankly.

We called our proposed spacecraft Cyclops because it had a big eye on it. It was designed to see small comets burst over such places as San Francisco and Germany and fall into the atmosphere. It was designed to obtain exciting pictures of the comets disintegrating in real time, at high resolution, and see what kind of simple organics they contain. It was called Cyclops because it would continually stare at the sky. If it saw an intruder, an object coming in its field of view, it would then automatically switch to high speed mode. It could sense and distinguish between cosmic rays and small comets and then set off a high-speed sequence of frames as the object went through its field of view and smashed into the atmosphere.

Cyclops was actually designed to carry two instruments. One was the camera for visible wavelengths, which we called Clio, which had such a high time resolution that it could produce a *movie* of an object coming in to its field of view. It also had great spatial resolution, so you would be able to see the pieces fly off as a small comet burst apart. The other instrument was an ultraviolet-light spectrometer from the Naval Research Laboratory that would look at the oxygen, methane, carbon, and sulfur emissions from the small comets, which are weak, but this was a big sensor. The two instruments would have allowed us to see dynamically just how these small comets behaved and what their composition was.

Cyclops was to orbit the Earth at about 600 miles, which is about a tenth the distance or less of Polar's orbital distance. Imagine what this would allow you to see. The camera would usually be pointed parallel to the atmosphere so it would have a big field of view as the small comets are coming in. And then when Cyclops went on the night side you could look off and detect the trails. For the price and for the effort, the scientific objectives were just out of sight. The price was a real bargain at just 30 to 35 million dollars, including the price of the launch vehicle. This would have been one of NASA's Small Explorer missions. It would also have been valuable in studying other near-Earth objects. It would have been just fantastic.

But the project got hung up in a battle over turf. There are four sections at NASA headquarters. One is Origins. Another is Solar Terrestrial Physics, of which the Polar project is a part. The third is Planetary, and the fourth is Astrophysics. We submitted the Cyclops project to the Origins section, but this disturbed them because they had their own program of instruments they wanted to fly. They were very unsettled with this unexpected proposal. So they tossed it back to George Withbroe, who heads the Solar Terrestrial Physics section.

External reviewers are recruited for evaluation of proposals to NASA. Then a debriefing of the strengths and weaknesses is provided by the NASA official in charge, usually via a telephone conference call. Withbroe debriefed me on the evaluation of Cyclops. I was astounded by the fact that our proposal was turned down on the basis of comments that made it clear that the reviewer had not read the proposal. Withbroe quietly agreed and noted that the review process sometimes fails and that this proposal was just "falling through the cracks."

But we have not given up on Cyclops yet. We can always resubmit the proposal. Often you have to go through a number of hurdles before these things are accepted by NASA, so what happened was really not too surprising. They treated it routinely and simply failed to realize its importance. So now we are looking for a mission of opportunity. At the moment we have repositioned the proposal as an instrument on another spacecraft, a communication satellite. The company, called Fast Analysis, is putting up a number of low altitude observation satellites. They have plans to launch a satellite every six months from the year 2000 to 2003 and beyond. So if someone else is paying for the launch, and if we can manage to pay for whatever modification needs to be done to the spacecraft to get our instrument on it, they are willing to give us that ride. That is why they call it "a mission of opportunity."

Who knows? Maybe one day in the future we will get a chance to send a major mission after these small comets and bring back some samples. What an exciting adventure that would be for everyone.

CHAPTER 31

Grand Slam

Our American Geophysical Union presentation set off a stampede for our small comet papers on the Polar data. Typical was the statement that Heinrich Holland, a planetary scientist at Harvard University, made to *The New York Times*: "Everybody is stirred up. But we need to slow down and look at the data." Unfortunately, the papers had not been published yet. In fact, they had not even been *accepted* at the time of the conference.

Early in 1997 we prepared a total of four papers for submission to *Geophysical Research Letters*, the journal that had published our original discovery papers with data from Dynamics Explorer. But James Burch, who had been our editor on our last small comet paper in 1989, was no longer the editor of that journal. He had been replaced by his associate, Hunter Waite, in 1994. Waite put all four of our Polar papers in the review process. One after the other, as they came in, he sent them out to reviewers. There were lots and lots of reviewers.

The first paper we submitted to *Geophysical Research Letters* was a revised version of "The Smallest Comet" that had been rejected by *Science* early in 1997. We sent the paper, now called "Detection of small comets in the vicinity of Earth," to Waite on the 17th of January. We had no idea how arduous a task we would face in getting this paper published. It underwent three grueling rounds of reviews over a six-month period. Involved in this marathon were three reviewers whose comments led Waite to repeatedly reject the original manuscript and its subsequent revisions.

Not surprisingly, the reviewers were concerned with my interpretation of the observations of the bright oxygen trails. They thought I did not draw a direct enough line between these trails and the small comets. The reviewers were bothered by the fact that I did not mention alternative explanations to the observations. They were troubled by the brief durations of the light trails, tens of seconds, and the relatively long lifetime of water molecules broken up into oxygen atoms, which is about a day. The reviewers also felt the paper was incomplete; they wanted to know whether Polar had imaged the atmospheric holes, for instance, because they were unaware that this was just one of four papers, and that a separate paper dealt entirely with atmospheric holes.

Some of these criticisms were constructive, some were not. Sometimes they criticized without making any constructive suggestions to improve it. (When I am a reviewer I provide both the criticism and a companion suggestion as to how to satisfy the criticism.) But I had no choice. Each time I revised the manuscript, implementing major changes in response to the criticisms of the reviewers. I answered the critical points and rewrote my interpretation in order to avoid the impression of circular arguments. I offered and dismissed various alternate interpretations. I considered debris in the vicinity of the spacecraft, for example, as well as an exploding dust ball. Of course, the shapes of the light curves are so uniform that they eliminate the possibility of these trails being due to debris in the vicinity of the spacecraft.

The reviewers also thought that the standard model for the disruption of cometary water to produce atomic oxygen did not satisfactorily explain our observations. They were correct. But a direct release of atomic oxygen in the interior of the comet, as was observed in Jupiter's moon, Ganymede, does. So we proposed that the trail in our image is actually the record of the release of atomic oxygen as the cometary water snow vaporizes, with the maximum brightness occurring near the end of the vaporization process. The released atomic oxygen is very reactive and will rapidly combine with such atoms as carbon and nitrogen in the cometary gases. This rapid removal of the atomic oxygen also explains why the trail is narrow and why there is not a large-scale diffuse glow of oxygen emissions across the sky.

We thought the review process would never end, but it finally did and the paper was accepted in July. Retitled "Detection of atomic oxygen trails of small comets in the vicinity of Earth," the paper centers around the remarkable September 23, 1996, image that shows an unexpected light trail positioned above the Earth's limb. This long light trail has two dark strips across it produced by the electronic shuttering of the camera. This shuttering is necessary to avoid a contamination of the image by light scattering from the large antenna booms on the spinning section of the platform and also to correct for the imbalance in the spacecraft. These shuttering sequences also guarantee that an object is being imaged and that the trails are not instrumental artifacts due to penetrating particles into the sensor. But as remarkable as this image is, it is by no means an isolated event. Approximately five to ten of these trails are recorded by the Polar camera each day.

We submitted our second paper to *Geophysical Research Letters*, "Transient decreases of Earth's far-ultraviolet dayglow," two months after the first, at the end of March 1997. It took two interminable months to get the reviews back, but thankfully the revisions necessary for its acceptance would prove minimal. Two of the referees recommended publication without significant modification. One of these lamented the fact that this confirmation of the atmospheric holes was not achieved by another research group, but realized that we could not be held responsible for this. The referee was well aware that the original Dynamics Explorer data were simply overpowered by the common belief that the atmospheric holes were due to a defect in the camera.

The third reviewer, however, raised a number of technical points that Waite felt should be addressed. The reviewer wanted an overview of the observing program on Polar, wanted to know why only portions of three images were presented in the paper, wondered if the other clusters of dark pixels in the images were holes or noise, and was concerned about the calibration of the Earth Camera, among other problems.

So we revised the paper, making significant additions in response to the critical reviewer. We clarified how the data processing was carried out, added the requested overview, and had to shorten the paper in the process to accommodate *Geophysical Research Letters*' four-page per paper limit. The changes, I admit, made the paper stronger than the original. It was accepted at the end of July.

This paper effectively laid to rest a decade of skepticism over the reality of transient atmospheric dark holes. It proves that the atmospheric holes seen in the far-ultraviolet dayglow with Dynamics Explorer were not due to defects in the camera. This was ten years later, we were using a different spacecraft and a totally different camera—and the results were the same. The paper shows one of the first global images of the dayglow that we obtained from Polar, an image from the 6th of April 1996. We established that the basis for the events was sound and that the observed events were real and could not be an artifact of our instrument or our analysis. We used several images to calculate the improbability of the coincidental clustering of dark spots, the "dumbbell" appearance of real events due to spacecraft wobble, and the continuity of moving dark trails in successive exposures. By the summer of 1997 we had acquired 14 months of data, totaling about 400,000 global images of the Earth. In each image there were typically one or two well-developed atmospheric holes. We estimated that there were about 500,000 holes in these images—a frequency of detection very similar to the average frequency previously observed with Dynamics Explorer. Only the foolhardy would now insist that the atmospheric holes are not a genuine geophysical phenomenon.

Our third paper sought to use another instrument aboard Polar to independently confirm the existence of atmospheric holes. As we mentioned previously, Polar carries two cameras capable of viewing Earth's far-ultraviolet dayglow. One is our Earth Camera. The other is the UVI instrument for which George Parks at the University of Washington is principal investigator. For about one hour on the 12th of April and less than 15 minutes on the 30th of July 1996, the UVI was programmed into a special operating mode in order to obtain simultaneous images with the Earth Camera. Would the two independent cameras aboard Polar be able to detect the same atmospheric holes at the same time?

We submitted our paper on this question to *Geophysical Research Letters* in early April 1997 and received the reviews six weeks later. The reaction, as Waite said, was "not unfavorable." One referee even felt that it settled the question of whether the UV dayglow depletions we originally reported on in 1986 are a geophysical phenomenon or not. The other referees had a variety of technical concerns that Waite requested we address in our revision. He also suggested that the title not use the words "atmospheric holes" in order to avoid further hysteria on the part of the scientific community. It was good advice and we would call the paper, "Simultaneous observations of transient decreases of Earth's far-ultraviolet dayglow with two cameras."

But there was another, rather obvious problem—and the reviewers quickly picked up on it. They wanted to know whether the UVI team endorsed our analysis. Waite also wanted to know why we had not secured the participation of the UVI team. Of course, we had requested their participation. For nine months we had repeatedly invited Parks to do the paper with us, but he never took us up on our offer, so we did it on our own. It would not be the first time we have had to analyze data obtained by others in an effort to confirm or deny the existence of these incoming comets. Bob Hoffman, the Polar project scientist, knew that we had extended every possible consideration to the Parks team and finally recommended that we proceed on our own and not delay the publication of this paper any further.

Before describing the results, it is important to understand the very great differences between the two instruments. The reviewers did not like the fact that in my original submission I had emphasized the UVI's "considerably lesser capabilities" in comparison with the Earth Camera, and I was forced to tone down my language in the revised version of the paper that was accepted at the end of July. But the truth is the truth. The UVI instrument is clearly inferior. During our analysis of the UVI images it became clear to us that without confirmation from an independent camera, the UVI was simply incapable of distinguishing a darkened cluster due to a real object from one due to electronic noise. The UVI does not have sufficient time resolution to detect the holes in consecutive images or to eliminate the effects of the platform's wobble.

But despite the poorer quality of the UVI images—including permanent small-scale dark regions—they reveal the remarkable, infrequent presence of very large clusters of darkened pixels with diameters of about 60 miles at a range of 300 miles above the Earth's surface, as expected. Because of the large fluctuations in the responses of the UVI pixels, only clusters with ten or more darkened pixels were identified as atmospheric holes. And, it should be noted, these darkened pixel clusters are not seen in the preceding or following images.

Subframes from the two cameras aboard the Polar spacecraft—VIS and UVI—that show the dayglow intensities at the same location in the Earth's upper atmosphere before, during, and after the occurrence of the atmospheric hole produced by the deposition of water vapor from a small comet. Other than one of two other such events, which were compromised by responses due to energetic charged particles in the camera sensors, there were no other examples for which a large atmospheric hole was detected in one camera's frame but not in the other camera's frame. This series of observations provide a conclusive identification of atmospheric holes as a geophysical phenomenon.

In 72 minutes of simultaneous observing time, UVI acquired a total of 61 UVI images in which we were able to identify 14 candidate atmospheric holes. During this period, we collected 127 images with the Earth Camera in which we were able to identify 13 atmospheric holes. The question was: were any of the atmospheric holes seen simultaneously with both imagers?

They were. Five atmospheric holes appear in near simultaneous frames of both the UVI and Earth Camera. The geographical positions of the five events from the two imagers were identical, after taking into account the effects of the platform wobble, the motion of the atmospheric holes, and optical distortions. The five pairs of detections are admittedly small, but that is simply due to the infrequency of the UVI images in comparison to the VIS. All in all, the simultaneous detections of the atmospheric holes with these two imagers is quite impressive considering how infrequently large holes occurred in the two sets of images. The simultaneous observations of atmospheric holes with the two Polar cameras provided further exciting confirmation that they are a genuine geophysical phenomenon. Parks, of course, would later argue otherwise.

Our fourth paper, which is pivotal to our interpretation of atmospheric holes as water vapor, was also submitted to *Geophysical Research Letters* in early April 1997. And again, it took the reviewers an interminable two months to reply. But at least they recognized the importance of the paper. "These new observations and their implications are startling..." wrote referee "A." Indeed the detection of comet trails with yet another Polar camera for visible wavelengths provided considerable insight into the nature of the small comets.

But the reviewers also thought my four manuscripts *to Geophysical Research Letters* should be combined into a single manuscript for publication in the *Journal of Geophysical Research*. I found this suggestion totally unacceptable. This diversion would delay the publication by at least a year and many scientists were already quite upset that these papers had not already been published. I urged instead that the editor of *Geophysical Research Letters* publish these four papers together as soon as possible. This is eventually what happened, but not until someone tried to stop the publication of the four papers, more about that later.

The referees for this fourth paper had a number of concerns. They wanted more information about the electronic shuttering of the camera, about why these trails with the camera for visible light differed from those seen by the Earth Camera in the ultraviolet light. We addressed all these concerns including the reason for the differences in the trails. These are simply due to the differences in electronic shuttering sequences caused by the different angular resolutions of the two cameras involved.

We had decided to include the special filter for the visible camera on Polar a decade earlier and the decision had paid off handsomely. It was one of 12 filters on the filter wheel that serviced this camera. Our paper, "Trails of OH emissions from small comets near Earth," presented the imaging results.

Water is hard to detect because it is unable to effectively scatter any wavelength in the visible or ultraviolet. So water vapor is difficult to see unless one goes to infrared wavelengths, which require very sophisticated instrumentation. Since there are no strong emissions from water at ultraviolet and visible wavelengths, there are emissions that act as proxy for the presence of the cometary water molecules. Essentially what happens is that the water is broken apart into H and OH by the ultraviolet radiation from the sun. The emissions from the OH fragment are very bright when exposed to sunlight. (The sodium in street lights produces an orange-colored light because these atoms are very efficient in radiating this light. In the case of the street light, the sodium atoms are not excited by solar radiation but are excited by passing an electrical current through the sodium gas.)

We used one of our two cameras for visible light on Polar to observe these emissions from the OH. The OH emissions of the small comets featured an intense core surrounded by a large region of dim luminosity. These OH emissions are a bright component of the host of emissions from the well-known large comets. This light from OH, noted referee A, "is a well studied, well understood emission which should be a very sensitive indication of anomalous concentrations of H_2O in the sunlit upper atmosphere. The detection by the VIS experiment of 'trails' of OH emissions must be viewed as persuasive evidence for the existence of 'small comets.'"

An independent camera for visible wavelengths on the Polar Spacecraft, the Low-Resolution Visible Camera, records the emissions from the OH molecules formed when the small comets plunging toward the upper atmosphere are disrupted and vaporized by sunlight. This stunning nighttime image of the light emitted by the breakup of water molecules from a small comet was taken on December 31, 1996. This multiple exposure is composed of three consecutive snapshots (in one image) of the event taken six seconds apart. These intense glows are created at an altitude of less than 2,000 miles above the Earth when sunlight illuminates the oxygen-hydrogen molecules formed after a comet's water molecules have been stripped of a hydrogen atom. A view of the Earth at the time of the event has been superimposed onto the image for reference as there are no detectable emissions from the Earth's nighttime atmosphere at the wavelength where these oxygen-hydrogen molecules glow.

Our paper focused largely on a sample image acquired on the last day of 1996. The image contains three exposures, or snapshots. For an object with constant apparent speed, the three detections must be in the same line, have equal separations, and be very similar in magnitude. These requirements separate these sightings from any source of image contamination, such as due to the penetration of cosmic rays in the sensor.

The same method was employed on February 22 and March 13 during which the OH filter was replaced with other filters, including those for sodium and cometary dust. Remarkably, no sodium or dust trails were detected. This finding led referee B to comment: "Certainly if the existence of a high influx of mini-comets can be confirmed, this result would be a very important advance in cometary science, especially if many of the unexpected chemical and physical properties envisioned by the authors are true."

By using the Earth itself to shadow the cometary water clouds from solar radiation, we were able to determine the approximate altitude range of the trails. The evidence we acquired placed a large fraction of the trails at altitudes of 2,000 miles or less. We also calculated the speed of these OH trails, and they fell in the range of three-to-ten miles per second, which is similar to the apparent speeds of the atmospheric holes. And when we calculated the diameter of diffuse emission of these trails, we came up with 30-to-40 miles, which again is similar to the diameter we had measured for the atmospheric holes. Likewise, the total amount of water in the dim diffuse region of OH emissions was similar to the mass estimates for the small comets we had calculated in 1993 from other considerations. Thus the amount of water vapor we found using the OH emissions as a proxy is sufficient to account for the atmospheric holes. The global rates of both these phenomena are in the range of five-to-thirty events per minute. The agreements between the independent observations of the OH light and the atmospheric holes were remarkably good.

Once I knew that our four Polar papers would be published, I submitted a fifth paper to *Geophysical Research Letters* at the end of September. This one was called "Observations of Transient Decreases of Earth's Dayglow with the Ultraviolet Imager on the Polar Spacecraft." The paper presented our further analysis of the UVI images. We obtained an independent confirmation of an atmospheric hole with UVI by identifying sets of three dark spots in the images, which are due to viewing a moving atmospheric hole under the effects of the back-and-forth motion of the spacecraft camera platform. One reviewer found it "entirely worthy of publication." The other called it "an interesting and intriguing manuscript."

But the reviewers also had a reservation. They both felt that it "did not carry the case much further" and that it "might form the beginning of a complete study" whose length would make it more appropriate for a journal other than *Geophysical Research Letters*. Based on these comments, Charles Kolb, the atmospheric sciences editor overseeing my submission at the time, turned down the paper in December on the basis of "its degree of novelty and definitiveness," adding that *Geophysical Research Letters* was "not the proper venue for a long series of incremental papers." I got the message. *Go away*, he was telling me, *we've had enough of you*. They were about to publish four papers from us on the subject and they just did not want a fifth.

In all, it took half a year for our four Polar papers to go through the review process. I think Waite just wanted to make sure that they were thoroughly reviewed so as to not be embarrassed by them. I understand that. I think he was objective and knew enough to ignore any reviewer who made ridiculous comments. That is what an editor should do and Waite was sufficiently knowledgeable in a large number of scientific fields that he could do that. But the review process took so many months that by the time they were due to appear he was no longer editor. Even though they had already been accepted, I knew this would spell trouble.

The new editor, Michael Prather, told Waite that he did not want four papers published in the same issue. In fact, he did not really want the papers published at all. But he was unable to stop the publication process.

Our four papers appeared in the October 1, 1997, issue of *Geophysical Research Letters*. Prather stuck them at the rear of the issue and attached a silly little note to them saying how he was not responsible for accepting them. The Editor's note, which ran at the bottom of the table of contents, stated in part: "A retiring editor made an earlier commitment to publish the four papers above as a 16-page group. *GRL* honors this commitment in spite of its being contrary to current editorial policy. Letters should represent independent, stand-alone research, and small groups of papers or special sections will be considered if they involve a greater research community." Prather's note needlessly scolded his predecessor in public and attempted to degrade our work. It was amazing and completely uncalled for.

But that, in a sense, was beside the point. Our four Polar papers were now published. The safe cocoon provided by the explanation that the Dynamics Explorer detections were simply noise had been stripped away. Everyone now had to face the stark truth.

CHAPTER 32

Dessler and His Pawns

Inevitably, once the shock of our Polar presentation wore off, the critics surfaced in force once again. They had not all been silenced. David Morrison and Alan Harris of the Jet Propulsion Laboratory were both urging skepticism over the Polar findings. Harris was insisting that my numbers were "a million or so too high," that we should be able to "see several in an evening scanning the sky with binoculars," and that a one centimeter rise in the ocean per decade was "measurable." Harris called my hypothesis "wacko." Donald Hunten of the University of Arizona granted the black spots in the images were real but did not think that my explanation for them was valid. Neither did Tom Gehrels apparently.

Gehrels sent me an email saying the Polar results could be explained by lightning. I thought he was joking and did not reply. Then he sent me a follow-up email asking if my lack of response could be interpreted as my believing his explanation to be true. So I replied: "I thought that your message concerning lightning was intended by you to be a joke. In order to clarify that this is the case I note the following facts. Apparently, you mean either regular lightning or 'lightning' in the sensor. I know of no reports of nature's lightning at altitudes of 5,000 km [3,000 miles] above Earth's surface. So you must mean 'lightning' in the sensor, otherwise known as a high-voltage discharge. First of all, a single electrical discharge of this type would totally destroy the delicate sensor. Secondly, the dramatic oxygen trails are stamped with a validation shutter sequence which cannot be attributed to an instrument artifact. Thirdly, the ultraviolet camera detects about 5 to 10 of these trails every day. The trails have various lengths and brightnesses but always bear the validation signature. We have records of approximately 3,000 of these events as gathered since camera activation early last year. In the face of this evidence I believe that you should begin to think about the source of the trails in terms of a real interplanetary object." Gehrels was right about only one thing though. In an email, he had signed off, saying: "Interesting times again!" It was.

Many of the objections to the Polar findings were again of "we would have seen them if they were real" variety. But Mike A'Hearn of the University of Maryland was one who realized just how false this common assumption really was. "I am simply unconvinced that asteroid searches even see the comets of the same size as the asteroids they detect. The coma inhibits most of the automatic detection schemes rather than helping. After lengthy discussions with Ted Bowell and Bruce Koehn during my last observing run in Flagstaff, they agreed that their planned detection scheme would not find comets. …The only comets detected by Spacewatch have been ones that were seen visually on the screen by the observer sitting in the control room. They were not found by the software." And later he went on to point out just how poor the asteroid surveys are at spotting comets: "Despite the existence of the Shoemaker Palomar survey and the Spacewatch survey in 1990-95 and the NEAT survey and the Spacewatch survey since then, two thirds of the known comets coming inside the Earth's orbit in both of those time periods have been discovered by amateurs when they were far, far brighter than the detection limits of the cited surveys. Part of this is, of course, the fact that the existing surveys have focused on the ecliptic where the asteroids are and most comets are not in the ecliptic. The point is that we are not looking at faint comets systematically."

A'Hearn's comment was a breath of fresh air amidst all the hot air emanating from the critics. And, of course, there was Dessler, who would once again head the opposition—or so the press made it seem. "The small comet hypothesis fails to agree with physical reality by factors that range from a thousand to a billion," he told Richard Kerr of *Science*. A doubting Dessler told Monastersky of *Science News* "there is too much evidence against the small comet theory." And to Corey Powell of *Scientific American*, he said: "Frank's ideas just can't be right." But Powell was nice enough to follow that awful statement with my rebuttal: "There are still some people who don't believe in continental drift."

I expected the backlash that followed our Polar announcement. Science is strange these days. New findings are usually well-received by the press—and then the naysayers come out of the woodwork. While I welcome criticism in science, I do not think that the harassment of individuals who simply propose something new is justified. What is wrong with trying to advance science into the frontiers? The energy expended by the naysayers, other than in writing a paper and criticizing it, is just wasted scientific manpower. It does not advance science. You are safe from the naysayers only as long as you regurgitate the things that we already know, perhaps with a little different arithmetic, a little different perspective. I think that tight funding has produced this kind of behavior. Unfortunately, it is forced by envy, too. People like their 15 minutes in the sun. It is almost embarrassing to watch.

The basic outline of the critics' arguments had changed little since the 1980s. There were no showstoppers. Nonetheless, the snowball fight now resumed. The next officially sanctioned battle would take place at the American Geophysical Union meeting on December 9, 1997, in San Francisco, a meeting which Dessler would not attend. He would be a guest at the Nobel Prize award and banquet ceremonies in Stockholm at the time. Then, in January, he would email me saying that if I was right about the small comets he would "personally nominate" me for a Nobel Prize. Dessler was now a foreign member of the Royal Academy of Sciences.

My critics blindsided me at the small comets press conference at this meeting of the American Geophysical Union. It seemed to me that there was a concerted effort by the editors, Robert Winglee and Michael Prather, and by Alex Dessler and the authors of the critical papers, to make sure that I did not see their results until the meeting. That showed a total lack of professional courtesy. How was I going to respond to them? When a reporter came up to me and asked me "Well, what about these recent pictures of the moon from Clementine?" "I don't know," I said. "Haven't seen the paper." And I only had an inkling of what George Parks would say. But each one of those papers would prove to have one fundamental flaw in it, fundamental in the sense that it just did not supply evidence against the small comets. Only one paper was of real significance, the rest of the critics were just blowing smoke rings.

Five papers attempting to undermine the small comet hypothesis were presented at this American Geophysical Union meeting and all five appeared in the December 15th issue of *Geophysical Research Letters*. In a "Highlights of This Issue" section, the editor, Michael Prather, claimed that the five studies together "refute [the small comet] hypothesis." This was obviously Prather's revenge for being obliged to publish my four papers in one issue six weeks previously.

The first critical paper came from George Parks, the space physicist at the University of Washington who was principal investigator for the UVI instrument aboard Polar. Parks made four points, all of them invalid. His first is that the small comets cannot be real because the black spots appeared in his calibration images before the spacecraft even got off the ground. Parks had taken laboratory images before the instrument was launched by shining a uniform ultraviolet light on his detector so that it could be calibrated. So Parks' interpretation is that any ultraviolet light source, even in a lab, will produce dark spots. He told reporters that "the white snow in space is no more than the black snow on the television screen." The fault, in other words, is with the ultraviolet camera.

But Parks only had five calibration images and his light curves do not extend to the realm where the real geophysical events are found. If you look at the chart in his paper and you look at the size of the black spots that would correspond to atmospheric holes, you will find that they are there in the data but not in the calibrations. He says that the noise profile looks the same in the flight data as in the calibration data. That is true. But the large events that are seen up there in space and are recorded in the flight data are *not* seen in the laboratory calibration data. His abstract says that the lab curves are exactly like the light curves and they are most definitely not. Hoffman pointed out the problem to him. Everyone told him.

Secondly, Parks' evaluation of the UVI images found that the dark holes in the images could not be positively determined to be a geophysical phenomenon, rather than as an instrumental artifact. This is in agreement with our conclusions in our previously published paper on the simultaneous measurements. Only when other features or data are used can this be resolved.

The third issue in the paper is Parks' claim that he could not find any smearing in the UVI images of the dark spots produced by the spacecraft's wobble. But they are there. What Parks forgot to take into account is that the atmospheric holes are in motion. It is something that many critics are prone to forget. These water vapor clouds are *not* motionless. By the way, we clearly identified the UVI spin signatures in our fifth Polar paper, which *Geophysical Research Letters* refused to publish.

The final item that really riled us was what Parks said about the oxygen trails in the Polar images. We had already published our paper that explained how the shuttering sequence insured that the oxygen trails are not cosmic rays. But he said that the cosmic ray events in his sensor can be mistaken in the VIS images as oxygen trails from small comets. I found that misrepresentation very irritating. With the UVI's small field-of-view, I estimate he would only see a real trail once every one or two months. He was told all these things, but his reply was just *I don't believe it*. What Parks is forgetting is that we are not starting a new religion. Belief has nothing to do with scientific investigation.

Alex Dessler, who seems to have nothing better to do than to round up small comet opposition at the University of Arizona and elsewhere, was behind the second paper not only in spirit but in name as well. He had co-authored the paper with Bashar Rizk of the University of Arizona's Lunar and Planetary Laboratory, though it was Rizk who actually presented it at the American Geophysical Union meeting. If the small comets were real, they claimed, our sky should sparkle like a Christmas tree. We should be seeing a continual display of bright clouds and flashes if these things were real. Of course, this was nothing new. Rizk does optical calculations on dust clouds and luminosities, but his paper with Dessler simply repeated what Dessler had brought up in his 1991 *Journal of Geophysical Research* paper and which I answered in a subsequent issue of that journal in 1993. They said that if you are in the dark at the right time, you should see an incoming ice cloud illuminated by the impending dawn or dusk and it would be brighter than Venus but not as bright as the full Moon. But if the sizes of the ice crystals are small, which you sort of expect, then you would not be able to see it at all. Only if you assume a maximum scattering brightness, as Dessler and Rizk do, would an incoming small comet be very bright. Rizk showed a curve with a variety of brightnesses and particle sizes and then he only took the very tip of it. How are ice crystals all going to end up at their favorite size? The brightness is much different depending on a large versus small crystal size. There is no disagreement there. There is no mystery.

The third paper was presented by Jennifer Grier and Alfred McEwen who are also from the University of Arizona. It makes me laugh. If the small comets are real, Grier and McEwen argued, then we should be seeing evidence of 400,000 comet hits on the Moon each year. They said that a small comet should jar the lunar surface almost once a minute and produce a crater 55 yards across or at least throw up a fan of bright lunar soil some 180 yards in diameter.

So what Grier and McEwen did was compare Apollo 17 images taken in late 1972 to Clementine images taken 22 years later. They covered an area about one fourth the size of the island of Great Britain or little more than one tenth of one percent of the lunar surface. Any crater and bright spot seen in the 1994 Clementine images but not visible in the 1972 Apollo photos might be the record of a small impactor hit, they reasoned. If the small comets theory was correct, they argued, Clementine imaging should have discovered more than 10,000 new bright impact spots over this area. But each of the 3,920 bright spots they found in the study area in 1994 by the Clementine spacecraft was also recorded by Apollo.

It was not a bad idea, but it was poorly executed. First of all, they claimed that when a small comet hits the surface of the Moon, it penetrates down into the white substrata, and puts out this enormous white plume of dust. That is what they *expected* to see, but that is not science—it's dreaming. The small comet crater would be dark and irregular. If you do not dig down into the bedrock, you are not going to get white stuff. You need a big rock to spread the white stuff over a large area. But that was not even the biggest problem with their paper.

While Apollo produced fabulous, high resolution images, Clementine, which was moving rapidly and had a poor camera in comparison, produced lower resolution images. I do not see how you can tell in the Clementine images whether anything as small as a small comet had impacted there. The resolution of the Clementine images is pathetically insufficient. The resolution for the Clementine images was only about 300 feet, while the diameter of a small comet impact is expected to be in the range of 30-to-50 feet, much less than the resolution of the camera. I agree that the Apollo images are fabulous, and if they had had that with Clementine, they would have had something.

What is really surprising is that Prather put this embarrassment on the cover of *Geophysi-*

cal Research Letters. The Apollo image is sharp, beautiful. The Clementine, in comparison, is simply horrible. How are you going to see a small crater in that smudgy blur? Perhaps the cover was just meant to be a cartoon.

Not so funny is the fact that I essentially lost access to *Geophysical Research Letters* over the publication of this paper. I was a reviewer on this paper and on the 4th of August I informed the authors that they had, presumably unintentionally, miscited the mass densities and speeds of the small comets in order to achieve crater sizes large enough to be visible in the Clementine images. I cited the literature for them and provided the reference. But they insisted otherwise.

Winglee accepted their paper on the 4th of November without asking them to correct this error. So in January, after their paper had appeared, I wrote a harsh "Comment" and submitted it for publication. In it, I accused the authors of misrepresenting the literature, and scolded Winglee for allowing such a misrepresentation to occur. On the 19th of February I received a letter informing me that my comment was being handled by editor-in-chief Michael Prather. Prather refused to consider the Comment for publication. Not even a short apology for the miscitation that had occurred ever appeared in *Geophysical Research Letters*.

The quality of the work coming from Dessler's Arizona groupies left a lot of be desired. So too with Timothy Swindle and David Kring, who are also from Arizona and presented the fourth critical paper at the December American Geophysical Union meeting. In their view, the small comets were formed in very cold regions of the solar system far from the Sun. Comets that formed far out in space condensed from the same dust and gas that accreted into planets, trapping "noble" gases in the same ratio as the sun and the rest of the solar system. Noble gases—which include argon, krypton, and xenon as well as the more common helium and neon—are inert, or non-reactive gases, not easily removed from the atmosphere. At the current rate of bombardment from the small comets, Swindle and Kring calculate that Earth should have 500 times as much krypton and xenon in its atmosphere as it does, and 30,000 times as much argon. So either the rate of supposed small comet bombardment is 30,000 times greater today than it has been over the 4.5 billion year lifetime of the solar system, or the comets formed nearer the Sun, at about the distance of Jupiter, for the theory to fit the observed noble gas inventory.

Swindle and Kring's attack was based on the underlying assumption that the small comets actually do bring in more noble gases. But how do we know that? We do not know how much noble gas was present when they were formed out there on the rim of the solar system. Noble gases are heavy, and maybe like iron and all the other heavy stuff, they all ended up closer. Swindle and Kring are being very presumptive. They must understand the entire history of the formation of the solar system in great detail, the formation of the planets, the distribution of all the elements, everything else that has happened in the solar system in the past four billion years or so. One would assume that we have knowledge like that, but we simply do not. I would never get trapped into an argument about something that is all assumptions.

The only one of the five critical papers to have real scientific merit came from Mark Boslough, a physicist at Sandia National Laboratories, and Randall Gladstone, who is with the Southwest Research Institute in San Antonio. They wondered if the atmospheric holes might not be produced by meteoroids. So they ran computer simulations that modeled the impact of a rock on the atmosphere, showing how as it channels down it can regurgitate some oxygen back through the channel and produce "black spots" in the atmosphere. It was a good solid calculation based upon the successful work of Boslough and his colleague David Crawford predicting the visible plumes from Comet Shoemaker-Levy 9's impact into Jupiter in 1994. That is the code they applied to the Earth's atmosphere and they showed that the atomic oxygen, which is the source of the dayglow, is momentarily displaced by the passage of meteoroids. They proposed that when a stony object as small as 20 inches across collides with the atmosphere and plunges into the lower layers, it ejects a very thin gas plume as high as 600 miles. It is these dark plumes, they concluded, that are being detected by Polar.

While it is true that a rock probably can do that, it has to be a kilogram or larger; this is the crux of the matter because there is simply not enough of them to account for all the atmospheric holes. And Boslough and Gladstone are well aware of that. Now, if by some miracle, you can get a one-gram dust particle to have the same effect on the atmosphere, that is another matter. But the one-gram particle just does not have enough mass to kick back enough stuff into the atmosphere and darken a large enough area. It would simply burn up. The idea is good a one, however, and perhaps some of the holes, a small fraction of them, may be due to this phenomenon—a few a month, maybe ten altogether. But this mechanism is just not going to account for the bulk of the atmospheric holes we see every day. It is an insignificant fraction of what we see. Maybe one part in 100,000. We record about one million atmospheric holes a year. The number of large stony or iron meteors is insufficient to account for all of the atmospheric holes observed by the Dynamics Explorer and Polar spacecraft. Gladstone and Boslough know that; they do not say that this accounts for all of the atmospheric holes we are seeing.

I countered my critics at the December meeting of the American Geophysical Union with a presentation of what I called the "ultimate test" of the reality of these objects. This ultimate test is based on one simple fact: if the atmospheric holes are real, then the number of holes detected in the images should decrease as the spacecraft's altitude increases. Why? Because as the satellite's altitude above the atmosphere increases, the apparent size of the atmospheric holes its camera sees should decrease, until most of the holes become just too small to be seen by the camera. On the other hand, if the data shows no decrease in the number of atmospheric holes as the spacecraft's altitude increases, then the atmospheric holes are an instrumental artifact.

Using Polar data from June 1, 1997—which is similar to those obtained on other dates—we compared the low altitude data, when the Polar spacecraft was between three and five times the Earth's radius above the surface, to the high-altitude data, when the spacecraft was between five and eight Earth radii above the surface. On the day in question, a total of 5,650 holes appeared in the images. For the same area in the images, the frequency of atmospheric holes in the high-altitude data was 80 percent less than in the low altitude data.

If the holes were an instrumental artifact, then obviously the frequency of the holes would not be dependent on the spacecraft's altitude and about the same number of holes should have been seen in the two altitude ranges. But more than five times as many holes were observed at the lower altitudes. This result is a marvelous confirmation of the reality of atmospheric holes.

This ultimate test could not be applied to the earlier Dynamics Explorer images because that spacecraft (with an apogee altitude of about 15,000 miles versus Polar's 30,000 miles apogee) did not get high enough above the atmosphere for the atmospheric holes to elude

Sighting of an atmospheric hole, taken on April 23, 1997, when the Polar spacecraft was at a very low altitude (7621 km). The number of pixels is large and the diameter of the hole is about 100 km.

detection by its camera. But as with the atmospheric holes found in the Dynamics Explorer images, those found in the Polar spacecraft's images also show a greater number of impacts during the morning hours as compared to the evening hours. This provides yet further confirmation that the holes are indeed real and not instrumental artifacts. If the holes were an artifact, then their appearance in the images should not be dependent either on the spacecraft's altitude or the time of day.

There is no way to avoid the fact that small comets are a genuine geophysical phenomenon—or that my critics would not go away anytime soon. The critics are argumentation's equivalent of the cockroach—squashable one at a time, but impossible to destroy once and for all.

Snowball fights are not fatal. Everyone would be back to fight another day.

CHAPTER 33

Steering Through Rough Waters

Bob Hoffman, the quiet-spoken project scientist on Polar, has somehow managed to stay impartial in the small comet fray. "People see me as being for Louis Frank," said Hoffman in a 1998 interview, "but that's because anyone who is not against him is regarded as being for him."

I first met Hoffman at a Midwest Cosmic Ray Conference, which was held at the University of Iowa in Iowa City every year. I was a junior at Iowa at the time in the late 1950s and he was a second-year graduate student at the University of Minnesota. We did not know each other very well, but we did share similar interests in science. Then, in the mid-1970s, Hoffman became project scientist on the Dynamics Explorer mission and we began working together.

The project scientist is responsible for the scientific success of the mission, but it is an ill-defined job. The project scientist basically makes of it what he or she wants. The hardest part of the job occurs before the launch. It is very important that the scientific community involved in spacecraft programs watch very closely how the spacecraft systems are developed so they do not interfere in any way in the ability to get clean measurements. Then, once the mission is underway, the project scientist has to coordinate the work of the principal investigators, each of whom is responsible for the development of his or her own instrument aboard the spacecraft. On Dynamics Explorer there were 15. On Polar there are 12. Hoffman interprets the job of project scientist very liberally: he basically sticks his nose into anything that affects the science.

So it was Hoffman who ran the science team meetings for Dynamics Explorer when we first presented the unusual characteristics of our data back in the early 1980s. He was responsible for keeping the meetings on an even keel, which sometimes took some doing. The top atmospheric and ionospheric scientists in the country were in those meetings, and we sought ideas and help from them regarding what the black spots in our images might be. But no one ever came up with an explanation for them. This went on for two years before we developed our own ideas about what they were and eventually published our first papers on the subject in 1986. Some of the heaviest criticism we would face came from those same people who were on the Dynamics Explorer science team.

Hoffman's initial reaction to the black spots in our images is the same as it is today: "Very interesting." He was intrigued by our first papers, too. "I guess at the time I was fairly convinced that Frank might be on to something," recalled Hoffman in 1998. "I don't think I was ever totally convinced, though I tried to keep an open mind on such things. It was such an unusual and revolutionary explanation, as was the set of observations. And, of course, immediately there were many objections which one certainly has to take into account. But in some ways I was very disappointed in the scientific community at the time. It's attitude in general seemed to be 'Let's try to find reasons why they can't exist' rather than asking the question as 'Well maybe they can exist, and what would be the impact on our concepts and theories if they actually did exist?' It was such a close-minded negative reaction by the scientific community."

This time, after the Polar presentation in May of 1997, Hoffman was offered a chance to channel the scientific community's response into a more productive direction. On June 6, 1997, NASA Goddard asked Hoffman to try to quickly resolve the small comets issue. The request was made by Joseph Rothenberg, then director of the Goddard Space Flight Center, and it came via Steve Holt, the Director of Space Sciences. "Holt wanted me to lead a small group to investigate whether currently available databases could be searched for the existence of small comets," recalled Hoffman. "He also wanted the group to investigate current instrumentation, either ground-based or in-orbit, that could be used to search for the existence of these objects.

"But there was a third item on the small comet group charter and that one bothered me the most. It called for a quick mission called Lunar Retrospective to be done in conjunction with the Air Force that would look for water on the moon. This was essentially a follow-up to the Clementine mission that had reported water on the moon. I almost immediately requested that we not be part of this effort for several reasons. First of all, I was not knowledgeable at all about that kind of science, and being a project scientist for a mission whose science I know nothing about would be a huge mistake. Then it quickly became clear that in terms of roles and responsibilities, it really was not Goddard's job to do this mission anyway, and NASA headquarters would probably not approve such a thing. So it was agreed that I wouldn't pursue the lunar project with the Air Force.

"But I did agree to attempt to pursue methods of confirmation. Although again because I am not a cometary scientist, I felt I needed advice, and through the contacts I did have and talking to various people, I asked a number of scientists in this area if they would act as an informal advisory group to me. I've always used the word group. I never called it a committee. It has never had any official position. It was never given any responsibility or authority by NASA. It was purely an ad hoc advisory group in my efforts to coordinate and identify activities that could be used in a confirmation or refutation of small comet activity. We would not make any pronouncements one way or another, however."

At a status meeting on the 13th of June 1977, Hoffman told Rothenberg and front office personnel that nothing would happen quickly. First of all, Hoffman noted that the scientific community was largely in the dark about the latest findings, as the four papers that I had submitted to *Geophysical Research Letters* had not yet been accepted and the various pre-prints we had circulated to people like Hoffman were still confidential. In order to work with other facilities in searching for the small comets, Hoffman felt it was necessary to provide these people with the essential small comet parameters. So we gave Hoffman permission to extract from the articles we had submitted to *Geophysical Research Letters* the various observational parameters that Hoffman could give to other scientists to help them in their search. By the 6th of August Hoffman had prepared a crib sheet with all the essential details. Everything was now set for the first Small Comet Steering Group meeting.

An impressive group of people attended that meeting on August 12th. Present were Michael A'Hearn, a University of Maryland astronomer knowledgeable about comets; Art Aiken, an atmospheric chemist at Goddard; Robert Robinson of the upper atmospheric research section of the National Science Foundation; Andrew Christensen from the Aerospace Corporation, a Department of Defense support organization with very close ties to Air Force facilities; Robert Meier of the Naval Research Laboratories who has been very interested in atmospheric holes; Ching Meng of the Applied Physics Laboratory at Johns Hopkins University and an investigator on Polar; Richard Vondrak, head of Goddard's Laboratory for Extraterrestrial Physics and also a scientist in atmospheric and ionospheric physics; Major Martin Whelan of the Pentagon, who has administrative responsibilities for facilities and spacecraft that appeared prime to verify the existence of the small comets; and finally Robert Carovillano, who attended for George Withbroe, director of the Sun-Earth Connections program at NASA headquarters of which Polar is a part.

The first meeting was essentially organizational in nature. "We discussed the concept of verification and defined three types," noted Hoffman. "One involved an independent analysis of the VIS and/or the UVI data. Another involved independent observations similar to the Polar observations, in other words, a verification of what Polar had seen. And the third type of confirmation involved independent observations of phenomena that were consistent with the interpretation of the Polar observations. We also recommended that the raw VIS and UVI data be made available with the necessary software to the scientific community for analysis and verification. And we recommended that a list of VIS atmospheric holes be developed and maintained for use by other observers. Finally, we recommended that plans should be made to coordinate more simultaneous VIS and UVI searches."

Hoffman's group also used their first meeting to develop a list of contacts and identify the facilities that might be approached to search for small comets. The Hoffman group came up with numerous candidates, including an ozone database at Goddard, some DOD spacecraft, various Air Force radars, and Sandia National Laboratories. They also learned that several searches appeared to be underway. Richard Goldstein at the Jet Propulsion Laboratory was going to do a search using the Goldstone radar beginning on August 17; he had obtained permission for six observation periods through September. A'Hearn had submitted a proposal to look for OH from the Moon by long-slit spectrograph. Ted Bowell of the Lowell Observatory intended to do a small comet search using the 60 cm Schmidt telescope. Eleanor Helin at the Jet Propulsion Laboratory intended to use the NEAT facilities for a search. John Mathews at Penn State requested support from the National Science Foundation for use of the Jicamarca radar in Peru, and Meier had requested data from NAVSPACECOM, a fence of eight radars at constant latitude across the United States. Finally, Ching Meng would look into using the MSX spacecraft for an atmospheric holes search. Things looked quite promising initially.

The next day Hoffman and Meng flew up to the Massachusetts Institute of Technology Lincoln Lab as Meng had identified and contacted people there who were responsible for a number of radars and optical facilities that might be of some help. Meng and Hoffman met with these people and discussed their use of the optical and radar space debris surveillance systems. They have extremely sensitive equipment and they were very much interested in a search. But Hoffman estimated that it would cost hundreds of thousands of dollars to fund their search, and that is money he simply did not have.

Eight days after the steering group meeting, on August 20th, Hoffman gave the Goddard director another briefing. A dozen people—front office administrative people, Holt and Rothenberg—attended the briefing, but Hoffman was never asked to give another briefing after that. "Maybe it was partly my fault," said Hoffman. "I didn't push it as it became obvious that there was no quick way of doing a verification. They wanted something quick, and it became obvious that it wasn't going to be quick. They had assumed initially that NASA probably had the resources to settle the matter in one of its programs, but it became obvious very quickly that this was not the case. It was more likely that the Department of Defense had facilities that could be used."

For Hoffman, the next stop was the Pentagon. Accompanied by George Withbroe, Hoffman met with Whelan and two other people at the Pentagon. They had a very lengthy discussion about the various Department of Defense facilities and determined that the most direct type of observation that could be done by a military surveillance system would be one designed to look for the infrared signal of hot water from rocket plumes. And the best such facility for the job, they concluded, was the MSX (Midcourse Space Experiment), a Ballistic Missile Defense system operated by the Applied Physics Laboratory at Johns Hopkins University.

This became one of the primary items on the agenda of the next steering group meeting, which took place on the 9th of October. But there were other items of note as well. One came from Mike A'Hearn who was developing a sensor he was about to test that could see water vapor being given off from one of these objects hitting the lunar surface. But out of three nights of viewing, the instrument was not operational on one night, on another night it was cloudy, and on the third night there was yet another problem, so he never did get any proper data. Another item of note came from Steve Knowles of the Naval Research Laboratories, who gave the first report on the radar signatures from the Navy radar fence. But his results were, for the moment at least, inconclusive.

Most of the session focused on the MSX. Larry Paxton of the Applied Physics Laboratory at John Hopkins presented the capabilities of the MSX spacecraft at the meeting. It turns out that APL was most interested in trying to have Hoffman's small comet group utilize the MSX satellite. The MSX is a very complex, very large spacecraft built by APL for the Department of Defense. On board is an infrared instrument and various imagers. Some APL people actually went down to NASA headquarters and gave Withbroe a presentation on the capabilities of the MSX before this meeting. The reason why APL was so interested is that the prime mission of the MSX had basically ended. One of the main instruments on board was no longer operational, so the Air Force was cutting back drastically on the money they were putting into the MSX operation. It was a very expensive spacecraft to operate. So APL was looking for sources of funding to continue operating the MSX.

Hoffman had also invited Drake Deming, a well-known infrared scientist at Goddard, to join the Small Comet Steering Group's second meeting. "It appeared at the time," notes Hoffman, "that the best way to look for the breakup signals of the small comets was in the infrared for water vapor. So I asked Deming to come up with what infrared signals would actually be seen, especially against the background Earth, based upon Frank's parameters for the small comets, and he came to the conclusion that it would be very difficult for any facility to see the water vapor in infrared against the sunlit Earth. The infrared signals from the Earth are too patchy, too irregular, and the infrared from such a breakup would be pretty much at the levels of a natural signal. I think the best way of identifying these things is with infrared but looking off the rim of the Earth into a background of space where there is essentially no water vapor. There they would stick out even better than a sore thumb."

The main problem, Hoffman realized, is that Department of Defense instruments like the MSX look down on the Earth, so their infrared sensors would have considerable difficulty seeing the large water vapor clouds produced by small comets. "I don't think they have the facilities to do scanning in an operational mode," Hoffman added. "And I don't know the details of how different the infrared signals would be for hot water versus cold water. If there is [a difference], they're certainly looking for hot water signals, whereas I believe that Lou Frank's comets after disruption would be not in that temperature range—although I think some people might disagree with me on this point."

It would not matter, however. APL had collected a little bit of data but it was not extensive enough to even put limits on the frequency of the small comets. The small comet group's assessment of the MSX was that it would be very expensive to utilize it in a proper search. The price tag was more than a million dollars. "I think we were all convinced that it could've done the job because of the narrow field of view," says Hoffman. But again, the price tag was too high.

Also at the second steering group meeting was Dee Pack, one of the management people from Aerospace, who talked about the capabilities of Air Force Space Command, or at least as much as he could because a lot of it is classified. To claim whether or not they can detect the small comets would essentially tell us what the sensitivities of their systems are and that is a security problem.

But I know that some Department of Defense spacecraft have already picked up these objects. Edward Tagliaferri of the Aerospace Corporation in Los Angeles told me he was interested in looking to see if Department of Defense spacecraft had picked up any atmospheric holes events. Tagliaferri works with infrared sensors on "star wars" type spacecraft that continuously monitor the Earth's atmosphere. It is all very secretive, but I suppose these spacecraft have sensors for rocket exhaust water emissions. So I gave him eleven atmospheric hole event times and their coordinates to search for in his data. Ten of those were big diffuse events, 50 to 100 miles across, and one was a compact event, meaning that it was much smaller, having been seen shortly after disruption.

Tagliaferri told Hoffman that they had found one of these eleven atmospheric holes events. They did not find the big diffuse ones, probably because the infrared signals from these large water vapor clouds were not bright enough to be picked up by their sensors. But the compact one, which was probably closer to what they are used to searching for in missile launches, they found. The signal is probably denser or much brighter when it is a compact object. They managed to locate a simultaneous signal in defense department data within 50 miles of the compact event we had imaged with the Polar spacecraft. They knew it was not a missile launch. I was very encouraged to learn that we were within 50 miles of the object because that is the accuracy for our cameras anyway.

That one hit was enough to intrigue Tagliaferri. He tried to get Aerospace and the Air Force to kick in some money, which means pay someone to develop the software needed to do a more thorough search of defense department databases. The amount of data they have is just so enormous. They have programs that look for certain things in the data, but their software probably automatically eliminates the small comet signatures. First of all because the water clouds from small comets are coming in quite a bit faster than a ballistic missile, and secondly because the small comets are not coming in on a ballistic missile trajectory. So these signals are probably getting thrown out.

As far as I know, Tagliaferri's efforts never panned out, or if they did, I was never told of the results. Actually, I was surprised that they even admitted to detecting one of these objects. The military is usually silent on such topics. But I think it is an important issue for them because obviously they should be seeing these things. How can there be ten million of these objects coming in every year, ten million rocket exhaust-like water trails, and the military not see them?

Dessler, however, kept insisting that defense department satellites could provide the information we were looking for. Hoffman tried to set him straight in an email message: "We have had both Andy Christensen from Aerospace and Dee Pack, a manager of space systems for the Air Force at Aerospace, at a Small Comets steering group meeting. I have also had conversations with people at Sandia about these systems, and met with three Pentagon Air Force officers involved in the management of these systems, as well as conversations with Tagliaferri, who sent me copies of his publications. None of these people claim that the infrared systems have acquired data bases that can be searched for the small comets. Much of the data are discarded after 90 days … The Air Force is looking for hot water vapor over the cold Earth. Calculations that I have heard about indicate that the snowy comet signals are about the same level as the background water vapor and would be difficult to see."

Hoffman did not realize the difficulties involved in doing a confirmation when he started out. Nobody did. Our skies are filled with advanced satellites packed with a variety of technologies to monitor the Earth's environment. We can now peer to the very edges of the known universe. So how could a phenomenon as flamboyant as a high frequency barrage of snowy objects be so darn elusive?

The reason is actually quite simple, as Hoffman came to realize. "The more we looked into these other systems," he explained, "the more we found problems with their use in a search. At first we thought that, yes, there might be some facilities out there that had the technical capability to look for these objects. In fact, there were a lot of comments from scientists in the community, which I found discouraging, who off the top of their head would quip, 'Well if these things existed, such and such a facility would have seen them.' And when we looked into it, we found that indeed, yes, they would have seen them, but maybe never identified them because of the nature of the operation and the different nature of Frank's comets compared to what they were designed to look for. In other words, you would either have to modify onboard software, or ground software, or have operators look for different types of signals than they were trained to look for."

The other major problem was that the small comet group was not exactly flush with funds. "I had effectively no money to carry out any of this work," noted Hoffman, "except for some Polar contingency money. Headquarters would not put up any money for the effort. I paid for some travel for the steering committee people, and I was willing to put in a small amount of money for some analysis. But I didn't have enough money available to conduct a major search. In general, NASA's approach was, let's find out what we could get essentially free before we consider spending any money." So NASA did not want to put up the money for small comet studies until there was absolute proof that the objects were real, but of course to secure such proof, more funding was needed.

A day before the third small comets steering group meeting Alex Dessler was giving a talk at Rice University. The title of his talk was "Small Comets: A Case History in Pathological Science." The only thing pathological about it was Dessler's obsession with destroying the small comets. While Dessler maintained this high ground, he sent his pawns into battle.

Hoffman invited me to the steering group meeting on April 17, 1998. He did so out of fairness, something that had been sorely lacking in this controversy. Tom Donahue was going to brief the group on the special session on the small comets for the upcoming spring meeting of the American Geophysical Union in Boston. So Hoffman very wisely called everyone several weeks before and said, "Why don't you present your findings at the steering meeting first?" Then, with the steering group's approval, he invited me so that I would not be blindsided by my critics as I had been in December. "I felt that it would be unfair to Lou Frank," recalls Hoffman, "to have these things kind of sprung on him at the next American Geophysical Union meeting. I thought that a dialogue at this steering group meeting between the various people would increase the quality of the presentations at the spring AGU."

Donahue was late to the meeting because he thought it was being held out at Goddard rather than in downtown Washington, D.C. Knowles presented the Navy Space Command radar fence analysis, which was nearly finished. James McFadden gave both his and Forrest Mozer's analysis. Larry Paxton presented the analysis that he had done on the atmospheric hole catalog data and the hot spots in the detector. Jim Scotti and Al Harris gave their presentation on the Spacewatch data, for which Hoffman had scheduled 50 minutes. It was considerably more time than anyone else was given, and for good reason. The presentation was a classic. I will comment on this and the other presentations in the subsequent chapters.

After lunch, Hoffman kicked out all the visitors and held a closed session about what the role of the steering committee should be. "We were being pressured by several people, like Dessler and Mozer, that this steering group should actually make a pronouncement on the small comets," recalled Hoffman. "But we all agreed that that was not the proper role of the steering group. We thought it really would be very improper for an ad hoc group to make any kind of a pronouncement. That's really the role of the scientific community. So we adamantly refused to be drawn into anything like that. We also discussed any future activities and we pretty much agreed that until there were new situations arising that there wasn't a need for the steering group to meet again."

Hoffman was terribly dismayed that some of my critics had tried to pressure the steering group "to declare the issue dead." But even if the steering group had felt that it was within its jurisdiction to do so, the small comets would not have gone away. They are here, right above our heads, right now, and nothing anyone says can change that fact.

CHAPTER 34

Contact Sport

The critics did their best to pour hot water on the small comets, but all they did was steam up the place. An Associated Press story previewing the Spring 1998 American Geophysical Union meeting in Boston noted that emotions about the small comet theory "have reached a red-in-the-face, fist shaking level." One of the speakers would call this "science as contact sport." Seven independent teams of researchers were prepared to assail the small comet hypothesis. The audience wanted blood—anyone's.

The chairman of the session was Tom Donahue. In his introductory remarks, he warned, only half in jest: "I don't want any hitting in the trenches. No fighting, no biting." It didn't help. The auditorium was filled with scientists eager to see this "intellectual slugfest," as the *Boston Herald* called it. "I've never seen anything like this," one NASA scientist told the *Herald* reporter. "It's not very constructive. They are not solving the problem. But people are curious."

Much of what happened at that American Geophysical Union meeting really had been stewing for months. And one of these squabbles, with Forrest Mozer, probably had roots dating back decades. My long-term relationship with Mozer has not exactly been sterling. About 20 years ago, I questioned his electric field measurements on the International Sun-Earth Explorer spacecraft and ever since then his measurement of electric fields has not been trusted. He had found enormous electric fields in space with his antennas on the spacecraft while my plasma instrument on the same spacecraft failed to observe these anticipated large plasma flows. I suppose that now he was trying to get me back.

It began when Mozer implied that our Polar data could not be trusted. "If it was really true," Mozer wrote to Hoffman on January 19th, "that RAW, UNTOUCHED, UNCORRECTED, UNEDITED data from Frank's instrument was available from NASA (not Iowa) in a decontaminated format giving the number of counts in each pixel for a large number of images, and if Frank's identification of the locations of cometesimals in these images was also available, I, for one, would be interested in examining that data." I felt that Mozer crossed the line with his implication that we were tampering with the VIS images and thus the data must come from the Goddard Space Flight Center. But I told Hoffman that we could supply Mozer a thousand or more consecutive raw images within a few days and that Mozer could take his pick.

Meanwhile Dessler was claiming that I was withholding the Polar data from such people as Ching Meng. In fact, Meng had not requested the data, and Dessler later admitted that his information was second hand. We were not withholding data from anyone. It is just that we have an enormous amount of it—every day about 1,000 images of Earth are acquired with the VIS—and managing all of it was no easy task. Subsets of our VIS data had been available to the scientific community on the web since 1997 and we were working hard at the time to provide the community with a complete set of images, together with the coordinates, engineering parameters, etc., on CDs. These became routinely available in the summer of 1998.

"Am I imagining things, or does it seem to be getting a bit tense?" Dessler noted in an email dated January 20th. He was not imagining it. Things were very tense. We supplied Mozer with the data and the support he needed to analyze them. I did not expect much as Mozer clearly had a closed mindset. "Although I have only had time to glance at your paper," he wrote to us, "I think that we will disagree with the results."

Soon afterward Mozer began claiming that his study was an "independent and blind" analysis of the VIS data. But it simply was not. Mozer's study was not "blind" because we had to lead him step by step through the analysis, which is acceptable as long as it is not advertised as "blind." He was actually "looking over our shoulders" in order to verify the software we used in the analysis. He thought that the small comets could be tested by a double-blind experiment with me and George Parks. But if Mozer had really been knowledgeable about the UVI and VIS data, he would have realized that this was impossible. For one thing, the UVI data is so poor that it is instantly recognizable when mixed in with VIS data. In any case, Mozer stuck his neck out by saying that one way to verify the small comet hypothesis "would be if seasonal and altitude dependencies were found" in an analysis of the data. We would do just that.

On the 20th of February we received abstracts submitted by Mozer and company for the spring meeting of the American Geophysical Union. The paper by Forrest S. Mozer and his cohorts at the Space Sciences Laboratory at University of California, was entitled: "On the Internal Inconsistency of the Interpretation of Dark Pixel Clustering Data as Small Comets." Of the approximately 700,000 dark pixel clustering events observed by the VIS Earth Camera on the Polar satellite during a 120-day interval in 1997, Mozer and company found that "the cluster occurrence rate, its variation with 7621 km, and the altitude dependence of clustering sizes are inconsistent with the small comet hypothesis and are fully explained as instrument noise. A portion of the dark pixel clustering in this data set is due to underestimating the standard deviation of the counts in a pixel. This underestimate is largest in the outer radiation belt where most of the apparent dark clustering events were found."

Mozer's paper essentially said the atmospheric holes are not real events but instrumental noise because the holes are the same size regardless of altitude. The problem is that he simply assumed the result. Razzle dazzle, I call it. If Polar can see a hole at eight Earth radii in a single pixel, Mozer argues, then the spacecraft camera should be able to see enormous holes when it is at low altitudes. He claims that the size of the holes seen at low altitude has be 500 to 1000 pixels. That is a big hole. He is talking about something that is 25 times the area of the 60-mile atmospheric holes we're talking about. But that did not bother him. So at the American Geophysical Union in Boston in May of 1998 he showed a chart with a whopping hole the size of Queensland, Australia, sitting there, and he said "I don't see anything like that in the data. I just see the same small hole when I go to perigee." But, in fact, you do find the large holes in the data; they are 25, 30, 35 pixels large, just as they should be. And they are there with the frequency one expects because the frequency is a lot less for large holes simply because when the spacecraft is down close to the Earth, the camera sees less of the atmosphere and therefore fewer atmospheric holes. Sure, sometimes there will be five-pixel holes at low altitude. But those are simply incipient holes in the process of expanding. I do not think he ever looked at the images, because if he had, he would have seen the 30, 40, 50 pixel holes just sitting there, grinning at him.

Mozer also played a supporting role in another critical paper entitled: "An Instrumental Source for the Dark Pixel Clusters Observed by the Polar VIS and UVI Imagers," where the primary author was J. P. McFadden. McFadden attempted to calculate how many dark holes appear in an image and he started out by analyzing the characteristics of the sensor used in our camera. His intent was good. At least he was making some basic calculations. But he was using the wrong sensor. He simply did not check what sensor we had used in building the VIS camera.

That was not the worst of it, however. The data McFadden used has all sorts of problems with it. He was using *all* the data without any of the restrictions that we had put on it. You cannot use the data produced when the satellite enters the radiation zone, for example, and he used it to show that there is no altitude dependence. His analysis criteria were completely arbitrary and based upon invalid data. We only made the atmospheric hole catalog available as an invitation to do collaborative work, not to have it abused that way.

When I saw what McFadden had done, I immediately wrote back to him: "I assume that you are not using the data in the hole for the period after 2035 UT on 1 June 1997. Any experienced scientist who is studying the images would quickly realize that this data is not usable because the spacecraft is inside of the radiation zones where the energetic particle contamination is so great that the images cannot be used for any purpose. We do not use such data in our papers, of course. I assume that you are not using this data either. If you are, you have made a grave and careless error. If this is the case, you are wasting our time."

He was. You cannot publish material like this, I told McFadden. And, since I was a reviewer on the paper, I told *Geophysical Research Letters* editor Robert Winglee this as well. Winglee replied saying that anybody can publish anything off the web; that it does not have to be approved by the principal investigators. When I sent that remark to Hoffman, he went through the roof. Hoffman said that is not the intent at all; they are supposed to work with the principal investigators. I sent Winglee a list of objections. But Winglee said that I was so negative, he could not follow our suggestions. He thought the authors had done enough work. He invited me to write a Comment for *Geophysical Research Letters* if I felt the paper was fundamentally flawed.

Hoffman was not happy with the situation either. "Mozer and McFadden were told several times that they really were using data they shouldn't be using to do their analysis," he explained in a 1998 interview. "And they went ahead and published their papers based on that data anyway. Now strictly, legally, you can argue that any data posted on the web is public domain, in spite even of qualifications on the data. The qualifications are only warnings, so there's the question of does the person have to abide by the warnings? I guess if you want to push it the answer is 'no.' It's more of a courtesy than anything. But the point that I made to Winglee is that if the paper is submitted using data in spite of some warnings, then my position is that the journal's reputation is at stake in publishing this paper without good reason. There was never any attempt by Mozer to get permission from Frank, or get certified data, or get Frank to agree that the way Mozer was using the radiation belt data was acceptable. There was never an attempt to essentially narrow down the magnitude of the disagreement, and obviously the Berkeley people felt justified in their mind to go ahead with the publications immediately."

This situation is really very sad because it reflects on all of science. That is why people are more and more afraid to do any science these days. This sort of thing just drags science under. It is inexcusable. But it is a good indication of how large a percentage of the work that appears in these journals is junk. It wastes people's time and money. It's politics, not science.

In mid-October, I saw the final version of the Mozer paper. It was the worst we had encountered in the 12-year debate on small comets. It was one thing for McFadden to make a fundamental error in his paper in using the wrong sensor for the Earth Camera. But Mozer's deliberate miscitation of the "Iowa method," the foundation of his paper, was simply an atrocity. I did not hesitate to let Mozer know how I felt about this. He replied that he was confident that "there are no errors in our papers." We then pointed out the serious, embarrassing error he had made. Simply put, in one table, Mozer showed an array of darkened pixels where no such array of darkened pixels exists in the raw image data we had provided him. He had essentially *created* a set of data.

The next day Mozer acknowledged his "oversight." He sent us a draft of their response to the "problem." Over the space of several paragraphs he mostly reiterated the conclusions of their original paper. He admitted his error in only two sentences, stating that the numbers in "Table 2 are incorrect" and that this material "was inadvertently included in the analyses associated with Table 2." Mozer was quick to note that "NONE of the conclusions reached from Table 2 are modified by correction of this oversight."

A reporter for UPI was quick to grasp the big picture when he asked me: "You mean they used noise to show it was noise?" "That's right," I replied. When a reporter for NHK television in Japan called to interview Mozer, he hung up the phone on them. Perhaps Mozer did not want his 15 minutes of fame.

Like Mozer, George Parks would argue that my analyses were misled by electrical noise in the imager. In this, his second small comet paper, which was written up in March, submitted to his close associate, editor Winglee, and rapidly approved by the month of May, Parks did a UVI-like analysis of the VIS data and concluded that the Earth Camera data is all noise, using the September and April data we gave him. And he said that the curves for the two months are the same. But if you go to those curves and you look at the larger holes with five, six, seven, eight pixels, they actually differed by a factor of 5 to 10 from noise. He says that they are the same, but they are not the same.

But again, that was not the worst of it. Parks says the reason that he did not analyze for comparison the UVI data for September 23rd was because it was not in the right observing mode. Of course, the reason that we gave him the September 23rd data in the first place was because it *was* in the right observing mode. We deliberately gave him this data and we sent him an email saying that the reason that we were giving him the September data, rather than November data, was because on that day it was in the correct mode. Parks insisted that it was not in the correct mode.

So we looked it up again, just to double check, and sure enough, it was in the correct mode. So in an email dated the 21st of July, I explained this to Parks and said: "Would you like us to send you a copy of your data?" He essentially ignored us. To this day we do not know whether he analyzed that data and found differences that he could not explain.

Everyone was looking for mundane explanations for the black spots in the Polar images—but no one could agree on what they were. Larry J. Paxton of Johns Hopkins University had his explanation for the atmospheric holes, and that gave him 15 minutes of fame at the American Geophysical Union audience in Boston. And he knew it. "Gosh," he told the audience, "I've never talked to this many people before—or at least to people who were awake." Paxton's presentation was entertaining. He was also the only one of the lot who understood anything about imagers.

Paxton made two basic claims, one worthy, the other worthless. His analysis of our imager was interesting because he identified localized "hot spots" in the CCD pixel array which exhibited dark clusters with a frequency greater than expected from the random distribution of real atmospheric holes. He had found, in other words, that the sensitivity of the sensor was not uniform and that these hills and valleys, you might say, favored certain portions of the sensor for sensing atmospheric holes. This was interesting to us because we naturally wanted to make sure that this cluster distribution was not actually overwhelming the real events.

If the holes are real, of course, they should be uniformly distributed over the sensor. To do a proper search for the atmospheric holes, you need to pick search criteria that gives you an even distribution. Unfortunately, Paxton picked search criteria that produced a lot of false atmospheric holes. Our search criteria for the holes was much more severe in order to eliminate the noise. When you do this, the holes are not associated with the "hot spots" in the sensor, but are evenly spaced across the face of the sensor, as expected for a real phenomenon.

John Sigwarth and I investigated Paxton's analysis of the VIS Earth Camera images following his presentation at the Small Comets Steering Group meeting on the 17th of April 1998 in Washington, DC. We then sent him the new calculations so he knew what the problem was before he gave his talk at the American Geophysical Union. But he refused to acknowledge our calculations before the audience by saying simply: "I didn't have time to look at them."

Paxton's other conclusion was interesting but had no merit. He claimed that the black spots in the Polar images were due to clouds in the lower atmosphere. Tests show that although ultraviolet imagers are designed to only detect ultraviolet radiation, they sometimes see some visible light. Paxton claimed that the cloud structures could mimic the atmospheric holes. Perhaps, he suggested, the breaks in the clouds could show up as dark spots in the images. He claimed that their numbers varied precisely with the daily and seasonal cycle of cloud cover.

Paxton said that must also be the reason for the appearance of atmospheric holes in the Dynamics Explorer images. But the problem is that *everyone*—our critics included—who has worked with the Dynamics Explorer data knows that the filters on that imager were pure. You simply do not see any clouds in the Dynamics Explorer data. So Paxton just shot himself in the foot on that one. His "cloud theory" does not hold for Polar either, even though the passband filter for the multiplier on Polar was not as good as the one on Dynamics Explorer. The first thing he should have done is check where the atmospheric holes were appearing. If he had then looked at the images that show atmospheric holes, he would have found that only five-to-ten percent of an image is covered with these clouds. When we examined the positions of the holes, we found that they do not lie in or near the clouds any more frequently than out of the clouds. End of topic.

If the black spots are only noise, as so many critics claimed, why then do they show a seasonal variation? That was the crux of my rebuttal at the spring 1998 meeting of the American Geophysical Union. We did what Mozer had said would be positive proof of the small comets—we found a strong seasonal variation in our data. Our analysis of the images indicates that the number of dark spots in the images vary by the season, just as meteors tracked by radar do. We found that the influx of small comets into the Earth's atmosphere is ten times greater in early November than in mid-January, when the small comet rate diminishes dramatically. This is the same seasonal variation we discovered nearly two decades before in the atmospheric hole data found in images from a different camera aboard a different spacecraft, Dynamics Explorer, which traveled a different orbit than the Polar spacecraft.

These new results also suggest that the small comet storms are at least 40 years old. While the Polar spacecraft provides data for the 1990s, and the Dynamics Explorer spacecraft gathered data during the 1980s, the oldest data set showing the influx of small comets into the Earth's atmosphere dates back to 1955. As mentioned before, two Canadian scientists, E. L. Vogan and L. L. Campbell, using forward scatter radar, had found exactly the same seasonal variation, a November high and January low, in their non-shower, or sporadic, radar meteor rate. Why the atmospheric hole rate should correlate so well with the meteor rate measured by forward scatter radar is no mystery. After all, small comets are just a part of the meteoric dust and debris that orbits the Sun and falls into the Earth's atmosphere on a daily basis. Because the weakly bound small comets and mantle debris are expected to produce ionization at higher altitudes than stony or iron meteoroids, forward scatter radar—which is much more sensitive to ionization at higher altitudes than backscatter radar—is ideally suited to reflect the infall of small comets. (Backscatter radar events, on the other hand, are dominated by the infall of iron and stony meteoroids.) So with this 1955 radar data set included, the current influx of small comets appears to have continued at least from 1955 to 1998, or more than 40 years.

Some people wondered why we only analyzed data for November through January. We have the Polar data for the whole year. But for Dynamics Explorer, the apogee moved so quickly that we did not get viewing after January. We only have it for November through January for Dynamics Explorer. But this period also happens to be the best part of the year for the radar studies. During the summer, radar data gets noisy from atmospheric irregularities driven by direct solar radiation. So, since we did not have Dynamics Explorer data or forward-scatter radar data for the whole year, the first order of business was to find the Polar data for the same period. It was just a matter of comparing the same number of apples, oranges, and bananas. We have since looked at the rest of the year with the Polar data and what we found is that the rates increase gradually from February into early May and then rise dramatically in June, with sustained high fluxes through early November.

Our analysis of the Polar images also indicates that the number of dark spots in the images vary with the time of day. The small comets strike the Earth more often in the morning than in the afternoon. When we presented this on the first of two days of small comet talks at the spring meeting of the American Geophysical Union, Al Harris of the Jet Propulsion Laboratory rose to challenge us during the question and answer period. Harris claimed that this finding failed to match the small comet's path in space, which we say moves in the same direction as Earth and overtakes it. But if that prograde motion were true, insisted Harris, the comets should be plowing into the Earth's trailing edge, hitting the Earth's atmosphere in the evening hours.

So we swapped insults in front of everybody. When Harris put up a chart that purported to shoot down my basic premise, I replied: "These cartoons are meaningless." Harris responded: "I have a PhD in celestial mechanics. I think I know what I'm talking about." The problem is that Harris went to one extreme and pointed right at the morning dawn terminator, right at the division point between night and day, and said, "Well, you won't have a concentration of them right at the dawn or very little relative to the ones that are coming directly at dusk." But there was something he was forgetting and when I tried to explain to him, I probably did not do it too gingerly. The actual maximum will not be at dawn, but in the late morning hours. Why? Because the small comets get focused by the Earth's gravitational field. So the small comets strike the atmosphere at 9-to-10 AM, rather than 6 AM, and that is what we measured with both Dynamics Explorer and Polar.

Harris must have lost a lot of sleep over this subject because a year later he submitted a paper on the subject to the *Journal of Geophysical Research*. The paper, which was subsequently published on August 1, 2000, and entitled "'Atmospheric holes' are not small comets," basically repeats his previous argument. But as we noted in our rebuttal in the same issue, the model he constructs of the situation assumes that the impacts of small comets into our atmosphere are the same as those for meteoric rocks (they are not). Harris fails to include one essential element in his calculations—the amount of time that a *disrupted* small comet can be observed above the atmosphere as an atmospheric hole. This error completely negates his conclusion.

Janet Luhmann, the senior editor of *Journal of Geophysical Research: Space Physics*, kept this "paper debate" between Harris, Sigwarth, and me civil, editing our respective papers so as to maintain the tone of impersonal professionalism—without altering the scientific content and scientific critique—that she adheres to for the journal. Her objectivity and sense of fair play were a welcome breath of fresh air in the steamy atmosphere of space research.

The small comets also eluded Stephen H. Knowles and his colleagues at the Naval Research Laboratory in Washington, D.C. They examined a six-week period of data collected by the Navy's string of radar installations spanning the southern United States, which is a backup to the Air Force's radar network. They call this Naval Space Surveillance System "the world's most powerful radar." According to Knowles, the system should have detected one comet every few minutes, but it failed to find any among the thousands of objects it did observe during this time period. They estimated that 800-to-5,000 small comets should have been detected by the radar during their 37-day search during the fall of 1997, and they concluded that either the impact rate of the small comets with Earth must be far less than we have postulated, or that the radar cross sections of the small comets must be much less than the values they had computed.

The lack of detection of small comets with their radar came as no surprise, however. The radar is poorly calibrated, and the cross section is at or below the threshold of detection of the small comets. Though they recognize the fact that the possibility of detection of water-snow comets is marginal at best in one part of their paper, they then introduce large quantities of dust into these objects in order to increase the cross section. I found this particularly annoying because Meier, who was a co-author on the paper, is well aware that we have shown that there is very little dust in these small comets. If the small comets were as rocky or dusty as

Meier would have it, these objects would have been discovered by radar shortly after the Second World War.

The Knowles presentation appeared on the second day of small comet talks at the spring 1998 American Geophysical Union. The second day did not draw as large a crowd as the first day. I suppose that after the virtual bloodbath, these talks were anti-climactic. David Deming's talk about the need for an extraterrestrial input of water to explain the volume of water on Earth was by far the most important of these, but few people seemed to realize it. I will have more to say about this important paper later. Tobias Owen of the Institute for Astronomy at the University of Hawaii spoke of the problem of the fingerprint of the ratio of heavy hydrogen to normal hydrogen in the large comets and the source of the Earth's water. Heavy hydrogen is also known as deuterium. He did not really address the small comet topic, however, because he could not. We just do not know this ratio for small comet water.

Most of the rest of the talks dealt with the presence of water in the upper atmosphere. James Russell and his colleagues at the Center for Atmospheric Sciences at Hampton University work on upper mesospheric water, and they realize just how complex an issue this is. They are working with something never really thought of before—a substantial amount of extraterrestrial water—and they are carefully going through and checking it. Right now, they do not eliminate the possibility of small comet water.

Bryan Hannegan of the University of California, Irvine, presented his paper that was being published in the May 1998 issue of *Geophysical Research Letters*. Using a stratospheric-mesospheric circulation model, Hannegan concluded that "current observations of HO in the stratosphere and mesosphere are consistent with an extraterrestrial source of HO..." but the flux is a hundred times less than the small comet hypothesis would suggest. Earth may be receiving two million tons of water a year from space at the most, he said, not 200 million tons. So there is not enough water in the mesosphere to account for the small comets.

But the fact that the cometary water vapor is injected deeply into the stratosphere is inconsistent with Hannegan's assumed injection at higher altitudes and hence renders his results inapplicable to the small comet situation. The paper shows no understanding of the comet's impact at all. Hannegan assumed that the small comets just stop dead cold up there in the upper atmosphere. But you just cannot lay the water up there like a blanket. Hannegan does not understand the dynamics; he is not a physicist. And he knew he had lost even before he got off the starting line because I had already done the calculations and showed him the graphs indicating that the small comets penetrate much deeper into the upper mesosphere. That is why, during the small comets session on that second day, Hannegan was so desperately asking the other researchers who were making presentations on upper atmospheric water vapor: "Can you say there is no water due to small comets up there?" And they would look at him and say: "We can't say that," which obviously upset Hannegan.

But there is more to this story. Hannegan was the primary author, but it was clearly Michael Prather's baby. His name is on the paper, and Prather is the senior author for all practical purposes. He was also the editor-in-chief of the journal in which this paper appeared. If you write a paper and you are the editor, you should ship it out to somebody else to publish, another journal. You should not have it published in your own journal. I was not sent a copy for review; I think Prather had a standing policy of not allowing me to see anything. If the critics cannot make small comets disappear with all these machinations, then they are in deep trouble because it will just come right back to haunt them.

By the end of this mad hatter's session, this two-day small comet symposium at the American Geophysical Union, Tom Donahue seemed to have changed his mind—again. "I think Frank's been seriously challenged," he told Richard Monastersky of *Science News*. "There's a strong likelihood that it is instrumental [noise]. That's the way the wind is blowing." Donahue has been pretty outspoken about the small comets since the beginning. But he has flipped back and forth on the subject, and when you reverse your position so many times, your opinion becomes worthless. I think the value of his dollar was now pretty low.

The main problem with the critics is their desperation. As I have said before, if we found something that shows that the small comets are not there, we would have reported it. I really would. If I did that I would probably be somebody's hero. Why waste people's time? But I cannot honestly do it for the same reason I could not hide it before. I will walk off this stage with my integrity intact and everyone else will have to worry about their own.

Meanwhile, during the first half of 1998, I was dealing with the publication of a major paper on the small comets for the *Journal of Geophysical Research*. It began as two papers. I had originally sent the journal a paper entitled "Atmospheric Holes: Altitude and local-time effects" in early January 1998. It essentially stated that if the atmospheric holes were due to instrumental artifacts, then neither the spacecraft's altitude nor local time should have a bearing on the matter. But they do. The frequency of atmospheric holes decreases as the spacecraft's altitude increases. The holes also show up more in the local morning hours relative to the local evening hours. These are things you would expect only if the holes were a genuine geophysical phenomenon.

The two reviewers of this paper were generally positive. One was quite impressed. "The strong points," the reviewer wrote "… are the automated detection of dark holes and the resulting observed dependencies on spacecraft attitude and local time. It is hard to imagine how these could result from any random process or artifact of the instrument." But both reviewers also wanted me to address the first Parks paper, and Janet Luhmann, the senior editor for the *Journal of Geophysical Research*, agreed, requesting that I address "the serious issues regarding detector backgrounds" that Parks had raised.

Almost simultaneously I had submitted a paper called "Observations of Transient Decrease of Earth's Dayglow with the Ultraviolet Imager on the Polar Spacecraft" to the *Journal of Geophysics Research* in early February. Two months later, Luhmann declined publication. One of the three reviewers thought it was more of a comment on Parks and should be submitted to *Geophysical Research Letters* instead. But the reviewers also thought that the noise that plagued UVI was insufferable and recommended that I not publish the paper at all. But I had gone ahead and analyzed the data anyway and the atmospheric hole signatures were there, although not persistently, I admit.

Our efforts were not wasted, however. I would essentially end up rolling the essential points of this second paper into the first. This new version of the first paper on the geophysical effects was greatly expanded and dealt with all the electronic bilge that Mozer, McFadden, Paxton, and company said was fooling me into thinking this small comet phenomena is real. To capture the all-encompassing focus of this new paper, I changed its title to "Atmospheric holes: Instrumental and geophysical effects." It was a monster, more than a hundred pages in manuscript form. Luhmann rightfully insisted that the paper be comprehensive on the topic to the extent that all reasonable comments by others be answered; she wanted to avoid a long

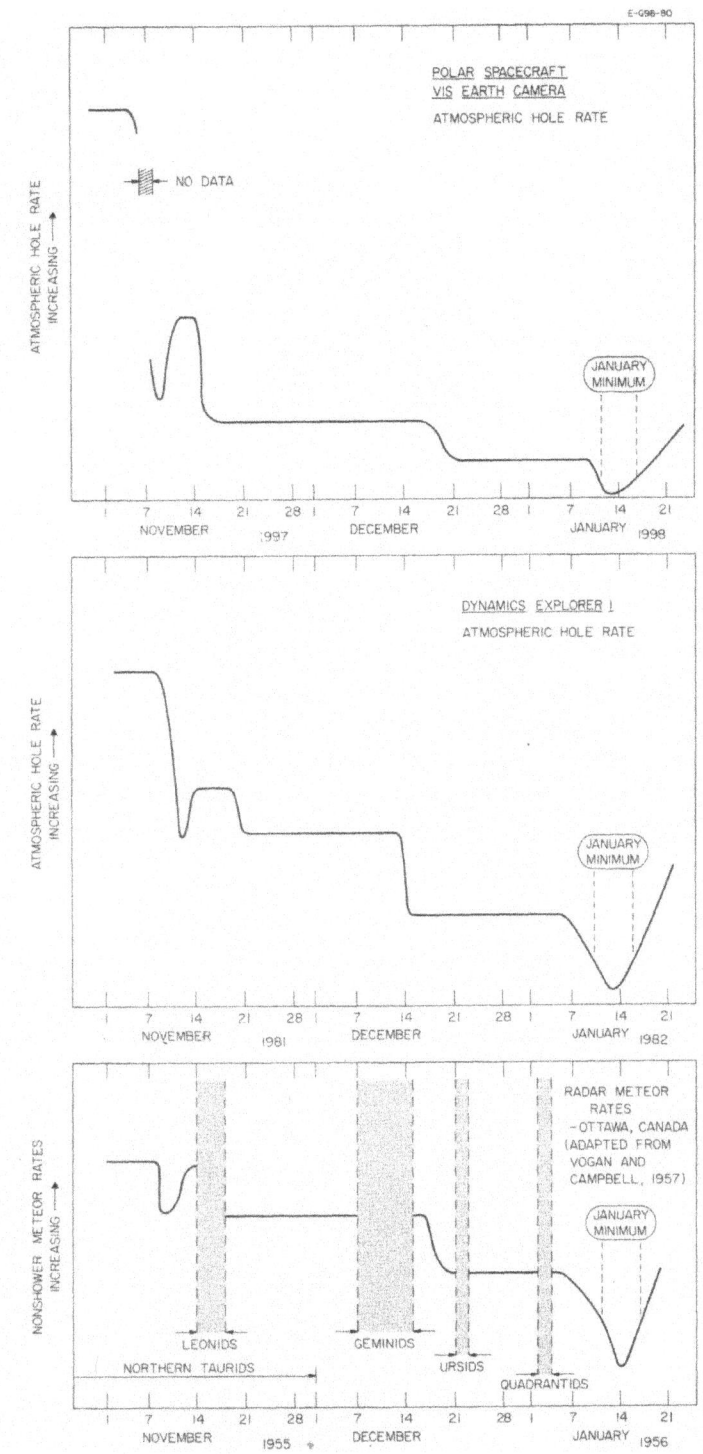

The atmospheric hole rates observed with the Polar spacecraft from November 1997 through late January 1998 (top) are remarkably similar to the rates observed during the same months but 16 years earlier with Dynamics Explorer 1 (middle), and with the non-shower, forward-scatter meteor rates observed in Ottawa Canada (bottom), some 42 years earlier.

chain of comments and replies in her journal.

This paper was a major milestone. In it we completely investigated all possible instrumental effects, including effects due to energetic electrons in the outer radiation zone, the non-uniform sensitivities of the pixels or "hot spots" in the sensor, and the contributions of longer wavelength radiation from the atmosphere to the camera's responses. We also evaluated the random rates for a period in mid-January during which there were no atmospheric holes and thus provided an excellent in-flight calibration of the instrument's noise. We showed that these instrumental effects were not a major contributor to the appearance of atmospheric holes.

Perhaps even more importantly we developed an automated system for detecting atmospheric holes using computers, completely eliminating any visual bias on our part. The computer searched for groups of dark pixels where the intensity of light varies significantly from the average. This "hands-off" search for these atmospheric holes revealed a strong altitude dependence of the frequency of atmospheric holes. It also showed a substantial local-time variation of rates, which favored the local morning sectors of the atmosphere relative to those in the evening, and the increasing sizes of the atmospheric holes as the spacecraft altitude decreases. Finally, this hands-off approach found large seasonal variations in the atmospheric holes rates during the period November 1997 through late January 1998; these were remarkably similar to those observed during the same months but 16 years earlier with Dynamics Explorer and in non-shower, forward-scatter meteor rates observed in Ottawa Canada, some 42 years earlier.

Luhmann accepted the paper for publication on August 31, 1998. Two days before Christmas, and a week before it actually appeared in print, we put the paper up on the web and notified Mozer, McFadden, Dessler, Parks, and Winglee where they could go find it.

And to all a good night.

CHAPTER 35

Spacewatch Redux

Our presentation of the Polar findings in the Spring of 1997 unleashed an avalanche of scientific activity. The topic that was once so unpalatable to the scientific community now drew people like a casino buffet. There was even a momentary scramble to search for the small comets. I think the people who specialize in tracking near-Earth objects were feeling faint.

The Near-Earth Asteroid Tracking, or NEAT, team headed by Eleanor Helin was one of those who went out and looked for the small comets. The NEAT telescope is located at the Ground-based Electro-Optical Deep Space Surveillance (GEODSS) site on Haleakala, Maui, Hawaii. This telescope is designed to do a comprehensive search of the sky for near-Earth asteroids and comets. The NEAT camera was designed, fabricated, and installed on a one-meter GEODSS telescope by the Jet Propulsion Laboratory, but the facility itself is run by the Air Force.

Helin, along with co-investigator David L. Rabinowitz, searched for the small comets but they did not have the sensitivity to lock on to the small comets because their field-of-view is so huge. That forced their sensitivity up and the only way to compensate for that is by actually moving with, and tracking, these objects. But that is not a mode used by the Air Force. And since this is part of our security network, it is very unlikely that the Air Force would let a bunch of people from JPL go in and fiddle with it.

The problem with the NEAT search is that most of the small comets are fainter than anything the telescope can detect. Because NEAT cannot lock on at the right speed, it is about a magnitude in brightness short of seeing the small comets, and the NEAT people know that. The threshold of the NEAT is a visual magnitude of 17.8. which means they could not see the brightest small comets, those with a maximum visual magnitude of 18.2 that Yeates had observed with the Spacewatch telescope. And it is important to point out that the number of small comets increases rapidly with decreasing brightness—by a factor of about a thousand when the threshold of the Spacewatch Camera is reached at 19.6. So the threshold of the NEAT was simply not sufficient for the detection of any small comets in the present mode of operation.

On January 12, 1998, David Rabinowitz reported to Hoffman that a preliminary search by NEAT had failed to find any small comets. It was not a surprising conclusion, and Rabinowitz was well aware of that. Their null result was "not inconsistent," he said, with the rate of Spacewatch detections we had reported. Nonetheless, Helin would later say that they had searched for the small comets and had not found them. It was just sour grapes.

I believe that Eugene Shoemaker of the U.S. Geological Survey also intended to look for the small comets. A leading astrogeologist, comet watcher, and the co-discoverer of the giant comet that plowed into Jupiter in 1994, Shoemaker admitted to being "deeply puzzled" by the Polar results. He was torn, on the one hand, by his conviction that the small comets hitting the Moon would blow holes in the loose, yards-thick layer of dust and rock that blankets the lunar surface. He believed this to be the single biggest problem for the small comet theory and compared these missing lunar impacts to dogs that fail to bark in the night. "On the other hand," he commented to Kathy Sawyer of the *Washington Post* about our Polar results, "these observations are extraordinary ... I'm casting around in desperation." I think Shoemaker was interested in doing a search for the small comets but fate intervened. Tragically, he died in a car accident in Australia in July of 1997.

The issue of using the Hubble Space Telescope for a small comet search was raised in March of 1998, when a NASA news release suggested that the Hubble Space Telescope should be able to spot the small comets. Over a period of three years a team of researchers headed by Robin Evans and Karl Stapelfeldt of NASA's Jet Propulsion Laboratory had searched through 28,000 Hubble images looking for the signature of asteroids. They had then made a statement to the effect that the small comets should also have left detectable trails in the Hubble images but had not.

We had looked into the possibility of observing small comets with the Hubble Space Telescope in 1996. In February of that year we had sent to Ralph Bohlin at the Space Telescope Science Institute a detailed description for a search for the small comet using the Hubble Space Telescope. I felt that the Wide Field and Planetary Camera aboard Hubble offered the possibility of decisively confirming or denying the existence of the small comets. Only about 40 minutes of observing time was needed. But in order to do the search the telescope's field-of-view would have to be moved properly across the celestial sphere. Otherwise Hubble would not see the small comets. And that was the problem. The people who control the telescope said that moving the field in this way was too risky. Because of that limitation the Hubble Space Telescope simply cannot detect small comets.

So the statement made by Robin Evans and Karl Stapelfeldt about small comets in the 1998 NASA press release was based upon an invalid assumption that the apparent motion of the small comets is more-or-less random, like that of asteroids, and occasionally one will dwell on the CCD pixels for a long time. This is a false assumption because the small comets are moving in a stream. It is too bad that astronomers cannot read the relevant literature. Anyway, I can do without the infamy if my special mode of operation ended up causing the catastrophic failure of the great Hubble Telescope.

Of all the telescopes out there, however, the Spacewatch Camera was still—a decade later—the best candidate for doing a small comet search. But after the controversy that had erupted in the 1980s over the first small comet search using the Spacewatch telescope, it was unlikely that the Arizona people would ever allow another search using their equipment. But I was wrong. Alan Harris played a large part in this not-very-glorious chapter in the history of small comets. Harris is a Senior Research Scientist in the Jet Propulsion Laboratory's Earth and Space Sciences Division. He has degrees from CalTech and UCLA, has studied asteroids with the telescopes at Mt. Wilson and Palomar, and specializes in the search for Near-Earth Objects, or NEOs. So Harris is obviously one of those people whose reputation would be on the line if the small comets turned out to be real.

Harris entered the picture arm-in-arm, so to speak, with Alex Dessler. Early in 1998 Harris offered to go out into the desert with Dessler to look for the small comets. He wanted to follow up Dessler's claim that you should be able to look up in the sky at dawn or dusk and see a flotilla of these things roll in with your naked eyes. Dessler had said as much in his 1991 *Reviews of Geophysics* paper. That is totally incorrect. When I did the calculation showing that it is not so easy and not really possible to see the small comets with the naked eye, Harris agreed. The calculation is straightforward and I think it surprised Harris because I did it correctly, despite the fact that I am not an astronomer and I do not have a PhD in astrometrics, as he does. But I can do these sorts of things because I am a physicist, which always raises the ire of all those with off-brand degrees. Anyway, the fact remains that you cannot see the small comets with the naked eye. You have to have binoculars at least.

Harris's email to Dessler on their proposed naked-eye/binocular search for what he calls LAFOs, (Louis A. Frank Objects), his derogatory term for the small comets, clearly shows he was anything but serious about the subject. His total proposed budget for $120.30 includes round-trip airfare from Los Angeles to Tucson, reimbursement expenses for the use of Dessler's car, no charge for a pair of binoculars and a pair of eyeglasses (personal property of the investigators), "optical cleaning tissue, to be sure any observed spots are real" for 10 cents, a bottle of Jack Daniels (Kentucky sour mash whiskey) "to sharpen the vision" for $20, and a postcard to submit the final results to NASA for 20 cents.

Harris's humorous work plan is worth quoting in its entirety, deliberate errors and all: "Harris flies to Tucson in the afternoon (after work). Dessler meets him at airport, and both drive out to the desert before dark. Sample the JD for quality and consistency, and to sharpen the vision. Conduct naked-eye search from dusk for a couple hours, or until increased thirst interferes. Have some more JD to prepare for the scopitel seech [slurred version of "telescope search"]. Check the JD once more to adapt eyes (and mind) for teleskeet shoot. Point barnicles at sky (not the bottle, stupid). Look for telescopish LAFOs ... try not to laugh. Have a bit more JD to regain composure. Look reeel hard, try to imagine ... Finish off the JD, pass out under the stars and forget it all until dawn. In the morning, Dessler drives back to the airport while Harris composes the final report on the postcard. Drop final report in the mail from the airport, and Harris flies home in time for work in the morning. Note that this work plan cleverly avoids any cost for lodging or manpower. One word of caution: may need to double budget and do it twice, in case weather interferes."

During this time, in early 1998, Jim Scotti of the Spacewatch Telescope joined the email discussion on small comets led, in a sense, by Bob Hoffman. Not surprisingly, Scotti disagreed with my summary of the Spacewatch search that had been conducted by Yeates in the late 1980s. Scotti thought that if the small comets were real, they should be showing up in the current Spacewatch survey data. They should, in fact, be showing up several times in each image, he insisted. No need for the Yeates "skeet" shooting technique either, he said, which was required by the older, noisier, and less efficient Spacewatch detector. In fact, he said, they didn't need any additional survey data to refute the small comets claim.

The Spacewatch Camera people were fixed on the concept that one looks in the direction opposite to Earth's orbital motion to search for small comets. In principle this is viable only if most of the small comet motions are ideally aligned so that their image dwells on a CCD pixel during an exposure. But the actual distribution of motions provides for relatively fast movement across the pixels and hence no detection of the small comets, except for an exceeding rare case. Yeates used this direction of viewing for his null observations. And indeed, he found no small comets in this mode.

To my great surprise, the Spacewatch people did finally do another search for the small comets. I think they must have been goaded into it. Harris must have told them: *Why don't we just put this issue to rest? Let's make the run. But not in the Gehrels mode of looking in the direction of motion. We'll do it according to Yeates, right to the letter.* I think Dessler probably had a hand in it, too. I think he must have said: *Look we get tromped on with the Yeates stuff all the time, let's just do it over, okay?* And they did.

News of just what they did trickled out slowly, however. The first mention suggesting that they were even planning a search came from an abstract that Jim Scotti and Al Harris submitted for the Spring 1998 meeting of the American Geophysical Union. The abstract concluded that the Yeates method was no longer required in order to observe the small comets. Upon seeing this abstract, Hoffman, at Dessler's suggestion, contacted Harris and Scotti on March 23, 1998. Hoffman suggested that they run the telescope in the Yeates skeet shooting mode anyway "to satisfy any doubters." He also added, "We would want the images to be made available to the scientific community," and invited them to present their findings before the Small Comet Steering Group.

Harris replied a couple of days later saying that Scotti had *already* successfully obtained two hours of data in the Yeates skeet shooting mode and expected him to obtain two more hours of data before the meeting on April 17th. Like Scotti, Harris stated that Spacewatch in its normal mode of operation should see the small comets if they exist in the numbers I claimed. Although Harris admitted that the "skeet shooting" mode does enhance the expected rate of detections by factors of 10 to 100, he went on to say that the 2-3 hours of operation in skeet shoot mode is actually less effective than the 10,000 or so hours of regular operation by the Spacewatch telescope.

Scotti also replied to Hoffman saying that it was important to note that the increased sensitivity and rate of sky coverage per frame of the Spacewatch telescope with the new larger format CCDs would allow them to duplicate the amount of sky coverage made by Yeates in less than an hour of observing time while increasing the limiting magnitude by at least 0.5 magnitudes. So two hours of operation would net about 7 square degrees of sky coverage, more than three times what Yeates and I had used in our analysis, according to Scotti. They now have so many pixels on their new CCD that they can no longer take a picture, read it out, and take another picture in a reasonable amount of time, as Yeates had done. Now they had to take a picture, close a shutter, take another picture, and then read it out. But that was fine. And that is just what they did.

The Spacewatch run took place in February of 1998. They took 17 big frames. That's all. They did it in one run. One night. There were 17 images, or so Harris and Scotti claimed at the small comets steering group meeting on April 17, 1998, where they showed one small segment of one frame. Since it was the Yeates mode and they were moving the telescope, Harris and Scotti thought the signature of a small comet would be a trail, just like a star trail, only shorter. But that is not true because the small comet signatures are at *angles* to the stars. They are not parallel. Yet Harris was saying, "There's nothing in there, no streaks parallel to the stars."

I was sitting at the head of the table, quite close to the image. "But look over here," I said.

"I think you have the wrong impression about what the signature should be, what the trail should look like." Then pointing to a corner of this segment of a frame they were showing us, I said: "It should look something like this streak that appears right here." Hoffman, who was sitting right next to me, says, "Yeah, and look up there in the upper left-hand corner at that strange signature of some object or something." So Harris and Scotti looked closely at what we had pointed out to them, and finally they realized what the small comet signature was supposed to look like, and Harris said, "Oh." And he pulled the viewgraph away. Just like that.

Trying to be as diplomatic as possible, I emailed Scotti the day after the Small Comet Steering Group meeting, saying how pleased I was that they had run the Spacewatch Telescope in the Yeates mode in a search for the small comets. I told him how tantalizing the viewgraph for part of a single image was but that inspection of the raw images in a high-resolution color format was necessary for a decisive analysis of these images. I tried to explain to Scotti what a small comet trail would look like. And I told him that during my brief, close-up examination of their viewgraph of a partial image I thought I may have identified a possible candidate. I also asked him when we might be able to obtain a copy of the new, raw, Spacewatch images on digital tape. But I knew I was wasting my time.

When Harris made the Spacewatch presentation at the American Geophysical Union meeting in Boston in May of 1998 he again showed just one image, but somehow it was even smaller than the fraction of an image he had presented with Scotti at the Small Comet Steering Group meeting a month earlier. They had cropped it. You could tell because the star trail sizes were longer, and there were fewer of them.

Harris told the American Geophysical Union audience that the detector on the present Spacewatch telescope was far more sensitive than the older version, so the small comets should show up much more clearly—if they exist. But they had found nothing in the Spacewatch images, no small comets at all, he said. They had done the search in the Yeates mode and found nothing. The trouble is they were not showing anyone the data, and they had reached their conclusion without doing any analysis. How could they come out with a conclusion like that without doing their analysis?

After the presentation Hoffman went up to Dessler and Harris and said: "Look, you don't have to worry about us swiping the data, all we want to do is examine the data. You have the right of first publication and everything else. We just want to look at them, to examine the 17 frames." Hoffman then tried to convince Dessler and Harris to go back and to persuade Scotti and the Spacewatch people to make the data available. Of course, I do not believe that Harris or Dessler would do anything of the sort, but they could hardly tell the head of the Small Comet Steering Group that. Harris informed Hoffman that the Spacewatch people would not release the data until they had done their own analysis and published it first. Hoffman did not have a problem with that. The question was: how long would it take for them to do this?

Later Hoffman called up Bob McMillan, who had taken over Gehrel's spot at Arizona in charge of the Spacewatch activities. Hoffman learned that no one at Spacewatch had pursued the matter. Finally, frustrated over McMillan's stalling, Hoffman said: "We'll pay for the images." But Hoffman got the cold shoulder from McMillan, who basically said, *We're too busy.* Hoffman just was dumbfounded.

Impatient with the Spacewatch people, Hoffman located three astronomers who were interested in doing an independent analysis of the new Spacewatch images. Since the Spacewatch people were not inclined to give this enough priority to work on it, Hoffman hoped they would agree to allow the three independent astronomers to analyze the 17 frames. These people had no relationship with the University of Arizona and they had no relationship with me. One group consisted of Jack Brandt and Ian Stewart at the University of Colorado. The second was Michael A'Hearn at the University of Maryland and the third was Ted Bowell at the Lowell Observatory. Hoffman planned to have the Spacewatch people then review the findings of the independent analysis by these three groups. Hoffman even dangled my oft-repeated promise in order to convince McMillan to take up the offer: I would accept a proper analysis of these 17 images as confirming or denying the existence of the small comets.

But it would not happen. Bob McMillan's reply to Hoffman regarding his proposal for an independent study of the 17 images was negative. McMillan asked for patience in a November 1998 email to Hoffman, saying that the Spacewatch results would eventually be submitted for publication, and only then was he willing to discuss the issue of third parties being allowed to examine the data. Hoffman found good news and bad news in this reply. "I am very disappointed that he wants to sit on the data for an undetermined amount of time," said Hoffman. "The good news is that he is willing to consider releasing the data for others to look at." I was not going to hold my breath.

CHAPTER 36

Deep Waters

"Over the past 10 to fifteen years, our knowledge of Earth's climate has been challenged by new data which cannot easily be explained by existing paradigms." So begins David Deming's landmark study that appeared in the refereed scientific journal *Palaeo* in 1999. Deming is a geologist in the School of Geology and Geophysics at the University of Oklahoma in Norman. The new data Deming refers to includes abrupt warming periods in the Earth history and a 100,000-year periodicity of the glacial cycle. To Deming these are clear signs of an astronomical influence on the Earth's climate. What could that influence be? Deming looked for clues in the Earth's volatiles—specifically carbon and water.

Deming pored through all existing estimates and studies, and found, as I had earlier, that the standard explanation—volcanic outgassing—was a grossly inadequate mechanism to explain the abundance of water and carbon at and near the Earth's surface. Recent investigations of the movement of oceanic continental plates into the mantle, known as subduction, show that the loss rates for water are very large as the plates carry the water deeply below the surface. Unless there has been a considerable influx of water to our planet on time scales much shorter than its age of 4.5 billion years, our planet would now be "dry as a bone." Deming found "… that the net rate at which water is added to the hydrosphere by volcanic outgassing is only 5-15% of the rate at which it is carried down into the mantle by subduction. Similarly, the net rate at which carbon is added to the atmosphere by volcanic outgassing is only 10-44% of the rate at which it is lost to the mantle."

How to explain all this? The Earth's oceans must have some source of replenishment. Deming found that everything falls neatly into place with a "new and unifying theory" that the Earth's supply of volatiles may have largely originated from "ongoing extraterrestrial accretion," in other words, via an influx of small comets. Deming was just fascinated by the fact that the supply rate necessary to account for the mass of the oceans today is exactly what I have inferred from what my critics call "instrumental artifacts."

Deming boldly raised the stakes in the small comet controversy by linking these objects to the well-known vagaries of life on Earth: "The extraterrestrial influx rate may also act as the pacemaker of terrestrial evolution, at times leading to mass extinctions through climatic shifts induced by changes in accretion rates with concomitant disruptions of the carbon and nitrogen cycles. Life on Earth may be balanced precariously between cosmic processes which deliver an intermittent stream of life-sustaining volatiles from the outer solar system or beyond, and biological and tectonic processes which remove these same volatiles from the atmosphere by sequestering water and carbon in the crust and mantle."

Deming had great difficulty getting this important paper published. He originally submitted the paper, then titled "Extraterrestrial Accretion and Earth's Climate," to *Geology* in May of 1997. The editor in charge of Deming's manuscript, Lee Kump of the University of Pennsylvania, sent it to two reviewers, both of whom wished to remain anonymous.

The first reviewer thought there were "enormous problems" with a theory which combines the climate dust connection of geologists R.A. Muller and G.J. MacDonald with the small comets hypothesis of L.A. Frank, but the reviewer never actually came out and stated a single problem. The reviewer also faulted Deming for taking the small comet hypothesis "at face value." Deming would reply that the small comet papers had been published in the peer-reviewed literature, and that he had also mentioned criticisms of my work in his paper. Deming would offer to change the wording to the effect that "Frank's ideas remain controversial." Besides, "the chief point of my paper," noted Deming in a letter to Kump, "is that there is a tremendous amount of circumstantial evidence from geology and meteorology which points to higher rates of volatile accretion." Finally, the reviewer faulted Deming for his many references to textbooks, rather than to primary sources, though his references, in fact, contain not a single textbook! The first reviewer thought the manuscript should be rejected.

The second reviewer said there was "little evidence" to support the "small comet hypothesis." Deming begged to differ, citing not only the global images of the Earth's ultraviolet dayglow, but also the microwave radiometer observations of Olivero, and the visual sightings by the Spacewatch Telescope. The reviewer also accused Deming of trying to "capitalize on the recent renewed interest in delivery of water and other volatiles to Earth by small comets." Deming found this remark personally insulting and replied: "My ideas were developed over a period of years. In fact, I was completely unaware of Lou Frank's work up until the time that about 90% of the present paper was written."

The reviewer also found "zero evidence" to support the hypothesis of extraterrestrial accretion, though, in fact, the evidence includes the asymmetry of the climate record; abrupt climate change; the temperature-volatiles conundrum; the presence of water on Earth's Moon, Mars and Mercury; a high correlation between mean global temperature and greenhouse gas concentrations; the 100,000 year periodicity of the Pleistocene glacial cycle; a 100,000 year periodicity in the flux of extraterrestrial He3; the decrease of mean terrestrial surface temperature over time; geothermal calculations which show that the oceans would be depleted if not replenished by ongoing high rates of volatile accretion; formation of noctilucent clouds; and the direct observational evidence of high volatile accretion rates. Apparently, this was not enough.

A number of the comments suggested to Deming that the second reviewer's knowledge of geology was considerably out-of-date. This suspicion led Deming to double check specific points with independent sources such as Paul Knauth, a geologist at Louisiana State University, and Isaac Winograd, a geologist with the U.S. Geological Survey in Reston, Virginia. Deming learned that the reviewer was indeed wrong.

The damage had already been done, however. The two negative reviews led to the rejection of Deming's paper on July 28, 1997. Deming sent *Geology* a detailed, nine-page rebuttal of the reviewers' comments, asking the editor to reconsider the manuscript. He would not. So in October Deming submitted the paper to *Palaeo*. While *Geology* had acted quickly, *Palaeo* acted with glacial slowness. Five months after his submission Deming still did not know the status of his manuscript. The editor had sent out Deming's manuscript to three reviewers. I was one of them. The second reviewer was Isaac Winograd, who found the paper "exciting," "timely," and worth publishing, as I did. But the third reviewer never returned the manuscript so the editor finally sent it out to a fourth reviewer.

This last reviewer, who remained anonymous, nearly proved to be the nail in Deming's coffin. The reviewer did not like Deming's review of the small comet hypothesis or the style in which he had written his paper, and recommended that Deming undertake a major redrafting of the manuscript. The reviewer's primary objection was that Deming had not discussed, for example, the George Parks criticism of the small comets. Of course, when Deming had first turned in his manuscript, the Parks paper was not even available! Nonetheless, the reviewer thought that "the idea of investigating the implications" of small comets was "a reasonable one" even "before they are substantiated."

After months of delays, the *Palaeo* editor finally decided to reject Deming's paper on the basis of this third review. This led Deming to comment to me in an email: "I guess the important thing is to find any excuse for rejection." The situation was not entirely hopeless, however, as the editor had invited Deming to submit a *revised* manuscript for a re-review. Deming was not sure if the editor meant a re-review by the original reviewers or by an entirely new set of reviewers, which Deming viewed as a waste of time. "No one can hit a moving target," Deming noted.

The *Palaeo* editor turned out to be a realist, and Deming took advantage of the opening offered by the last reviewer. Deming's total revision of the manuscript, submitted in August of 1998, resulted in a paper that was more than twice the length of the original. Deming noted to the editor that the hypothesis he presented in the paper was "not identical to the small comet hypothesis, merely related. I originally developed my ideas with no knowledge of the small-comet hypothesis. The small comet hypothesis and the satellite data which support it are included as additional pieces of evidence." Because of its strengthened arguments and cautious, arms-length review of the small comet material, Deming thought the manuscript needed a new title and changed it to "On the possible influence of extraterrestrial volatiles on Earth's climate and the origin of the oceans." The editor agreed, accepting the manuscript just a week later.

Deming is not alone in upsetting the geological applecart of late. In October of 1992, William Fyfe, a geologist at the University of Western Ontario and the president of the International Union of Geological Sciences, sent me a book chapter he had written called "Crust Formation and Destruction" and wrote in his letter on October 1, 1992: "We have a problem—Earth today is in-gassing, not out-gassing. We need a water source!"

There has been increasing acknowledgement from geologists that the standard model of the origins of the Earth's water just does not hold water. The topic has generally been treated as something that you need not worry about, yet it is so important as far as the history of our Earth is concerned. The standard view is expressed by Frank Press and Raymond Siever in their 1994 college text, *Understanding Earth*, published by Freeman. "Most geologists believe that the origin of the oceans and atmosphere can be tracked to Earth itself," they write. "The amount of water vapor released by currently active volcanoes indicates that over several billion years, volcanic activity could easily have filled the oceans."

But this belief is "blind extrapolation of the present recycling," according to Armand

Delsemme, a University of Toledo astronomer, writing in *Comets and the Origin and Evolution of Life*, a set of conference papers published by Springer-Verlag in 1997. And Delsemme is by no means the only space scientist to have attacked the standard view. Like him, they argue that most volcanic water is not primordial in origin but recycled during the slow upheavals of plate tectonics.

In a 1993 issue of *The Sciences,* Raymond Jeanloz, a professor of geology and geophysics at the University of California at Berkeley, stated: "Just as many astronomers think most of the mass of the universe is made up of invisible dark matter, a growing number of geologists, and I am among them, are becoming convinced that most of the water on Earth may lie unseen, deep below the surface, dissolved into the rocks of the mantle and the core." How much water? "More than ten times the amount present in all the oceans and the atmosphere combined," notes Jeanloz.

Ever so slowly, the small comet theory has been making inroads into the mainstream literature. *Fresh Water*, a 1998 work by naturalist E. C. Pielou, notes the growing acceptance of the notion that the Earth's water supply is increasing all the time through an influx of small comets. And Dale E. Ingmanson and William J. Wallace in the 5th edition of *Oceanography: An Introduction,* published by Wadsworth in 1995 write: "Although this seemingly radical theory of the ocean's origin has yet to gain acceptance by the scientific community as a whole, more and more data seem to offer additional support."

As it happens, the influx of extraterrestrial water brought in by the small comets is *not* inconsistent with the water budget in the atmosphere. Many people were curious about this, but I did not tackle it in any great detail until early in February 1998 when Daniel Kirk-Davidoff of the Atmospheric Research Project at Harvard University asked my opinion about any major inconsistency in the flux of small comets into the Earth's upper atmosphere. Specifically, he wanted to know, is such a cometary water input inconsistent with our current knowledge of the upflow and downflow of water vapor at the tropopause? I spent a considerable amount of time reading the literature on the subject, in particular the major observational advances published in the five years previous concerning the upflow of water into the stratosphere at tropical latitudes, the return of water to the troposphere in the polar vortex, the conversion of methane into water vapor, and the cycling time for the air mass above the tropopause.

The cometary water input that I used was an influx of 10 million small comets per year with individual masses in the range of 20 to 40 tons. It is interesting to note that the cycling time of the entire air mass above the tropopause is rather long, about two years. This implies that two years of small comet influxes are stored above the tropopause. It also implies that the small comet water input is generally well mixed across the tropopause, the only exception being the small amount of water that is entrained at higher altitudes in the mesosphere during the small comets' passages to lower altitudes. I found that the small comet influxes I had calculated are not inconsistent with our current knowledge of the water vapor transport across the tropopause.

To illustrate this, I prepared a summary diagram of water transport across the tropopause that accommodated the input from the small comets. It consists of four elements, three of which make up the influx, and one the outflux. The first element of influx is the water from the tropics. A 1993 paper led by K.K. Kelley of the NOAA Aeronomy Laboratory in Denver found that this ranged from 1.8 to 3.2 ppmv (parts per million of volume), with an average of 2.4 ppmv. The second element of the influx is the methane conversion, in other words, the methane that comes up from our atmosphere and is photochemically converted into water. A 1996 paper led by Andreas Engel of the Institute of Stratospheric Chemistry in Julich, Germany, determined this factor to be about 1.1 ppmv. The third and final element of the influx are the small comets, which I calculate bring in anywhere from 0.7 to 1.4 ppmv, with an average of 1.0 ppmv. For the outflux, or the observed return flux into the polar vortex, I took the numbers from a 1995 paper led by Andy Dessler. Andy Dessler, who is Alex Dessler's son and works at the Goddard Space Flight Center, calculated that the outflux, or the observed return flux into the polar vortex, ranged from 3.7 to 4.7 ppmv. I used the intermediary figure of 4.2 ppmv, which matches pretty well the total upflow into the upper atmosphere I had calculated of 4.5 ppmv, including the input from the small comets.

My chart, which I called "H2O Cycling Above Tropopause with Inclusion of Small Comet Fluxes," really upset people. Bob Meier wrote that it "generated a lot of discussion in our hallways" at the Naval Research Laboratory. The critical points are net inflow and outflow. Andy Dessler believes science understands the water budget of the stratosphere so well that it provides a severe constraint for any source of water vapor. But the fact is we just do not know that much about it. He insists that science knows these numbers to a tenths of a part per million but that is simply not true. The uncertainties are not so much on the return flux into the polar vortex, which is pretty smooth coming in. And the methane conversion number is not controversial. The problem is that there are atmospheric structures in the tropics that just punch right through the tropopause. So spatially, at any given instant in time, there is a tremendous variation in the upflux. That is where the great uncertainty lies. And as long as we have that great uncertainty about what the upflux is through the tropopause, then we cannot make a definitive statement about the small comets existing or not because the small comets influx is just not that large compared to the uncertainties in the tropical upflux. The standard version, which obviously does not consider the small comet influx, has the influx and upflux matching by simply taking the high part of this upflux range or 3.7-4.7 and not using the small comet influx of .7 to 1.4 that we have calculated.

There is a tremendous amount of uncertainty surrounding these numbers. But it all boils down to this issue of water in the mesosphere, and the fact remains that small comets, even as clouds of water, will penetrate down just above the tropopause. At an altitude of about 20 miles, which is the top of the stratosphere, the downward speed of most of the cometary water vapor is an impressive 200 miles per hour. This slows down to the tens of miles per hour at the tropopause at airplane altitudes of 10 miles, where the mixing of cometary water vapor and atmospheric flows should be in full swing.

There is, of course, hard data to back up the passage of this cometary water through the atmosphere. In August of 1997 an ozone-monitoring satellite trailing the space shuttle Discovery detected much more water vapor in the upper atmosphere at northern latitudes than current theories predict. This region is supposedly drier than the Sahara Desert. While water vapor is abundant in the lower atmosphere, very little of it manages to drift upward in the polar regions, so the stratosphere and mesosphere remain dry. Only in the tropics, where thunderheads push moisture into the stratosphere, does water vapor manage to reach the upper atmosphere. With the Polar confirmation still on everyone's mind when this discovery made the news, the press quickly touted this preliminary finding as supporting the small comet theory.

The high readings came from an instrument called the Middle Atmosphere High Resolution Spectrograph Investigation (MAHRSI) operated by Robert Conway, a planetary physicist at the Naval Research Laboratory in Washington, D.C. MAHRSI found a surprising amount of our old friends, the hydroxyl ions, in a layer of the mesosphere at altitudes from 43 to 56 miles. The hydroxyl ions form at high altitudes when ultraviolet light splits apart molecules of water vapor. These results confirmed an earlier set of MAHRSI measurements taken during a 1994 shuttle flight, but Conway had questioned them since they contradicted standard atmospheric theory.

Conway, however, was less than convinced that the small comets were responsible for the presence of water vapor in the polar atmosphere. On the one hand, he told Richard Kerr of *Science*: "You have to give the man [Frank] credit for predicting something we're now seeing." But he then told Richard Monastersky of *Science News*: "… I can't believe that his explanation is the right answer." This comment led to a heated reply from a *Science News* reader in a subsequent issue, and rightly so. "Who cares what Conway believes unless he can propose a reasonable alternative?" wrote in Dana Thibault of Denver, Colorado. "This kind of thinking is a disservice to the ideals of scientific inquiry."

This was not the first report of excess water in the mesosphere. Nor was it the first time that such results had encountered disbelief. Not long before the surprising MAHRSI results, James Russell of Hampton University in Virginia had also found a water spike in the upper atmosphere in their data from the Halogen Occultation Experiment (HALOE), which has been flying on a satellite since 1991. The HALOE instrument measures water vapor directly by peering through the atmosphere while the Sun rises and sets behind the Earth. While reprocessing their data late in 1996, Russell and his colleagues were shocked to find that their instrument had discovered 50 percent more water vapor than they expected to find at an altitude of about 45 miles above the sunlit poles.

Of course, the critics replied that even if the mesosphere is wet, it was not nearly as wet as it should be if the small comets were bombarding the Earth. But they forgot one thing. You just cannot stop this incoming cometary water. There is so little atmosphere up there, the water vapor comes in like a freight train. It travels at Mach 10 as it passes through the upper atmosphere. How could it possibly stop there? Sure, some of the "skin" gets left up there, but who knows what percentage that is? It could be one percent or ten percent.

The important thing is that an excess of water has been measured up there which no one can account for as being from an Earthly source. If they could totally account for the water vapor up there, that would make a strong statement about there being no significant extraterrestrial influx. Sorting out water fluxes and why the water abundance varies with latitude and season will need close monitoring of the variation in the small comet influx. "It's when we get challenging observations like this that we start to rethink all of our assumptions," John Olivero told *Science* reporter Richard Kerr. Olivero, whose Penn State microwave measurements of water bursts in the upper atmosphere have still not been published, is now at Embry-Riddle Aeronautical University in Daytona Beach, Florida. "That's what science is supposed to be all about."

The amount of water involved with these objects surely signifies that they are indeed comets. But we have now learned that they are clearly a different class of comets than the well-known larger comets. Using the James Clerk Maxwell submillimeter radio telescope atop Hawaii's Mauna Kea, a group of researchers in the United States, France, and Switzerland headed by Tobias Owen of the University of Hawaii's Institute for Astronomy reported in August of 1997 that the ratio of deuterium atoms to hydrogen atoms in Comet Hale-Bopp was three to 10,000, which is the same as the ratio found in other large comets such as Halley and Hyakutake. This is about twice the ratio in the Earth's seawater, however. The significance of this finding is that all three of these water-rich comets could not have been the main source of water for the Earth's oceans. The hydrogen "fingerprint" simply does not match.

This came as something of a shock to some scientists, as the possibility that the water in our oceans is due to an influx of large comets during the early history of our planet has been quite popular during the past decade or so. But many small comet critics were quite thrilled with the finding. They thought that the small comets—if they exist at all—must simply be miniature versions of the large comets. So they assumed that the hydrogen fingerprint of the small comets would be similarly inconsistent as the source of water in our oceans.

This conclusion is not correct, however, because we have shown that the small comets are quite different beasts. Other than in water content, their composition happens to be quite different from the large comets. There is little dust and sodium in these small comets compared to the large comets. Since the contents of the small comets greatly differ from those of large comets, there is no reason to conclude that the hydrogen fingerprints of these two classes of solar system objects would be the same. Back in 1992 a study by Uwe Fink of the Lunar and Planetary Laboratory of the University of Arizona had already upset theories of comet uniformity. His analysis of the spectra of a comet known as Yanaka, discovered in 1988, showed that it was drastically depleted in two charged carbon molecules in comparison to other comets. "The comet is really totally different from others," Fink told *The New York Times*. We can't expect all comets to be identical.

There is lots of water on Earth, in comets, and everywhere in the cosmos. Water has even been found on the Sun. In 1995 an international team of scientists led by Oleg Polyansky, a theoretician from Russia's Institute of Applied Physics, discovered water—not in liquid form because the Sun is too hot, but as water vapor or steam—on the Sun. The finding was confirmed two years later when they compared the infrared spectrum of very hot water in the laboratory with that of a dark sunspot. Water vapor apparently forms in the sunspots because of their relatively low temperature.

Not only are there snowballs in hell but on its front door as well. Although Mercury bakes in temperatures hot enough to melt lead, radar observations discovered in 1991, and confirmed in 1994, the presence of vast sheets of water ice at the planet's south pole, which because of constant shadow reaches temperatures as cold as minus 235 degrees Fahrenheit. The original experiments, which were conceived by Duane Muhleman, a planetary scientist at Caltech, found distinctive bright reflective patches characteristic of water near its poles utilizing the Goldstone antenna in the Mojave Desert and the Very Large Array of radio telescopes near Socorro, New Mexico. The follow-up study by JPL scientists in 1994 used the radio telescope at Arecibo in Puerto Rico and confirmed that the patches were indeed water ice cupped in Mercury's polar craters.

Earth's twin planet, Venus, once had abundant water, also. In 1993 Tom Donahue headed a team of researchers at the NASA Ames Research Center in Mountain View, California, who analyzed data from the Pioneer Venus spacecraft. They estimated how much water existed on Venus billions of years ago by measuring the amount of hydrogen in the atmosphere and the current ratio of hydrogen-to-deuterium. Calculating backward they determined that Venus was covered by as much as 75 feet of water several billion years ago. Venus, dry now, was a very wet planet.

The presence of water on the Moon is still controversial. But in 1998, just two months after its launch, the Lunar Prospector spacecraft found solid evidence for the existence of water ice at both lunar poles, vindicating the results of the Clementine probe four years earlier. Estimates of the amount of water ice present on the Moon ranged from 11 million to 330 million tons. Twice as much of the water ice mix was detected at the north pole than at the south pole.

Then in July of 1999 the Lunar Prospector spacecraft was deliberately crashed into a permanently-shadowed crater where water was thought to exist. Scientists hoped the impact might liberate water vapor along with a plume of dust, but none of the hundreds of professional and amateur astronomers watching the event saw or photographed any such thing. Actually, scientists did not expect to actually see the water, but only detect the vaporous haze and water fragments by means of sensitive spectrometers tuned to the ultraviolet and infrared. Scientists are still searching their data for such a signature. But even if a water signal is never found, it does not mean that water is necessarily absent from the Moon. Prospector's impact may not have been powerful enough to separate the water from the minerals, or it may simply have hit a dry spot among the patchy ice deposits at the bottom of the crater.

Water exists beyond the Moon, as well. In 1997 water vapor was found in the atmospheres of Saturn, Uranus, and Neptune by the European Space Agency's Infrared Space Observatory. Saturn's moon Titan was also shown to have water. We said a decade ago that there would be water vapor in the atmospheres of the giant planets. Not surprisingly, Donald Hunten of the University of Arizona pointed out that the amount of water found was only one ten-thousandth of the concentration that would be expected if the small comets were real. But how does he know? The water vapor is not going to lay up there. Only a small fraction of the water vapor that the small comets bring in gets entrained in the thin mist of the atmosphere. He must think that somehow this supersonic cloud gets stopped in the thinnest of vapors. The reason they have only found a trace of water is because most of the water plummets deeper into the atmosphere than they can see.

These reports are very bothersome for most people. So scientists generally explain the water vapor as coming from meteors. A micrometeorite will be stopped very high if it is tiny. But what always puzzles me is, if you have a tiny little micrometeorite, how do you get the water in? It will be destroyed as it comes into the atmosphere.

Water is showing up all over the universe. Scientists have found the presence of massive amounts of water vapor within a cloud of interstellar gas a trillion miles across close to the Orion nebula, which is known to be a center of star formation. This region in the sword of Orion is a veritable "water factory" that churns out enough water molecules to fill Earth's oceans 60 times each day, according to Martin Harwit, an astrophysicist at Cornell University, whose report was published in the April 20, 1998, issue of *Astrophysical Journal Letters*. These new observations from the Infrared Space Observatory confirm predictions by astrophysicists over the past 25 years that whenever the temperature exceeds 200 degrees Fahrenheit, chemical reactions will convert most of the oxygen atoms in interstellar gas into water. These findings could have implications for the origin of water in the solar system, says David

Neufeld, a professor of physics and astronomy at Johns Hopkins University. Eventually the water vapor will freeze, becoming small ice particles. Similar ice particles are thought to have been present within the gas cloud from which the solar system originally formed. "It seems quite plausible that much of the water in the solar system was originally produced in a giant water vapor factory like the one we have observed in Orion," Neufeld said.

Astronomers using a radio telescope have also detected water 200 million light years from Earth in a galaxy called Markarian 1, located in the constellation Pisces, or fish, appropriately enough. Normally, the signature for the water molecule is too weak to be detected by radio telescopes, but so much energy spews forth from the center of this galaxy that it creates a very powerful beam of concentrated radio signal known as a MASER. The water in Markarian 1 is contained in clouds that are rotating at thousands of miles an hour around the galaxy's center, which is thought to house a black hole. Interstellar water was found in our own Milky Way galaxy in 1969.

Science once believed that Earth was an isolated little well of water, that Earth was a unique circumstance. All the water was here and there was next to nothing anywhere else. But everywhere we are looking now, we are seeing water. Why? Because our instrumentation has improved. Water just does not smile at you like a sodium lamp does. It's dim, it's gray, it's hard to detect. And that is why the water vapor from the small comets was hard to detect.

CHAPTER 37

Last Dance

The ultimate irony, perhaps, is that the University of Iowa now has a new telescope, known as the Iowa Robotic Observatory (IRO), that is capable of conducting a search for the small comets. This 20-inch telescope is located in the high-altitude Sonoran Desert at the Winer Mobile Observatory, which is about three miles outside of Sonoita, Arizona, and not far from Tombstone. The sky glow at this position is significantly greater than for the observatories at the Kitt Peak National Observatories, but it is adequate for a small comet search during the darkest nights.

The Iowa Robotic Observatory is the project of Robert Mutel, who has been a member of our faculty in the Department of Physics and Astronomy at the University of Iowa since 1975. Mutel, whose specialty is radio astronomy, supervised the design and construction of the Iowa Robotic Observatory, which is shared among the astronomy departments at the three Iowa Regents Universities. The IRO was designed primarily as an educational tool that could also do a little research.

The IRO telescope was due to open for business at the beginning of 1998, but a welder at the site in Arizona started a fire that did considerable damage to the camera proper, the cables, and other equipment. When Mutel and his colleagues arrived at the site after almost nine months of absence, they found that mice had damaged the computer and cables. By working day and night, they managed to operate the telescope in a limited manner in September of that year. But even then, the software that would have allowed us to operate the telescope remotely from the University of Iowa was not functional.

In August of 1998, John Sigwarth and I approached Mutel about the possibility of using the IRO to verify the existence of small comets. Mutel and his friend in the astronomy department, John Fix, agreed to conduct the study. Fix spent three decades on the University of Iowa faculty. His research focused on observational astronomy and ranged from the structure of circumstellar clouds to an investigation of zodiacal light using the Dynamics Explorer satellite. But in the summer of 1999, Fix was busy moving to Huntsville where he would become the dean of the College of Science at the University of Alabama. To my knowledge he did not participate in the subsequent observations, do any analysis, or write any part of their eventual paper.

John Sigwarth and I designed the actual search. We had to. There are only three people who are familiar with the way to operate a small telescope to detect small comets: John Sigwarth, Clayne Yeates, and myself. Unfortunately, Clayne is deceased. So we would design the search, Mutel would do the observations, and the data would be shared. We also agreed that our analysis of the data would be done independently.

The small comets are small, dark, and moving with high apparent angular rates at small ranges from Earth. For this reason, the viewing mode chosen for IRO had to meet several criteria. First of all, the IRO had to be capable of tracking the small comets, and it can because the telescope drive is completely flexible and the software can drive the telescope field-of-view along any path across the celestial sphere. The small comet search also had to obtain its images during periods that were not too short (when the total signal is small and too faint) and not too long (when the signal is swamped out by the sky background). We calculated that the IRO should be able to track the small comets at distances of about 45,000 miles, as compared to the Yeates' Spacewatch Telescope search at 90,000 miles with the telescope drives stopped. Although the number of events would be less, the small comets should be four times brighter than in Yeates' case. This would compensate for the smaller collecting area of the IRO.

In order to distinguish a real comet signal from "camera noise," we asked that two exposures be taken with a mechanical shutter. Taking two exposures of a small comet in the same picture should prevent any further claims of "camera noise." That way, one of the small comet exposures should be twice as long as the other and these exposures would be separated by a shutter closed time that is equal to the short exposure time. The result for the detection of a single small comet should be two trails, one about twice as long as the other and separated by a gap with no trail.

The observing campaign for the small comets covered the New Moon periods, when the sky is darkest, from September 24, 1998, to June 11, 1999. The candidate periods were approximately 10-to-14 nights of observations centered on New Moon, with the requirement that moonlight was not illuminating the atmosphere above the telescope and obscuring the dim small comets.

Unfortunately, many circumstances severely reduced the actual body of observations with the IRO, including rain, clouds, atmospheric haze, and high winds, as well as a variety of telescope malfunctions, shutter, loose bolts, and faulty software commands. One of the requirements of the observations was the availability of at least 30-to-60 minutes of continuous good viewing time in order to provide a measure of their rates. Otherwise the detection of a single event or two could not be put into the perspective of the previous Spacewatch Telescope measurements. This requirement of a reasonable continuous search period, together with the typically poor atmospheric clarity at the site and frequent malfunctions of the telescope, took a terrible toll on the available images. As a result, there were no useful observations for two of the months, December and May. The best observations were obtained during October, January, February, and April, with two of those months, January and February, being a known low point in small comet activity.

As we had agreed, Mutel gave us copies of the raw observational data as they became available, and we began our analysis almost immediately. Our careful analysis revealed that the small comets trails were there, just as we knew they would be. I planned to use one of the IRO images, taken on October 19, 1998, for a prestigious talk I was asked to give at the University of Iowa's Sixteenth Annual Presidential Lecture on February 21, 1999. The image created quite a stir. When Mutel examined the image, he concluded, without any rigorous analysis of the obvious trail, that the evidence supporting the detection of a small comet was not significant; he posted his analysis on the web. But later, just before Fix left Iowa City in July for his new job in Alabama, I showed him examples of other successful small comet detections on the computer screen. He then congratulated me on the successful search.

Fix's cordial reaction left us unprepared for the events of December 1999. Sigwarth and I knew that Mutel and Fix were presenting the results of their analysis at the December meeting of the American Geophysical Union in San Francisco, but we certainly did not suspect that members of our own department would engage in the sort of "surprise attack" they pulled on us at that meeting.

On Thursday morning, December 16, Sigwarth and I attended Mutel's presentation of the results of their search for the small comets. Their paper was placed on a board in the presence of hundreds of other "poster papers" describing a myriad of scientific results. Mutel had not told us of the results of his analysis before the meeting; like everyone else, all the information I had was his abstract printed in the meeting program, the last sentence of which read: "We will discuss the results of this search and possible implications for the small comet hypothesis."

We knew, of course, that the easy way out for Mutel and Fix would be to present a negative report, and they did not disappoint us on that score. They concluded that there was no evidence of small comets in the images taken with the Iowa Robotic Telescope. We questioned Mutel about the method used in his analyses and our discussion was civil. But that was not to last.

According to the poster presentation, the IRO had searched for objects moving in the same direction and orbital plane as the Earth at a distance of about 33,000 miles from the Earth. During their nearly nine month observing campaign, they had acquired a total of 6,148 images. Of these, 2,713 images with a limiting magnitude of 16.5 or fainter had been "visually inspected." This magnitude is about 10,000 times fainter than the faintest stars visible with the naked eye. According to Mutel, this is easily bright enough to detect the small comets. He claimed that their search using the 20-inch IRO reflector telescope had been designed so that expected trails would be easily seen without the need for elaborate statistical analysis of faint trails. Not so; we designed the search and never made any such claims. Based on the atmospheric hole detections, Mutel estimated that about 65 small comets should have shown up in the survey images. But their "visual inspection" of the images, according to Mutel, had found no trails consistent with small comets.

"We got zero," Mutel told the reporter for *Sky & Telescope*. If the small comets exist at all, Mutel estimated that their number is only about five percent of what I claim.

We were naturally very disappointed that Mutel did not communicate his results to us prior to the meeting. After all, our offices are in the same building at the university. But that was not all. The University of Iowa's Press Office had prepared a press release, written by Gary Galluzzo, on the Mutel and Fix presentation. "We failed to find evidence that small comets exist," Mutel was quoted as saying. I had no knowledge that the University News Service was issuing this news release; they never even had the courtesy of calling me for a comment. The Director of the University Communications and Outreach would later state: "it would seem fair, as a matter of common courtesy, to notify Professor Frank of Professor Mutel's findings."

But their poster paper made one thing perfectly clear: Mutel's and Fix's analysis of the images was simply faulty. It was so obviously subjective. What they did was draw what they thought the small comet trails would look like on some images and then asked a few undergraduate students to find such trails in the image set. Of course, there were no such actual trails because Mutel assumed that the trail images were very bright, without width due to atmospheric effect and small motions of the telescope, etc. In actuality, the trails are fuzzy in appearance and yield smudgy trails that must be evaluated for validity with careful statistical evaluation—not "the eyeball." Their entire analysis technique had no solid foundation for determining whether or not the trails were significant.

Various news media attempted to reach me for my comments on the Mutel and Fix paper. I returned some of these calls—specifically to John Kenyon of the *The Gazette* of Cedar Rapids, Iowa, and Ann McGlynn of the *Des Moines Register*—from the San Francisco airport, while waiting for my plane back to Iowa. John Sigwarth and Bill Paterson, another senior research physicist in my group, were with me when I returned these calls. I communicated the same basic information to both of them. I discussed the techniques used by Mutel in an attempt to enhance the images, and I denied the claim that the images failed to support the small comet theory. Following my return to Iowa City, I was again contacted by Kenyon and reiterated the statements I had made from the San Francisco airport.

I also called Gary Galluzzo at the University News Office and asked why he had never been in touch with me for a comment on the Mutel and Fix press release. This entire messy situation would probably have been avoided if I had had the chance to make a statement in the press release. When I asked Galluzzo by telephone about the way he had handled this, he told me that he had just been informed that he had been diagnosed with a fatal disease.

Then on December 19, 1999, *The Gazette* ran a front-page story with the explosive headline: "Frank alleges fraud." The first three sentences read:

"What at first seemed like a simple disagreement between two University of Iowa researchers has escalated into an accusation of fraud committed for revenge.

"Louis A. Frank, a James A. Van Allen professor of physics at the UI, says his colleague, Robert Mutel, doctored photographs from a telescope so the results would disprove a theory Frank has about ice comets that bombard the Earth's atmosphere.

"'They've been very angry at me for casting the final vote against tenure for a friend of theirs,' Frank said by way of explanation Friday."

I never said that Mutel had "doctored" any images. I never accused Mutel of fraud. Sigwarth and Patterson are my witnesses. Kenyon simply got it wrong, as none of the other reporters I spoke to on the subject ran such a story. But the reverberations of Kenyon's story would run deep; over the year that followed the university would conduct a "witch hunt" against one its of most honored faculty members.

CHAPTER 38

The Witch Hunt

I was ill with a vicious virus as the new year 2000 approached. During that week, Kenyon's inflammatory story appeared in *The Gazette* of Cedar Rapids, Iowa. A few days later David Skorton, Vice President for Research at the University of Iowa, advised me that it was his obligation to evaluate any serious allegations of scientific misconduct. Shortly thereafter I met with Skorton and Steve Parrott, who was then Director of University Relations, and we decided that the best way to handle the matter would be for me to write a letter-to-the-editor to the *The Gazette*. The entire text of my letter read as follows:

"I write to set the record straight regarding the debate over my 13-year-old theory that small comets composed of snow are continually bombarding the Earth and that those small comets could have, over billions of years, provided enough water to fill the Earth's oceans.

"I disagree strongly with the recently announced results of research conducted by Professor Robert Mutel of the University of Iowa and Professor John Fix of the University of Alabama at Huntsville, which seemed to find no evidence of the small comets.

"However, I emphatically do not believe that my colleagues' research is in any way tainted by fraud. This is an issue that will be decided on the basis of further research. That is how science works."

This letter-to-the-editor should have ended the matter. Instead the reporter used it to fan the smoldering coals of this unfortunate incident. On Christmas day, *The Gazette* ran a story by John Kenyon with the headline "Allegation of fraud retracted." Citing my letter to the paper, Kenyon stated that I had "backed away from those assertions" of fraud. But I did no such thing. There was no retraction. I did not back away from any such assertions because I never made them in the first place. There was nothing to retract.

Incredibly, my letter-to-the-editor, which Kenyon had based his story on, did not actually appear in the paper until two days later. This is shameful conduct. I decided to make no further objections to his articles as they probably would have ended up in the used car section of the newspaper. Since that time, I have refused to have any contact with either the *Gazette* or other news media on the matter.

At the end of December, I sent an email to Fix and copied Mutel regarding the *Gazette* articles. I reiterated that I had not used the word "fraud" in my conversations with Kenyon. I pointed out that Kenyon had not put the word in quotation marks in the article, which would have indicated that it was a direct quote from me. Fix never replied to my email and neither did Mutel. They had other plans for me.

Their silence was deceptive. They were actually busy obtaining a Notice of Charges against me from the provost of the university, Jon Whitmore. In cases of complaints against other faculty members, accusers normally communicate their concerns to the department chair in an attempt to resolve the issue before it becomes a university-wide problem. Unfortunately, Mutel and Fix apparently persuaded the physics department chair, Wayne Polyzou, to keep their complaint secret from me. I would later learn that on January 10, 2000, Mutel and Fix had asked the provost to initiate a formal investigation of the facts and circumstances behind the *Gazette* article.

In the meantime, Mutel had sent me a copy of the paper he had submitted to the *Journal of Geophysical Research* on his small comet search. I told him that I didn't find it to be much improved over the poster paper he had given at the American Geophysical Union meeting. I found his analysis to be fundamentally flawed. His determinations of the telescope's sensitivities in the search were incorrect. So was his estimate of the limit on the small comet fluxes. And, unfortunately, part of his image set was taken in January and February, when it is known that the small comet fluxes are very low. Consequently, I strongly objected to Mutel's inclusion of my name in the Acknowledgments and requested that he remove my name immediately. To a reviewer or a reader, it implies that I was supportive of Mutel's work, and that was, of course, not the case at all.

At the same time, I informed Mutel that our preliminary analysis had revealed the presence of a number of small comet detections in the 400 or so usable IRO images we had looked at so far. The range of brightnesses for these small comets happened to be the same as that reported previously by Yeates' Spacewatch survey. I told Mutel that we were preparing a paper on our analysis and ended my email with these words: "I strongly support your right to publish any results which you desire. On the other hand, it is my right to point out that you have fundamental flaws in your work." Mutel never replied.

I later had to reiterate my request for the removal of our names from the acknowledgments directly to Janet Luhmann at the *Journal of Geophysical Research*. At the same time, I also pointed out the fundamental flaws in their paper. One was that Mutel had run the images through a filter that increased the noise in the images and obscured the trails of the small comets. Two, that even though the authors stated that the tuning distance for the images was set at 55,000 km, none of the images Mutel and Fix listed in their "Table 1, Log of Observations," was taken at this tuning distance. In fact, the tuning distance in Table 1 after October 1998 was 47,000 km. This is a very important point because the volume of space is an extremely strong function of tuning distance. I reacted with dismay that Mutel and Fix could not even determine the distances at which the small comets were to be detected with the telescope.

On January 26 I received a surprise letter notifying me that Mutel and Fix had filed charges that accused me of ethics violations. In order to protect the rights of the accused faculty member, a preliminary investigation normally takes place before such charges are issued, but no such initial probe ever took place in my case. Not one university administrator checked to see whether the *Gazette* article had any merit. The sad fact of the matter is that if Whitmore had contacted me about this matter, he would have found out that there were witnesses to my telephone call with the reporter and the matter would have been dropped. Whitmore, by the way, had just been involved in the attempted removal of the very popular International Writers Workshop from the University of Iowa campus. Such behavior left no doubt about the rise of incivility on the Iowa campus. In this environment I suppose that it is best to burrow into one's office and spend more time "smelling the roses," as my 90-year-old mother would say.

In the complaint, Mutel and Fix claimed that the *Gazette* article had been "extremely damaging" to them "both personally and professionally," and they felt that my actions constituted "a serious violation of the university code of conduct for faculty." Of course, since I never used the word "fraud," there had never been a violation of the faculty code of conduct. If scientific results are in error, then this must be reported to the public. That's all I had tried to do. Science is a hard taskmaster; it is not a tea party.

At this point I had no choice but to obtain a lawyer in order to defend myself. Whether I was guilty or not, the university had put into gear its vast resources in order to prosecute me. Very fortunately, one of my best friends is a lawyer whom I have known since my days as a teenager. His name is Robert Downer and he is regarded as one of the best lawyers in the state of Iowa. He was amazed at the university's crude treatment of me in view of my contributions, which Downer summarized as follows in one of his briefs: "Dr. Frank is a 36-year member of the faculty at the U of I where he holds an endowed Carver professorship ... He has authored or co-authored 382 publications over 41 years, of which 141 were published since 1993. ... In the seven years ending December 31, 1999, Dr. Frank was principal investigator on projects for which grants totaling over 15.7 million were received. Over the past five years, while performing all of his research, writing and speaking, Dr. Frank has taught between 316 and 376 undergraduate students annually."

When the university made no serious attempt to mediate the dispute, it became clear that the university would continue to openly support the efforts against me by Mutel and Fix. A full-scale hearing was inevitable. Unfortunately, the President of the University, Mary Sue Coleman, remained silent on the matter. I felt despair in the face of the awesome resources that the university had available to prosecute me.

In the meantime, we continued our effort to resolve our scientific disagreement with Mutel and Fix. In April, Sigwarth and I had a brief correspondence with them, requesting that they correct the flaws in their paper. Despite my lengthy and repeated explanations, Mutel and Fix remained "confident that the paper is correct as it was originally submitted." When their paper was accepted for publication in the *Journal of Geophysical Research*, I had no choice but to submit a Comment in an attempt to address the damage done to the body of work on small comets by Mutel's and Fix's research.

The wheels of justice at the university shifted into gear on July 11th, when a letter to my attorney from the Faculty Judicial Commission announced that my hearing was scheduled for August. Amazingly, this would be the first opportunity that the university had given me to address the matter. A story in *The Gazette* seemed to relish my predicament; the headline read, "Panel to scrutinize professor's ethics: UI physicist could be fired over accusations leveled at colleague."

The "hearing," which to me had all the trappings of a "trial," occurred on August 17, 2000, and was presided over by a panel of three volunteer faculty members. This was an eight-hour ordeal during which the legal counsel for the university, Mark Schantz, portrayed me as a "recidivist." I thought I knew the meaning of this word but to make sure I looked it up in the dictionary. And, yes, the meaning is: "A tendency to relapse into a former pattern of behavior; especially, a tendency to return to criminal habits."

These are dark days indeed for the university when it has to stoop this low in the proceedings of a kangaroo court. The only crimes that I have been convicted of are a traffic violation as a teenager when my brother took the license plates off the car I was driving, and during the rest of my life, one speeding ticket and two parking tickets. I have also had a couple of divorces. Do they count?

So what are my "criminal" habits? One of the prominent instances offered by Schantz came from Forrest Mozer of the University of California at Berkeley who provided email messages stating that I was a bad guy because I had criticized one of his scientific papers. I clearly explained during the hearing that my problem with Mozer was that he tried to publish a false table of numbers in a scientific journal. I knew this data was incorrect because it was obtained with one of my instruments orbiting the Earth. Mozer had no choice but to correct the numbers in his paper. Is pointing out the error to Mozer a crime? Obviously Schantz was unfamiliar with the world of scientific debate and discussion. To him it was a crime.

For me, the most painful moment of the hearing came when comments made by John Craven, who had co-authored our original small comet papers in 1986, were offered in support of Mutel and Fix. I had a close, long-term scientific association with Craven while we worked together at Iowa. I was also his thesis advisor. Some years ago, he wanted to leave my "shadow" and go elsewhere. I understood how he felt and helped him obtain a professorship at the University of Alaska. So it surprised me when Schantz used testimony of Craven (presented as Exhibit 6 at the hearing) who had sent Fix an email saying, "Email is now equally occupied with Lou's subsequent poor behavior. Some things never change. Nice to be reading about it from 4,000 miles away vs. in the local newspaper." When I subsequently asked Craven about these comments, he told me that he had intended them to be confidential and that he hoped our relationship would continue as before. But that's just not possible anymore, not when one gets stabbed in the back by a trusted friend. *Et tu, Brute?*

It also came to light during the hearing that on July 26th, *The Gazette* had refused the university's request for the reporter's notes, or tape recordings, of the interview Kenyon had with me in which he claims I used the word "fraud." In a letter, the editor wrote: "The reporter, John Kenyon, had no tape recording and no longer has his hand-written notes or other contemporaneous record of the details of the first interview and a follow-up interview he had with Professor Frank before writing his stories. Having said that, *The Gazette* has a consistent position of declining to turn over notes used in its news gathering process. We believe such notes are irrelevant when determining what was said in an interview because we published what we consider to be an accurate depiction of what was said." I was amazed by their position—what we print is the truth. Their arrogance defies any logic.

The hearing, which was recorded on video, was not lacking in melodrama, but we won't go into that. On December 18, 2000, the panel members released their report. They found that "there is not sufficient evidence to conclude that Frank used the word 'fraud' when speaking with the *Gazette* reporter." They concluded the incident had probably been due to a misunderstanding by the reporter: "Disagreements between Frank and Mutel centered around the methods used to analyze the digital images and the definition of what was statistically significant for an apparent comet trail. Mutel had introduced a false signal, a comet trail, into the images to see if it was visible and [to] determine the detection threshold for a comet. Frank believed that the detection threshold had been set too high. A person not familiar with scientific and technical jargon may have interpreted the 'introduction of a false signal' as 'doctoring of data.' The apparent accusation of fraud may have been an honest mistake by

[the] reporter, a misunderstanding arising from the reporter's lack of familiarity with the scientific principles in question."

The panel also found that little harm had been done to Mutel's and Fix's scientific reputation. "Other than a few embarrassing inquiries," the report stated, "no evidence of damage is apparent." In fact, the panel recognized that Mutel and Fix were not entirely blameless in the whole affair: "We understand that Frank may have felt a great deal of resentment and a complete lack of kinship with Mutel and Fix, not without good cause."

The panel report explains: "Frank was frustrated because he believed that Mutel was interfering with his efforts to identify small comets. About 50% of [the] images that Mutel shared with Frank had been processed to reduce background noise and Frank contended that such image processing may have erased some of the comet images of interest. Mutel conceded that the image processing may have distorted the very signals that Frank was trying to detect, yet Mutel did not provide Frank with the unprocessed images that Frank required. Frank may have felt that his own analysis had been 'sabotaged.'

"In addition," the report continues, "Mutel and Fix may have provoked Frank by not extending to him certain professional courtesies that a scientist would expect of colleagues, especially those at the same institution…"

However, because Schantz, as the university's lawyer, claimed I was some kind of habitual criminal, the panel decided they had to find me guilty of something anyway, even if the charges they found me guilty of were beyond those directed against me in the original complaint. It does not take a legal expert to note that the accused must be informed of the charges *before* the trial.

The panel found that I had committed two violations. "Frank accused Mutel and Fix of research fraud," according to the report, "by asserting that their professional conduct was clouded by anger over a tenure decision." My assertion about the tenure decision, of course, has nothing to do with fraud; I was simply speculating on the reasons for the ill feelings between us.

The second "violation" held no water either. "Although Frank claims he did not make other accusations attributed to him," the report states, "he made no efforts to deny those statements publicity or correct the record, thus failing to fulfill his obligations as a scholar." Not only did I send a letter-to-the-editor of the *Gazette* correcting their error, but I had also emailed Mutel and Fix personally about the matter. What more did the university want me to do? Shout it from the rooftops?

The behavior of the university in this matter was disgraceful. If any good is to be gained by this unpleasant experience with the university, it is amply stated by my lawyer Downer in one of his briefs. "This type of response to the unfortunate circumstances involved, set in motion by [Mutel's and Fix's] failure to afford professional courtesy to [Frank], sends a chilling message to other senior faculty members who have devoted their professional lives to this university and, regardless of the outcome, inflicts great professional damage upon Frank." And for those young scientists who might consider a faculty position at the University of Iowa, beware.

CHAPTER 39

Snowballs in Hell

Snowballs, as it turns out, do stand a chance in hell.

The overheated university atmosphere did nothing to stop the small comets from falling—and a few from reaching the ground it seems—nor my scientific work on them. On the 8th of January 2000, a chunk of ice fell from clear skies and supposedly hit a car in Tocina, Spain, which is near Seville. The chunk measured about eight inches and weighed nine pounds. In the two weeks that followed at least 50 other ice falls, some as big as basketballs, were reported in Spain. Others reportedly fell in Italy, Switzerland, and Holland.

A team of investigators from the Consejo Superior de Investigaciones Científicas (CSIC), headed by geologist Jesús Martínez Frías from the National Museum of Natural Sciences, collected as many as they could locate, urging people to wrap the ice chunks in plastic and refrigerate them. But as the reports continued, the team stopped collecting the ice chunks, as some of the later reports had been practical jokes perpetrated after the initial reports. The scientists later judged that nine falls of more than 22 pounds of ice had been genuine, all occurring between January 8th and January 17th.

As is usually the case, the first suspect was human excrement ejected from aircraft. But the ice chunks were nearly 100 percent frozen water and lacked the coloring of aircraft wastewater. Meteorologists then ruled out weather as a possible explanation. Others mentioned comets, but they weren't thinking small comets, but bits and pieces from the tail of a large comet. A subsequent analysis of five recovered ice chunks showed that they were chemically and isotopically varied, with large densities of ions corresponding to solutions of such materials as quartz or feldspar aerosols. Another test indicated that there had been either a variation in condensation temperature or perhaps isotopic exchanges during the formation of each ice chunk. This data, along with the high frequency of the events, suggested to the investigators that the chunks didn't come from aircraft, nor were they "minicomets." I'm not so sure.

Their official explanation was that global warming had affected the classic processes of ice formation in the atmosphere. Their report, published in June, states: "Ozone distribution maps from NASA show that, on Jan. 5, a thin jet of ozone depression passed through all the areas in Spain where the ice falls took place. It is commonly posited that global warming and ozone depression are linked. Despite the fact that the greenhouse effect leads to an increase in the global mean surface temperature, it leads to cooling in the stratosphere." They ventured that the ice chunks may have formed in the upper atmosphere at altitudes of about 10 to 20 miles, where humidity is more abundant. And they wondered if these ice chunks could be an indication that dramatic changes are taking place in the atmosphere.

No one was thinking about the effects of small comets on the atmosphere or about what might be expected if a larger than average small comet entered the atmosphere. Or if a few larger-than-normal small comets fell upon the Earth in the span of a couple of weeks. Certainly, if the small comets are falling through our atmosphere at a rate of twenty per minute, these ice fall numbers are not large.

But whatever the real cause of this ice fall, they certainly served to focus public interest on the small comets again. A newspaper column in the *Bangor Daily News* was headlined: "Spanish ice chunks of interest to Iowa astrophysicist." Of course, I'm not an astrophysicist, and the columnist never bothered to call me to see if I was indeed interested. He assumed it and so did many who emailed their inquiries to the Small Comets website. The ice falls report, noted one, "makes a fine retort to those who myopically insist that the comets are 'instrumental noise.'" Even if the details are wrong, the spirit was right.

In the meantime, Sigwarth and I were trying to right the wrong caused by Mutel's and Fix's faulty analysis of the IRO search for the small comets. Unfortunately, Mutel and Fix severely overestimated the volume of space viewed by the IRO during their search. A correct calculation of this volume shows that the expected number of sightings for their search criteria is only about two, which in effect means that the results of their search are insignificant, one way or the other.

But all is not lost for the efforts of Mutel and Fix. There is a usable "window of opportunity" in the set of images taken with the IRO. Small comet trails with a detectable brightness should be present, if the small comets exist. We encouraged Mutel and Fix to follow up on this opportunity, but they did not. So once again we did it ourselves, proving again the truth of the old adage: if you want to do it right, you have to do it yourself.

Approximately 6,000 images were acquired during the search which took place from the Fall 1998 through Spring 1999 viewing season. First of all, I should say that Mutel did carry out a heroic effort to obtain these images. If only he had done the same in looking at these images. Greg Pickett and I spent a combined 400 hours in the examination of these images. Because of eyestrain only about 50 images could be carefully examined by an experienced person each day in a three-to-four-hour period.

One of the requirements of the observations was the availability of at least 30 to 60 minutes of continuous good viewing time in order to provide a measure of the occurrence rates. Otherwise the detection of a single event or two could not be put into the perspective of the previous Spacewatch Telescope measurements. This requirement for a reasonable continuous search period, together with the typically poor atmospheric clarity at the site and frequent malfunctions of the telescope, took a terrible toll on the available images.

The surviving images numbered only about 1,500, or about 25 hours of viewing. Some of these were taken in January when the small comet fluxes are known to be very low. In addition, there were no good observations in the most desirable viewing directions for small comets; that is, when sunlight is reflected from their surfaces at small angles nearly equal to those when you view yourself in a mirror. This resulted in a further loss of sensitivity for the small IRO telescope, which was already often degraded by the atmospheric conditions. It would have been wonderful to have had the telescope located at Kitt Peak and/or equipped with a larger primary mirror, but this was not the case.

In the set of 1,500 usable images from the IRO search, we managed to find nine unmistakable small comet sightings. The small comets are there but they are too faint to be discerned by the naked eye in the images. We used a set of very strict standards in order to show that the trails were not noise in the sensors. Of course, it is very important that the reality of each of the events be established without question. It took John Sigwarth three months of computations on our fastest computers to eliminate the possibility of noise. Though the small comet trails were few, given the small diameter of the IRO primary mirror in comparison to the Spacewatch Telescope, the numbers were within a factor or two or three of what the Yeates search found using the Spacewatch Telescope. This was very encouraging.

We could not take consecutive images of the small comets with the IRO, as Yeates had using the Spacewatch Telescope, because of the IRO's slow readout of images. To compensate for this, we used a mechanical shutter to acquire either two or three trails of the same small comet in a single frame as the comet's image moved across the CCD. From October 19, 1998, to February 11, 1999, the telescope was operated in the mode for acquisition of the same small comet as two trails in an image. That way, one of the small comet exposures should be twice as long as the other and these exposures would be separated by a shutter closed time that is equal to the short exposure time. The result for the detection of a single small comet should be two trails, one about twice as long as the other and separated by a gap with no trail. From February 14 through May 1999, the operation of the telescope was changed to acquire three trails of the same comet in an image. A strong confirmation that the trails seen were not due to random events is that we found no three trail events in the two-trail campaign and no two-trail events in the three-trail campaign.

On the 21st of February 2000, we submitted a paper to the *Journal of Geophysical Research: Space Physics* with our own analysis of the IRO results. The number of reviewers is usually two, but we were blessed with three. Two of the three essentially recommended publication without much comment. But the third reviewer seemed to have a direct and detailed knowledge of the IRO and questioned our analysis of the signal because of the accuracy limitations of the IRO. His 18-page critique, half of which were color copies of IRO images, said that a paper that "once again purports to prove the existence of these comets by what amounts to a statistical analysis of noise is not likely to be accepted by the community." He also claimed to find three trails in the two trails images. But an examination of the actual values of the pixels he identifies as a three-trail event shows that the values are buried deeply in the noise levels of the camera. The event suggested by the reviewer does not even remotely satisfy the requirements of a valid small comet trail.

The reviewer also believed that the small comet identified in one of the images in our paper could be more reasonably interpreted as the result of a faint star at the detection limit of the IRO. But it is not possible that this two-trail event was a star trail because it is oriented at an angle of 13 degrees to all the other star trails in the image.

In any case we revised the paper and resubmitted it. The critical reviewer found it "significantly improved," but he wanted us to do even more work on it. Unfortunately, he failed to quantitatively justify his criticisms. He kept wanting us to test our "skeet shooting" mode of operation to see if it worked with a known dim satellite. The problem is that a dim satellite has a variable light curve due to its rotation in sunlight. It is much more effective to verify the proper operation of the shutter by viewing star trails. The positions of the stars are well known and a large range of brightnesses are available. And, in fact, the stars were frequently used as verification during the search.

So we made some further small revisions to the paper and submitted it again. And again, we heard from the referee. It seems that we had finally pleased him—almost. "You have done a commendable and thorough job in addressing my issues," he wrote, and then proceeded to offer a final set of suggestions, in effect recommending that we incorporate our responses to his objections in the paper in order to "strengthen the arguments." He still had qualms about our experimental design, but admitted: "I would rather see the paper published in something close to its present form than prolong this impasse to no apparent gain."

This exchange took quite a long time and effort but the paper is well worth it. The suggestions made by the reviewer were useful and much was gained in the exchange—on both sides: for us to improve the paper with respect to skeptics, and for the referee to understand the details of the search itself. This is the way science is supposed to work. Occasionally we luck out.

A small comet appears in this image taken on January 20, 1999, with the Iowa Robotic Telescope. The large streaks seen in these images are star trails; these are longer than they normally would be due to the motion of the telescope that attempted to track the small comets across the sky. The trails of a small comet—indicated by the yellow box in the upper left-hand corner of the image—are not parallel to the star trails, and were acquired in the two-trail search mode. The long trail is on the right of the short trail. Note that the small comet trails are not single pixel events.

On August 11, 2000, our paper, "The Detection of Small Comets with a Ground-Based Telescope," was accepted for publication by the *Journal of Geophysical Research: Space Physics*. It was published in the March 2001 issue. I didn't expect it would change the minds of my critics; it was just another brick in an increasingly solid foundation supporting the existence of the small comets.

A month later I traveled to Copenhagen, Denmark, to deliver a Niels Bohr International Lecture. This prestigious lecture series is named for Niels Bohr, the Danish physicist and 1922 Nobel Laureate in physics. These invited lectures are given by prominent scientists three or four times each year and are directed to a broad spectrum of physicists. The lectures are sponsored by the Danish Space Research Institute and the Nordic Institute for Theoretical Physics.

My topic? Small comets, of course.

The three-trail search mode was used to acquire this image of a small comet on February 16, 1999, with the Iowa Robotic Telescope. The left-hand member of the three-trail set is slightly out of line with the two trails to the right. This displacement is due to the effects of motor drives and wind motion on the telescope's field of view. These effects are also seen in the star trails. Viewers may think they see other small comet trails in the images, but the identification of valid trails must be evaluated quantitatively since visual impressions can be misleading. Real small comet events have a signal-to-noise ratio of 6 or better.

CHAPTER 40

Small Comets and the Future

I am a very hard man to deal with. At times I may even be scathing and insulting. Some people live just to catch me in a big error. But I do not make mistakes. I know that is arrogant, but I have dedicated my life to my work and when I go, I will know that everything I did was of the highest quality. I am not out to win a Mr. Popularity contest. Paul Feldman calls my determination and perseverance "extraordinary," but then again he views my behavior as unscientific. Since when is standing up for what you think is true unscientific? I came from a strict family. I was raised to have integrity and honesty. So I have had no choice.

People complain that I have changed the numbers on the small comets. But the number of objects has stayed the same—and that alone is remarkable given that the instruments on Dynamics Explorer and Polar are so different. The speed of the objects has not changed either. The only thing that we have changed is their mass. Our first estimates of the mass of these objects was 100 tons. Now we have lowered our estimate. We are saying these objects are more on the order of 20 to 40 tons. The change comes simply because we acquired measurements that allowed us to improve our estimates. Once we had the Yeates measurements, which gave us an idea of the size of these objects and their visual magnitude, we could come up with a more accurate picture of their mass, based on a reasonable value for dark reflecting objects. And then, looking at the atmospheric holes and particularly the emissions from water fragments from small comets with Polar, we were able to verify that 20 to 40 tons was the best value.

We did what every scientist does; we used additional information to refine our numbers. It is just a progression. We did not go from a hundred tons to one ton or a thousand tons. It is just a matter of trying to refine what the best values are. Besides, as I have said before, we are talking about an average. Some of the small comets are going to be 80 tons, or 100 tons, and some of them are going to be 10 tons. And looking at the telescope measurements and at some of the very large holes that we see from high altitudes, some will even be up in the 150-to-200 ton range. These objects were not stamped out with a cookie cutter.

The atmospheric holes, and the small comets that produce them, are there. Nothing anyone says can make them disappear. And it is all going on right now, 100, 200 miles over our heads. We must come to terms with the thought that as our planet twirls around the Sun, as the Earth's tectonic plates heave and dive, the cosmos is bathing us in a gentle cosmic rain.

The small comets have always existed. What we did was isolate a set of data that allowed us to give them a name. Looking back, I now think that the discovery of the small comets was inevitable. After nearly three decades of space exploration, scientists began to aim their satellites, their eyes in space, towards the Earth. Our home is finally getting the attention it deserves. If we had not discovered the evidence for the small comets with the camera on the Dynamics Explorer satellite and then confirmed the discovery with our instruments on the Polar spacecraft, someone else with another camera like these would have done so sooner or later. Ours was just the first of its kind to fly on an Earth satellite. The future will see many more satellites geared exclusively to the study of the Earth and its environment in space. There will be other surprises. The hole in the ozone layer was one surprise. The small comets were another. This is just the beginning.

The timing was right for this discovery to be made. The technology was available. Someone had only to use it and we did just that. Yet, despite our use of state-of-the-art instruments, the small comet observations took place at the limits of detection. This is not a place to shy away from. This is the frontier where many scientific discoveries are made. It is where new particles are found in physics, where unsuspected lifeforms are discovered in the deepest seas, where the clues to earthquake prediction are being sought. The satellite images of the atmospheric holes, the measurements of water bursts in our atmosphere, and the telescope observations of the small comets were all made at this frontier.

A decade ago I said that I thought it would take another five, ten, or more years before the idea of small comets was finally accepted. This is essentially what happened, but we still have a way to go before *everyone* gets on the bus. I have done my part. People will no longer listen to me on this subject, as science does not take too kindly to those who plow through its well-tended gardens. Most likely, the process of accepting these objects into the scientific fold will fall to the new generation of scientists, for as the physicist Max Planck once drolly remarked: Scientists don't change their minds, they just die. Today's young scientists will be responsible for changing our mistaken impressions about the role of these objects in our past, present, and future.

This brings me to the end of my account and to a final note on final things. The number of small comets, if not their size, tends to strike a note of terror in people. In this sense, at least, the small comets are the latest chapter in a long book. From the earliest times, large comets have been regarded as objects of fear and superstition. They were seen as symbols and harbingers of wars and famines and other great natural and human disasters. Some even saw them as signs of retribution by the gods for the sins of men. This view led one 16th century rationalist to quip: "If comets were caused by the sins of mortals, they would never be absent from the sky."

In keeping, then, with proper tradition, I will end with a brief tale of cometary peril. If the present influx of small comets continues, and there is no reason to think that it is likely to end anytime soon, then eight billion years from now, when the Earth's age has tripled, our planet will have gained twice more water than it has today. All during this time there will be a little evaporation from the top of our atmosphere but it will never keep up with the thousand times greater rainfall from the small comets. The Earth will just continue to accumulate water. The doubling of the water on the Earth will no doubt affect the delicate balance that we live in. But we do not know what this will mean for us. Will we just have twice the water and less land? Or will this be enough water to turn the Earth into an ice planet, or a complete water planet? All we know is that something is going to happen. Fortunately, it is not going to happen next year.

It will be a very gradual filling-up, like the steady drip of a faucet into a barrel. There is not going to be any severely rapid change from the fall of one extra inch of cosmic rain on Earth every 20,000 years. We are safe. We will not have an ocean beach here in Iowa for a while yet.

NOTES

The Big Splash (1981-1990)

1 A Radical Departure

A radical departure. L.A. Frank, J.B. Sigwarth, and J.D. Craven, 1986. "On the Influx of Small Comets Into the Earth's Upper Atmosphere, II. Interpretation." *Geophysical Research Letters*, 13: 307 (April).

"Astronomers should leave to astrologers…" Anonymous, 1985. "Miscasting the Dinosaur's Horoscope." *The New York Times*, April 2.

Ongoing planet formation. Nils Aall Barricelli, 1985. "Preferential Pehrihelion and Aphelion Distances, and Planetary Formation." *Theoretic Papers* Vol.3 No.6, Oslo.

Large comets exhaustively studied. Carl Sagan and Ann Druyan, 1985. *Comet*. New York, Random House.

Previous work on small comets. Herbert A. Zook, J.A. Fernández and E. Grün, 1985. "Selection Effects Against Small Comets." In R.H. Giese and P. Lamy, eds. *Properties and Interactions of Interplanetary Dust*. D. Reidel, p. 287.

2 Cosmic Rain

One story bore the mark. Philip M. Boffey, 1986. "Did Ocean Waters Arrive in Rain of Icy Comets?" *The New York Times*, April 1.

A pair of papers. L.A. Frank, J.B. Sigwarth, and J.D. Craven, 1986. "On the Influx of Small Comets Into the Earth's Upper Atmosphere, I. Observations." *Geophysical Research Letters*, 13: 303 (April). L.A. Frank, J.B. Sigwarth, and J.D. Craven, 1986. "On the Influx of Small Comets Into the Earth's Upper Atmosphere, II. Interpretation." *Geophysical Research Letters*, 13: 307 (April).

Gleeful critics. Mark Washburn, 1988. "The Waters Above, the Storm Below." *Sky & Telescope,* December, p. 628.

"Crazy as they come…" Patrick Huyghe, 1986. "Origin of the Ocean." *Oceans*, July-August, p. 10.

"I've never heard anything like it…" *Ibid.*

"There is nothing like…" *Ibid*.

The circuit. Gina Kolata, 1990. "To Make the Big Time of Science, Better Take Your Show on the Road." *The New York Times*, January 2.

3 The Black Spot Mystery

It all began in 1981. L.A. Frank and J.D. Craven, 1988. "Imaging Results from Dynamics Explorer 1." *Reviews of Geophysics*, 26: 249 (May).

"How can we get rid…" James Ehmann, 1986. "Cosmic Comets of the Sea." *Omni,* July, p. 72.

In public for the first time. J.B. Sigwarth, L.A. Frank, J.D. Craven, 1983. "Atmospheric Holes Possibly Associated with Meteors." *EOS*, May 3, p. 281, abstract.

Press picks up story. Stefi Weisburd, 1985. "Atmospheric Footprints of Icy Meteors." *Science News*, Dec. 21 & 28, p. 391.

Fourth presentation. J.B. Sigwarth, L.A. Frank, J.D. Craven, 1985. "Latitudinal and Longitudinal Distribution of Atmospheric Holes Associated with Meteors." *EOS*, November 12, p. 1005, abstract.

Description and interpretation papers. L.A. Frank, J.B. Sigwarth, and J.D. Craven, 1986. "On the Influx of Small Comets Into the Earth's Upper Atmosphere, I. Observations." *Geophysical Research Letters*, 13: 303 (April). L.A. Frank, J.B. Sigwarth, and J.D. Craven, 1986. "On the Influx of Small Comets Into the Earth's Upper Atmosphere, II. Interpretation." *Geophysical Research Letters*, 13: 307 (April).

"Uncomfortable ramifications." A.J. Dessler letter to L.A. Frank, Feb. 7, 1986.

"If you restrict the journal…" A.J. Dessler interview by P. Huyghe, Oct. 14, 1987.

4 True Confessions

Co-author of two scientific papers. James A. Van Allen and Louis A. Frank, 1959. "Survey of Radiation Around the Earth to a Radial Distance of 107,400 Kilometers." *Nature*, 183: 430. James A. Van Allen and Louis A. Frank, 1959. "Radiation Measurements to 658,300 Kilometers with Pioneer IV." *Nature*, 184: 219.

Hannes Alfvén. John Noble Wilford, 1989. "Novel Theory Challenges Big Bang." *The New York Times,* Feb. 23.

Hannes Alfvén. Stephen G. Brush, 1990. "Prediction and Theory Evaluation, Alfvén on Space Plasma Phenomena." *EOS*, Jan. 9., p. 19.

Alfred Wegener. Martin Schwarzbach, 1986. *Alfred Wegener: The Father of*

Continental Drift. Madison, Wisconsin, Science Tech.

5 A Storm of Controversy

"A terse distillation..." A.J. Dessler, 1986. "A Turbulent Interface." *Geophysical Research Letters*, 13: 1 (January).

Comments received. A.J. Dessler letter to L.A. Frank, August 13, 1986.

First Comment. Thomas M. Donahue, 1986. "Comment on the Paper 'On the Influx of Small Comets Into the Earth's Upper Atmosphere, II. Interpretation'." *Geophysical Research Letters*, 13: 555 (June).

"Pretty unavoidable." Philip M. Boffey, 1986. "Did Ocean Waters Arrive in Rain of Icy Comets?" *The New York Times*, April 1.

Angry letter. T.M. Donahue letter to J. A. Van Allen, April 1, 1986.

"Bitter." Stephen Cain, 1988. "Comet Theory Data Misused, U-M Man Says." *Ann Arbor News,* May 18.

Donahue denies it. T.M. Donahue letter to L.A. Frank, May 24, 1988.

"Most important finding in space science." Philip M. Boffey, *Ibid*.

"We regret the public contention..." Unpublished letter to *The New York Times* from T.M. Donahue and L.A. Frank, April 22, 1986.

"A large number of problems." Thomas M. Donahue, 1986. *Ibid*, p. 555.

How Venus lost its oceans. Thomas M. Donahue, J.H. Hoffman, R.R. Hodges, Jr., and A.J. Watson, 1982. "Venus Was Wet: A Measurement of the Ratio of Deuterium to Hydrogen." *Science*, 216: 630 (May 7).

My Reply to Donahue. L.A. Frank, J.B. Sigwarth, and J.D. Craven, 1986. "Reply [to Donahue]." *Geophysical Research Letters*, 13: 559 (June).

Others had questioned water on Venus. Sig J. Bauer, 1983. "Water on Venus: Lack or Loss?" *Annales Geophysicae*, 1: 477.

"Your observations and interpretations..." S.J. Bauer letter to L.A. Frank, May 28, 1986.

"Venus's present hydrogen loss rate..." John S. Lewis, 1974. "Volatile Element Influx on Venus from Cometary Impacts." *Earth and Planetary Science Letters*, 22: 239.

Microwave measurements. John Olivero, J.J. Tsou, C.L. Hale, and R.G. Joiner, 1986. "Solar Absorption Microwave Measurement of the Upper Atmospheric Water Vapor." *Geophysical Research Letters*, 13: 197.

"Ingenious." D. Hunten letter to L.A. Frank, April 14, 1986.

6 The Creation of the Oceans

Bathtubs full of water. Fred Powledge, 1982. *Water*. New York, Farrar Straus Giroux, p. 23.

Alternate view. Carl Sagan and Ann Druyan, 1985. *Comet*. New York, Random House, p. 317.

"Trickle down theory..." Patrick Huyghe, 1986. "The Origin of the Oceans." *Oceans*, July-August, p. 11.

Lack of evidence to the contrary. *Ibid*. p. 12.

"Data concerning ocean volume..." Thomas M. Donahue, 1986. "Comment on the Paper 'On the Influx of Small Comets into the Earth's Upper Atmosphere II. Interpretation'." *Geophysical Research Letters*, 13: 555 (June).

"The exceedingly slow rate..." William W. Rubey, 1951. "Geologic History of Seawater: An Attempt to State the Problem." *Bulletin of the Geological Society of America*, 62: 1135 (September).

"By small increments..." *Ibid*. p. 1117.

Problem of excess volatiles not solved. P. Wilde letter to L.A. Frank, June 29, 1988.

"Some of the excess water..." P. Wilde interview by P. Huyghe, Nov. 29, 1989.

"A startling geological revelation..." P. Wilde letter, *Ibid*.

A study of Halley's Comet. D. Krankowsky, P. Lämmerzahl, I. Herrwerth, J. Woweris, P. Eberhardt, U. Dolder, U. Herrmann, W Schulte, J.J. Berthelier, J.M. Illiano, R.R. Hodges and J.H. Hoffman, 1986. "In Situ Gas and Ion Measurements at Comet Halley." *Nature*, 321: 326 (May 15).

Halley's fingerprint. P. Eberhardt, U. Dolder, W. Schulte, D. Krankowsky, P. Lämmerzahl, J. H. Hoffman, R.R. Hodges, J.J. Berthelier and J.M. Illiano, 1987. "The D/H Ratio in Water from Comet P/Halley." *Astronomy and Astrophysics*, 187: 435. P. Eberhardt, D. Krankowsky, W. Schulte, U. Dolder, P. Lämmerzahl, J.J. Berthelier, J. Woweries, U. Stubbemann, R.R. Hodges, J.H. Hoffman, and J.M. Illiano, 1987. "On the CO and N_2 Abundance in Comet P/Halley." *Astronomy and Astrophysics*, 187: 481.

7 Heat, Dust, and the Origin of Life

Comet watchers flock to Alice Springs. Peter H. Lewis, 1986. "Disappointed Comet-Gazers Bidding Halley's Farewell." *The New York Times*, April 13.

Critic on composition. David Parry Rubincam, 1986. "Comment on the Paper

'On the Influx of Small Comets into the Earth's Upper Atmosphere, II. Interpretation'." *Geophysical Research Letters*, 13: 701 (July).

Critic on composition. Christopher P. McKay, 1986. "Comment [on the Paper 'On the Influx of Small Comets into the Earth's Upper Atmosphere, II. Interpretation']." *Geophysical Research Letters,* 13: 976 (September).

Critic on composition. Donald E. Morris, 1986. "Comment on 'On the Influx of Small Comets into the Earth 's Upper Atmosphere II. Interpretation'." *Geophysical Research Letters*, 13: 1482 (December).

Critics on composition. John T. Wasson and Frank T. Kyte, 1987. "Comment on the Letter 'On the Influx of Small Comets into the Earth's Upper Atmosphere, II. Interpretation'." *Geophysical Research Letters*, 14: 779 (July).

My Replies. L.A. Frank, J.B. Sigwarth, and J.D. Craven, 1986. "Reply [to Rubincam]." *Geophysical Research Letters*, 13: 703 (July). L.A. Frank, J.B. Sigwarth, and J.D. Craven, 1986. "Reply [to McKay]." *Geophysical Research Letters*, 13: 979 (September). L.A. Frank, J.B. Sigwarth, and J.D. Craven, 1986. "Reply to Morris." *Geophysical Research Letters*, 13: 1484 (December). L.A. Frank, J.B. Sigwarth, and J.D. Craven, 1987. "Reply to Wasson and Kyte." *Geophysical Research Letters*, 14: 781 (July).

Laboratory experiments. L.J. Lanzerotti, W.L. Brown and R.E. Johnson, 1985. "Laboratory Studies of Ion Irradiations of Water, Sulfur Dioxide, and Methane Ices." In J. Klinger, D. Benest, A. Dollfus, R. Smoluchowski, eds. *Ices in the Solar System*. Boston, D. Reidel. R.E. Johnson, J.E Cooper, L.J. Lanzerotti, G. Strazzulla, 1987. "Radiation Formation of a Non-Volatile Crust." *Astronomy and Astrophysics*, 187: 889.

Halley's Comet. Hans Balsiger et al., 1986. "Ion Composition and Dynamics at Comet Halley." *Nature*, 321: 330.

Carbon mantle. Donald E. Morris, 1986. "Comment on 'On the Influx of Small Comets into the Earth's Upper Atmosphere, II. Interpretation'." *Geophysical Research Letters*, 13: 1482 (December).

Examination of carbon on Earth. Heinrich D. Holland, 1984. *The Chemical Evolution of the Atmosphere and Oceans*. Princeton, Princeton University Press, p. 82.

Biochemical molecules from comets. J. Oró, 1961. "Comets and the Formation of Biochemical Compounds on the Primitive Earth." *Nature*, 190: 389.

Study of comets and life. William Ward Maggs, 1989. "Comets and Life." *EOS*, March 28.

8 The Atmosphere and the Ice Ages

Noctilucent clouds. George C. Reid, 1975. "Ice Clouds at the Summer Polar Mesopause." *Journal of Atmospheric Science*, 32: 523. M. Gadsden, 1982. "Noctilucent Clouds." *Space Science Reviews*, 33: 280.

"Greatly exceed those that have been observed." George C. Reid and Susan Solomon, 1986. "On the Existence of an Extraterrestrial Source of Water Vapor in the Middle Atmosphere." *Geophysical Research Letters*, 13: 1129.

Solomon's earlier point. Susan Solomon et al., 1982. "On the Chemistry of H_2O, H_2, and Meteoritic Ions in the Mesosphere and Lower Thermosphere." *Planetary and Space Science*, 30: 1117.

Results from intercosmos satellites. R. Knuth, G. Sonnemann, D. Felske, L. Martini and B. Stark, 1976. "Some Problems and Results of Solar Occultation Measurements in the Thermosphere. Part II: Neutral Gas Density Variations." *COSPAR Space Research XVI*, Berlin, Akademie-Verlag. D. Felske, R. Knuth, G. Sonnemann, L. Martini and B. Stark, 1977. " On the Additional Lyman Alpha Absorber in the Winter-Thermosphere." *Journal of Atmospheric and Terrestrial Physics*, 39: 1423. G. Sonnemann, D. Felske, R. Knuth, L. Martini and B. Stark, 1977. "How Dry is the Thermosphere?" *COSPAR Space Research XVII*, New York, Pergamon Press.

Rocket-borne measurements of water vapor. K.U. Grossmann, W.G. Frings, D. Offermann, L. André, E. Kopp, and D. Krankowsky, 1985. "Concentrations of H_2O and NO in the Mesosphere and the Lower Thermosphere at High Latitudes." *Journal of Atmospheric and Terrestrial Physics*, 47: 291.

Three launches, three positive results. K.U. Grossmann interview by P. Huyghe, Oct. 14, 1987.

Water vapor concentrations in the middle atmosphere. J.J. Olivero, J.J. Tsou, C.L. Hale, and R.G. Joiner, 1986. "Solar Absorption Microwave Measurement of the Upper Atmospheric Water Vapor." *Geophysical Research Letters*, 13: 197.

"We thought Frank's hypothesis…" J. Olivero interview by P. Huyghe, Oct. 13, 1987.

Master's thesis. Dennis M. Adams, 1988. "Extreme Short Term Variability in Upper Atmospheric Water Vapor as Measured by Ground-Based Microwave Radiometry." Master's Thesis, Pennsylvania State University, University Park, August.

Olivero's paper. Dennis M. Adams, John J. Olivero, and Charles L. Croskey, 1987. "A Search for Extra-Terrestrial Water in the Upper Atmosphere by Ground-Based Microwave Radiometry." *EOS*, 68: 372 (April 21), abstract.

"Observed, short-term perturbations…" D.M. Adams, *Ibid* p. 90.

Not let down. J.J. Olivero letter to L.A. Frank, March 2, 1990.

Four years of data analyzed. M.E Bonadonna, J.J. Olivero and C.L. Croskey, 1990. "In Search of Small Comets: H_2O Bursts Observed in the Mesosphere." *EOS*, 71: 570 (April 24), abstract.

9 A Masquerade for Radar

"If his comets have all this dust..." Patrick Huyghe, 1986. "Origin of the Ocean." *Oceans*, July-August, p. 12.

"Our radar is infrared..." *Ibid.*

"NORAD has a grouping of radars..." *Ibid.*

"A target strength of enough ice..." *Ibid.*

Several billion meteors. Malcolm W Browne, 1989. "Radio System Uses Fiery Meteor Trails to Transmit Data." *The New York Times*, Aug. 22.

A study of radar meteor rates. E.L. Vogan and L.L. Campbell, 1957. "Meteor Signal Rates Observed in Forward-Scatter." *Canadian Journal of Physics*, 35: 1176.

Paper prepared in June 1986. L.A. Frank, J.B. Sigwarth, and J.D. Craven, 1986. "Temporal Variations of the Influxes of Small Comets Into the Earth's Atmosphere." University of Iowa Research Report, 86-33.

Dessler was inflexible. A.J. Dessler letter to L.A. Frank, Aug. 6, 1986.

Meteor trails enlisted by Pentagon. Malcolm W Browne, *Ibid.*

Experiment of meteor burst communications. Billy P. Ficklin, 1986. "Possible Comets." SRI International Working Paper, Project 8055, July 8.

Small comet search by radar. Gerard Caudal, 1989. "EISCAT Finds No Small Comet." Centre de Recherche en Physique de l'Environnement, CNET-CNRS, Paris.

10 How to Spot a Small Comet

Öpik's Tables. Ernst J. Öpik, 1963. "Tables of Meteor Luminosities." *Irish Astronomy Journal*, 6: 3.

We had our answer. L.A. Frank, J.B. Sigwarth, and J.D. Craven, 1987. "Estimate of the Luminosity of the Impact of Small Comets with the Earth's Atmosphere." University of Iowa Research Report, 87-18.

Dustballs. Ernst J. Öpik, 1958. *Physics of Meteor Flight in the Atmosphere*. New York, Interscience.

11 Flying Saucers and Other Strange Events

"This meteor is not characteristic..." J. West letter to L.A. Frank, Aug. 21, 1986.

"Unlike a trailed meteor..." E. Kurczewski letter to L.A. Frank, April 5, 1986.

Nebulous meteors. William Corliss, ed., 1986. *The Sun and Solar System Debris*. Glen Arm, Maryland, Sourcebook Project, p. 232.

"Direct evidence." K. Bigg letter to L.A. Frank, Dec. 8, 1986.

Bigg's account. E.K. Bigg and W.J. Thompson, 1969. "Daytime Photograph of a Group of Meteor Trails." *Nature*, 222: 156 (April 12).

Mystery object photographed. Bruce Maccabee, 1987. "Analysis of an Anomalous Image Found in a DSMP Weather Satellite Photograph." Fund for UFO Research, Mt. Rainier, Maryland.

"The most likely explanation yet" for UFOs. A.H. Lawson letter to L.A. Frank, July 12, 1986.

"There is water..." Charles Fort, 1974. *The Complete Books of Charles Fort*. New York, Dover Publications, p. 180 and p. 188.

Study of ice falls. James E. McDonald, 1960. "The Ice-Fall Problem." *Weatherwise*, June, p. 110.

Griffiths' icefall. Simon Welfare, 1980. *Arthur C. Clarke's Mysterious World*. New York, A & W, p. 40.

"Ties in beautifully." A.C. Clarke letter to K. Keeton, cc L.A. Frank, June 26, 1986.

Chinese ice falls. Hu Zhong-Wei letter to L.A. Frank, April 24, 1989.

Tunguska. John D. O'Keefe and Thomas J. Ahrens, 1982. "Cometary and Meteorite Swarm Impact on Planetary Surfaces." *Journal of Geophysical Research*, 87: 6669 (August 10).

Flynn Creek. *Ibid*. p. 6668.

Bright flashes of light. William Corliss, ed., 1982. *Lightning, Auroras, Nocturnal Lights and Related Luminous Phenomena*. Glen Arm, Maryland, Sourcebook Project, p. 180.

Odd halos and strange rainbows. William Corliss, ed., 1984. *Rare Halos, Mirages, Anomalous Rainbows, and Related Electromagnetic Phenomena*. Glen Arm, Maryland, Sourcebook Project, p. 6.

Expanding ball of light. Richard F. Haines, 1988. "Expanding Ball of Light (EBL) Phenomenon." *Journal of Scientific Exploration*, 2: 83.

Clear Air Turbulence. H. Blake letter to A.J. Dessler, cc L.A. Frank, April 15,

1986.

A rare event. Frank W. Gibson, 1976. "A Rare Event in the Stratosphere." *Nature*, 263: 487 (Oct. 7).

Sodium clouds. T.J. Beatty, R.E. Bills, K.H. Kwon and C.S. Gardner, 1988. "CEDAR Lidar Observations of Sporadic Na Layers at Urbana, IL." *Geophysical Research Letters*, 15: 1137 (September).

12 A Scramble for Satellite Data

OG0-4. Robert R. Meier and Talbot A. Chubb, 1987. "Theoretical and Observational Problems with 'Holes' in the Far UV Dayglow." Naval Research Laboratory Report 9078, November 5.

Dynamics Explorer 2. William B. Hanson, 1986. "Comment [on the Paper 'On the Influx of Small Comets into the Earth's Upper Atmosphere, II. Interpretation']." *Geophysical Research Letters,* 13: 981 (September).

Re-examination of data. L.A. Frank, J.B. Sigwarth, and J.D. Craven, 1986. "Reply [to Hanson]." *Geophysical Research Letters*, 13: 985 (September).

A satellite called Viking. C.D. Anger, S.K. Babey, A.L. Broadfoot, R.G. Brown, L.L. Cogger, R. Gattinger, J.W. Haslett, R.A. King, D.J. McEwen, J.S. Murphree, E.H. Richardson, B.R. Sandel, K. Smith and A. Vallance Jones, 1987. "An Ultraviolet Auroral Imager for the Viking Spacecraft." *Geophysical Research Letters*, 14: 387.

Joint paper. L.A. Frank, J.B. Sigwarth, J.D. Craven, J.S. Murphree and L.L. Cogger, 1988. "A Search for Atmospheric Holes in Viking Images of Earth's Ultraviolet Dayglow." *EOS*, April 19, p. 413, abstract.

Viking confirms findings. L.A. Frank, J.B. Sigwarth and J.D. Craven, 1989. "Search for Atmospheric Holes with the Viking Cameras." *Geophysical Research Letters*, 16: 1457 (December).

13 Extraordinary Evidence

Size of atmospheric hole as function of altitude. Talbot A. Chubb, 1986. "Comment on the Paper 'On the Influx of Small Comets into the Earth's Upper Atmosphere, I. Observations'." *Geophysical Research Letters*, 13: 1075 (October).

Our Reply. L.A. Frank, J.B. Sigwarth, and J.D. Craven, 1986. "Reply [to Chubb]." *Geophysical Research Letters*, 13: 1079 (October).

Revised Comment. B.L. Cragin, W.B. Hanson, R.R. Hodges, and D. Zuccaro, 1987. "Comment on the Papers 'On the Influx of Small Comets into the Earth's

Upper Atmosphere, I. Observations and II. Interpretation'." *Geophysical Research Letters*, 14: 573 (May).

When we did the same analysis. L.A. Frank, J.B. Sigwarth, and J.D. Craven, 1987. "Reply to Cragin et al." *Geophysical Research Letters*, 14: 557 (May).

Master's thesis. John B. Sigwarth, 1988. "Imaging of Absorption and Emission Features of Water-Vapor Clouds Associated with Small Comets as Observed with Dynamics Explorer 1." Master's Thesis, University of Iowa, Iowa City, May.

Sigwarth presentation. J.B. Sigwarth, L.A. Frank and J.D. Craven, 1988. "Imaging of Absorption and Emission Features of Water-Vapor Clouds Associated with Small Comets as observed with Dynamics Explorer 1." *EOS*, 69: 1350 (Nov. 1), abstract.

14 Launch of the Artificial Comets

Third launch. M. Mendillo et al., 1989. "Preliminary Results from ERIC-3: Attempts to Create Atmospheric Signatures of Comet-Like Objects." *EOS*, 70: 405 (April 11), abstract. Mark Washburn, 1988. "The Waters Above, the Storm Below." *Sky & Telescope*, Dec., p. 628.

Space shuttle water-vapor. M. Mendillo, J. Baumgardner, D.P. Allen, J. Foster, J. Holt, G.R.A. Ellis, A. Klekociuk and G. Reber, 1987. "Spacelab-2 Plasma Depletion Experiments for Ionospheric and Radio Astronomical Studies." *Science*, 238: 1260.

First launch. M. Mendillo interview by P. Huyghe, Dec. 3, 1987.

Polar BEAR. Anonymous, 1986. "Museum-Piece Satellite Goes Into Space." *Science News,* 130: 361, Dec. 6.

Second launch. Michael Mendillo, John B. Sigwarth, John D. Craven, Louis A. Frank, John Holt and David Tetenbaum, 1990. "Project ERIC: The Search for Environmental Reactions Induced by Comets." *Advances in Space Research*, 10: (7)83.

15 Some Cometary Competition

Donahue's last shot. Thomas M. Donahue, 1987. "Small Comets: Implications for Interplanetary Lyman Alpha." *Geophysical Research Letters*, 14: 213.

Discrepancies in Lyman alpha. Ralph C. Bohlin, 1973. "Mariner 9 Ultraviolet Spectrometer Experiment: Measurements of the Lyman Alpha Sky Background." *Astronomy and Astrophysics*, 28: 323.

Donahue's cometesimals. Thomas M. Donahue, Tamas I. Gombosi and Bill R. Sandel, 1987. "Cometesimals in the Inner Solar System." *Nature*, 330: 548 (Dec.

10).

His baby comets. Patrick Huyghe, 1988. "Oceans from Comets—New Evidence." *Oceans*, April, p. 9.

Donahue's detailed paper for *Icarus*. Thomas M. Donahue, Tamas I. Gombosi and Bill R. Sandel, 1987. "Evidence for Cometesimals in the Inner Solar System: Interplanetary Lyman Alpha and Lunar Cratering." University of Michigan.

Wetherhill's reaction. G.W. Wetherhill letter to J.A. Burns, January 28, 1988.

"I think that Donahue's work..." A.J. Dessler interview by P. Huyghe, Oct. 14, 1987.

Hydrogen gas around Halley's. J.M. Ajello, A.l. Stewart, G.E. Thomas and A. Graps, 1987. "Solar Cycle Study of Interplanetary Lyman-Alpha Variations: Pioneer Venus Orbiter Sky Background Results." *The Astrophysical Journal*, 317: 964.

Dynamics Explorer looks at Halley's. J.D. Craven et al., 1986. "The Hydrogen Coma of Comet Halley Before Perihelion: Preliminary Observations with Dynamics Explorer 1." *Geophysical Research Letters*, 13: 873.

"But who's right?" T.M. Donahue letter to L.A. Frank, Dec. 23, 1987.

16 The Verdict

Donahue cometesimals. Thomas M. Donahue, Tamas I. Gombosi and Bill R. Sandel, 1987. "Cometesimals in the Inner Solar System." *Nature*, 330: 548 (Dec. 10).

Published in *Nature*. James A. Van Allen and Louis A. Frank, 1959. "Survey of Radiation Around the Earth to a Radial Distance of 107,400 Kilometers." *Nature*, 183: 430. James A. Van Allen and Louis A. Frank, 1959. "Radiation Measurements to 658,300 Kilometers with Pioneer IV." *Nature*, 184: 219.

My interpretation of Donahue's Voyager results. L.A. Frank, J.B. Sigwarth, and J.D. Craven, 1988. "An Atomic Hydrogen Torus Around the Sun from a Large Population of Small Comets." University of Iowa Research Report, 88-9.

"At this stage..." D. Lindley letter to L.A. Frank, Feb. 22, 1988.

News and views. Paul D. Feldman, 1987. "Encounters of the Second Kind." *Nature*, 330: 518 (Dec. 10).

Three stories. Richard A. Kerr, 1988. "In Search of Elusive Little Comets." *Science*, 240: 1403 (June 10). Richard A. Kerr, 1988. "Comets Were A Clerical Error." *Science*, 241: 532 (July 29). Richard A. Kerr, 1989. "Double Exposures Reveal Mini-Comets." *Science*, 243: 170 (Jan. 13).

17 The Competition Fizzles

Torus of hydrogen. L.A. Frank, J.B. Sigwarth, and J.D. Craven, 1988. "An Atomic Hydrogen Torus Around the Sun from a Large Population of Small Comets." University of Iowa Research Report, 88-9.

Galileo spacecraft. John Noble Wilford, 1987. "Swarms of Miniature Comets May Elude Detection." *The New York Times*, Dec. 10.

Bombshell. D.T. Hall and D.E. Shemansky, 1988. "No Cometesimals in the Inner Solar System." *Nature*, 335: 417 (Sept. 29).

"The primary objective..." Donald Hunten, 1988. "The Hunten Report." The University of Arizona.

"I blew it." Richard A. Kerr, 1988. "Comets Were A Clerical Error." *Science*, 241: 532 (July 29).

Final version. D.T. Hall and D.E. Shemansky. *Ibid*.

Mistake snowballed. Richard A. Kerr, *Ibid*.

18 Why the Moon Doesn't Ring Like a Bell

One response on the Moon problem. R.B. Baldwin, 1987. "On the Current Rate of Formation of Impact Craters of Varying Sizes on the Earth and Moon." *Geophysical Research Letters*, 14: 216.

Making the same mistake. P.M. Davis, 1986. "Comment on the Letter 'On the Influx of Small Comets into the Earth's Upper Atmosphere'." *Geophysical Research Letters*, 13: 1181 (Nov.)

"To ring like a bell." *Ibid*.

The most difficult problem. L.A. Frank, J.B. Sigwarth, and J.D. Craven, 1986. "Reply to Davis and Nakamura et al." *Geophysical Research Letters*, 13: 1186 (November).

Dessler wrote me. A.J. Dessler letter to L.A. Frank, Sept. 26, 1986.

Nakamura and Oberst Comment. Yosio Nakamura, Jiirgen Oberst, Stephen M. Clifford, and Bruce G. Bills, 1986. "Comment on the Letter 'On the Influx of Small Comets into the Earth's Upper Atmosphere, II. Interpretation'." *Geophysical Research Letters*, 13: 1184 (November).

Those who worked on the physics of lunar impacts. Gary V. Latham, Maurice Ewing, Frank Press, George Sutton, James Dorman, Yosio Nakamura, Nafi Toksoz, David Lammlein, and Fred Duennebier, 1972. "Passive Seismic Experiment." *Apollo 16 Preliminary Science Report*, NASA SP-315, p. 9-1.

My Reply to the Moon problem. L.A. Frank, J.B. Sigwarth, and J.D. Craven, *Ibid*.

Brilliant work. Yakov B. Zel'dovich and Yu.P. Raizer, 1967. In W.D. Hayes and R.F. Probstein, eds. *Physics of Shock Waves and High-Temperature Hydrodynamic Phenome*na, Vol. II. New York, Academic Press.

"Splattered into the Moon..." Mark Washburn, 1988. "The Waters Above, the Storm Below." *Sky & Telescope*, Dec. p. 629.

"Most interesting..." J. Oberst letter to A.J. Dessler, cc L.A. Frank, Sept. 28, 1986.

"A flat-floored crater..." John D. O'Keefe and Thomas J. Ahrens, 1982. "Cometary and Meteorite Swarm Impact on Planetary Surfaces." *Journal of Geophysical Research*, 87: 6668 (Aug. 10).

Bright flash on Moon. Gary V. Latham et al., *Ibid*. p. 9-1.

History of strange glows on Moon. William Corliss, ed., 1985. *The Moon and the Planets*. Glen Arm, Maryland, Sourcebook Project, p. 126.

Sigwarth runs in with article. G. Kolovos, J.H. Seiradakis, H. Varvoglis, and S. Avgoloupis, 1988. "Photographic Evidence of a Short Duration: Strong Flash from the Surface of the Moon." *Icarus*, 76: 525 (December).

"From a point slightly..." *Ibid*. p. 528.

Uruguayan astronomer. Julio A. Fernández, 1988. "End-States of ShortPeriod Comets and their Role in Maintaining the Zodiacal Dust Cloud." *Earth, Moon, and Planets*, 41: 155.

"To the actual absence..." *Ibid*. p. 159.

Ice trapped at lunar poles. James R. Arnold, 1979. "Ice in the Lunar Polar Regions." *Journal of Geophysical Research*, 84: 5659 (Sept. 10).

Burst of water vapor reported. J.W. Freeman Jr., H.K. Hills, R.A. Lindeman, and R.R. Vondrak, 1973. "Observations of Water Vapor Ions at the Lunar Surface." *The Moon*, 8: 115.

"Firmly ruled out…" *Ibid*. p. 125.

19 Mars and the Outer Planets

"These networks resemble…" Harold Masursky, J.M. Boyce, A.L. Dial, G.G. Schaber, and M.E. Strobell, 1977. "Classification and Time of Formation of Martian Channels Based on Viking Data." *Journal of Geophysical Research*, 82: 4022 (September 30).

"May have formed..." *Ibid*. p. 4027.

"The water ice deposits…" H.H. Kieffer, S.C. Chase, T.Z. Martin, E.D. Miner, and F.D. Palluconi, 1976. "Martian North Polar Summer Temperature: Dirty Water Ice." *Science*, 194: 1343 (Dec. 17).

Cyclical Martian climate. B.M. Jakosky and E.S. Barker, 1984. "Comparison of Ground-Based and Viking Orbiter Measurements of Martian Water Vapor: Variability of the Seasonal Cycle." *Icarus*, 57: 322.

Warm period on Mars. Jonathan Eberhart, 1989. "Signs of Old Mars: Written in the Dust." *Science News*, 135: 173 (March 18).

Disposing of water on Mars. Fraser P. Fanale, James R. Salvail, Aaron P. Zent, and Susan E. Postawko, 1986. "Global Distribution and Migration of Subsurface Ice on Mars." *Icarus*, 67: 1.

Similar thoughts. S.H. Zisk and P.J. Mouginis-Mark, 1980. "Anomalous Region on Mars: Implications for Near-Surface Liquid Water." *Nature*, 288: 126 (Nov. 13).

Water ice structures. Jonathan Eberhart, 1989. "Neptune: A New Page in the Book of Worlds." *Science News*, 136: 300 (Nov. 4).

Planetary rings. John Noble Wilford, 1989. "New Clues Emerge in Mystery of Planetary Rings." *The New York Times*, June 27.

Heating of the outer planets. L.A. Frank, J.B. Sigwarth, and J.D. Craven, 1986. "On the Influx of Small Comets into the Earth's Upper Atmosphere, II. Interpretation." *Geophysical Research Letters*, 13: 309 (April).

Atmospheric temperatures of Jupiter, Saturn, Uranus, and Neptune. R.A. Hanel, B.J. Conrath, L.W. Herath, V.G. Kunde and J.A. Pirraglia, 1981. "Albedo, Internal Heat, and Energy Balance of Jupiter: Preliminary Results of the Voyager Infrared Investigation." *Journal of Geophysical Research*, 86: 8705. R.A. Hanel, B.J. Conrath, V.G. Kunde, J.C. Pearl, and J.A Pirraglia, 1983. "Albedo, Internal Heat Flux, and Energy Balance of Saturn." *Icarus*, 53: 262. H. Moseley, B. Conrath and R.F. Silverberg, 1985. "Atmospheric Temperature Profiles of Uranus and Neptune." *The Astrophysical Journal*, 292: L83.

20 Where the Small Comets Come From

Oort's paper. Jan. H. Oort, 1950. "The Structure of the Cloud of Comets Surrounding the Solar System, and a Hypothesis Concerning Its Origin." *Bulletin of the Astronomical Institutes of the Netherlands*, 11: 91 (Jan. 13).

Oort Cloud supplied by Oort Disk. J.G. Hills, 1981. "Comet Showers and the Steady-State Infall of Comets from the Oort Cloud." *The Astronomical Journal*, 86: 1730 (November).

Photograph of Beta Pictoris. Bradford A. Smith and Richard J. Terrile, 1984. "A

Circumstellar Disk Around Beta Pictoris." *Science*, 226: 1421 (Dec. 21).

Likely origin of large comets. Martin Duncan, Thomas Quinn and Scott Tremaine, 1988. "The Origin of Short-Period Comets." *The Astrophysical Journal*, 328: L69 (May 15).

Likely candidate for the task. L.A. Frank, J.B. Sigwarth, and J.D. Craven, 1986. "Reply to Morris." *Geophysical Research Letters*, 13: 1484 (December). L.A. Frank, 1989. "Atmospheric Holes and the Small Comet Hypothesis." *Australian Physicist*, 26: 19 (January-February).

Work of physicists. Edgar Everhart, 1968. "Change in Total Energy of Comets Passing through the Solar System." *The Astronomical Journal*, 73: 1039 (December). Edgar Everhart, 1969. "Close Encounters of Comets and Planets. "*The Astronomical Journal*, 74: 735 (June). Julio A. Fernández, 1980. "On the Existence of a Comet Belt Beyond Neptune." *Monthly Notices of the Royal Astronomical Society*, 192: 481.

21 Death of the Dinosaurs

Alvarez scenario. Luis W. Alvarez, W. Alvarez, F. Asaro and H.V. Michel, 1980. "Extraterrestrial Cause for the Cretaceous-Tertiary Extinction." *Science*, 208: 1095. Luis W. Alvarez, 1987. "Mass Extinctions Caused by Large Bolide Impacts." *Physics Today*, 40: 24 (July).

Similar periodicity in work of paleontologists. David M. Raup and John J. Sepkoski, Jr., 1986. "Periodic Extinction of Families and Genera." *Science*, 231: 833 (Feb. 21).

Additional evidence. M.R. Rampino and R.B. Stothers, 1984. "Terrestrial Mass Extinctions, Cometary Impacts and the Sun's Motion Perpendicular to the Galactic Plane." *Nature*, 308: 709.

Nemesis. Richard Muller, 1988. *Nemesis: The Death Star*. New York, Weidenfeld & Nicolson.

Planet X. Daniel P. Whitmire and Albert A. Jackson IV, 1984. "Are Periodic Mass Extinctions Driven by a Distant Solar Companion?" *Nature*, 308: 713. D.P. Whitmire and J.J. Matese, 1985. "Periodic Comet Showers and Planet X." *Nature*, 313: 36. J.J. Matese and D.P. Whitmire, 1986. "Planet X and the Origins of the Shower and the Steady State Flux of Short-Period Comets." *Icarus*, 65: 37.

Dark Planet. L.A. Frank, J.B. Sigwarth, and J.D. Craven, 1988. "A Hypothesis Concerning the Inner Oort Disk and Its Relationship to Comet Showers and Extinction of Species." University of Iowa Research Report, 88-8.

Carbon at geological boundary. W.S. Wolbach, R.S. Lewis, and E. Anders, 1985.

"Cretaceous Extinctions: Evidence for Wildfires and Search for Meteoritic Material." *Science*, 230: 167.

Shocked quartz. B.F. Bohor, P.J. Modreski and E.E. Foard, 1987. "Shocked Quartz in the Cretaceous-Tertiary Boundary Clays: Evidence for a Global Distribution." *Science*, 236: 705.

Ages of large craters on Earth. W. Alvarez and R.A. Muller, 1984. "Evidence from Crater Ages for Periodic Impacts on the Earth." *Nature*, 308: 718.

Paper not well received. L.A. Frank et al. *Ibid*.

22 The Debate Comes to an End

"The importance for scientific progress..." A.J. Dessler, 1986. "Controversial Publications: The Role of Comments and Replies." *Geophysical Research Letters*, 13: 1363 (December).

Paper on cometesimals. Thomas M. Donahue, Tamas I. Gombosi, and Bill R. Sandel, 1987. "Cometesimals in the Inner Solar System." *Nature*, 330: 548.

"Scientists would like to think..." Patrick Huyghe, 1988. "Oceans from Comets—New Evidence." *Oceans*, April, p. 11.

"Only if the small-comet hypothesis..." A.J. Dessler letter to L.A. Frank, Dec. 5, 1986.

Last Comment. John T. Wasson and Frank T. Kyte, 1987. "Comment on the Letter 'On the Influx of Small Comets into the Earth's Upper Atmosphere, II. Interpretation'." *Geophysical Research Letters*, 14: 779 (July).

Comets coated with viruses. Fred Hoyle and Chandra Wickramasinghe, 1979. *Diseases from Space*. New York, Harper & Row. Fred Hoyle and Chandra Wickramasinghe, 1981. *Evolution From Space*. New York, Simon & Schuster. Fred Hoyle and Chandra Wickramasinghe, 1985. *Living Comets*. Cardiff, University College Cardiff Press.

"A nice, appropriate response." A.J. Dessler letter to L.A. Frank, Jan. 27, 1987.

"As I see it..." A.J. Dessler letter to L.A. Frank, March 24, 1987.

Explaining to Dessler. L.A. Frank letter to A.J. Dessler, April 6, 1987.

Poem for Dessler's birthday. L.A. Frank letter to A.J. Dessler, Oct. 24, 1988.

"I'm still convinced..." Tex (A.J. Dessler) letter to L.A. Frank, April 5, 1989.

"Dear Mr. Excitement..." Tex (A.J. Dessler) letter to L.A. Frank, April 20, 1989.

23 Astronomers Enter the Fray

Soter's Comment. Steven Soter, 1987. "Comment on the Paper 'On the Influx of Small Comets into the Earth's Upper Atmosphere.'" *Geophysical Research Letters*, 14: 162 (February).

GEODSS Network. Jesse Hall, 1989. "Deep Space Surveillance Keeps Track of Objects From Gloves to Satellites." *Space Trace*, December, p. 6.

Taft papers. Laurence G. Taft, 1981. "A New Asteroid Observation and Search Technique." *Publication of the Astronomical Society of the Pacific*, 93: 658. Laurence G. Taft, 1986. "Satellite Debris: Recent Measurements." *Journal of Spacecraft*, 23: 342 (May-June).

"Frank is wrong…" L.G. Taft interview by P. Huyghe, Oct. 15, 1987.

Time variations in occurrence of atmospheric holes. L.A. Frank, J.B. Sigwarth, and J.D. Craven, 1986. "Temporal Variations of the Influxes of Small Comets Into the Earth's Atmosphere," University of Iowa Research Report, 86-33.

I redid Soter's work. L.A. Frank, J.B. Sigwarth, and J.D. Craven, 1987. "Reply to Soter." *Geophysical Research Letters*, 14: 164 (February).

24 Where Are You Now, Galileo?

"I could more easily believe…" Maurice Ebison, ed., 1977. *The Harvest of a Quiet Eye*. London, The Institute of Physics.

Gehrels' review. T. Gehrels letter to L.A. Frank, July 17, 1986.

Soter convinced small comets should have been seen. Steven Soter, 1987. "Comment on the Paper 'On the Influx of Small Comets into the Earth's Upper Atmosphere.'" *Geophysical Research Letters*, 14: 162 (February).

Gehrels' paper. Tom Gehrels, 1985. "Asteroids and Comets." *Physics Today*, 38(2): 33.

"Several times the energy..." T. Gehrels letter *Ibid*.

Gehrels may have detected them. Tom Gehrels and R.S. McMillan, 1982. "CCD Scanning for Asteroids and Comets." In W. Fricke and G. Teleki, eds. *Sun and Planetary Systems*. Dordrecht, Holland, Reidel. T. Gehrels, B.G. Marsden, R.S. McMillan and J.V. Scotti, 1986. "Astrometry with a Scanning CCD." *Journal of Astronomy*, 91: 1242. T. Gehrels and F. Vilas, 1986. "A CCD Search for Geosynchronous Debris." *Icarus*, 68: 412.

25 The Great Search Begins

"I thought it was pretty wild…" C.M. Yeates interview by P. Huyghe, Nov. 14, 1989.

"We were sitting…" *Ibid*.

"A lot of astronomers…" Patrick Huyghe, 1988. "Oceans from Comets—New Evidence." *Oceans*, April, p. 11.

Yeates search method. C.M. Yeates, 1989. "Initial Findings from a Telescopic Search for Small Comets Near Earth." *Planetary and Space Science*, 37: 1185 (October).

"I went through my own calculation…" E.M. Shoemaker letter to A.J. Dessler, cc C.M. Yeates, Jan. 4, 1987 (sic).

"The failure to find Frank's microcomets..." *Ibid*.

"It was a pretty crude system…" C.M. Yeates interview by P. Huyghe, Nov. 14, 1989.

The runs on the telescope. C.M. Yeates letter to T. Gehrels, July 19, 1988.

Unidentified streaks appear in 36 images. C.M. Yeates, 1989. *Planetary and Space Science*, *Ibid*.

"The whole thing…" C.M. Yeates interview, 1989. *Ibid*.

26 Double Trouble

"I was shocked…" C.M. Yeates interview by P. Huyghe, Nov. 14, 1989.

"Gehrels was telling the press…" *Ibid*.

"Apparently confirmed…" Jet Propulsion Laboratory Public Information Office Release #1191, March 30, 1988.

Yeates' findings made public. Patrick Huyghe, 1988. "Oceans from Comets—New Evidence." *Oceans*, April, p. 11. Richard Monastersky, 1988. "Comet Controversy Caught on Film." *Science News*, 133: 340 (May 28).

"What I was left with…" C.M. Yeates, *Ibid*.

Consecutive images first suggested. C.M. Yeates letter to T. Gehrels, July 19, 1988.

"If you see them…" T. Gehrels note to C.M. Yeates, April, 1988.

Analysis of consecutive images. L.A. Frank, J.B. Sigwarth, and C.M. Yeates, 1990. "A Search for Small Solar-System Bodies Near the Earth Using a Ground-Based Telescope: Technique and Observations." *Astronomy and Astrophysics*, 228:

522 (February).

Short summary. L.A. Frank, J.B. Sigwarth, J.D. Craven, C.M. Yeates, and T. Gehrels, 1988. "Telescopic Search for Small Comets in Consecutive Images with the Spacewatch Camera." *EOS*, 69: 1293, Nov. 1, abstract.

"If only you could..." T. Gehrels letter to C.M. Yeates, July 11, 1988.

"We have been in this business..." T. Gehrels letter to L.A. Frank, Oct. 17, 1988.

My reply. L.A. Frank letter to T. Gehrels, Oct. 25, 1988.

"The complete absence..." C.M. Yeates letter to T. Gehrels, Nov. 2, 1988.

"Your reasonings are impressive..." T. Gehrels letter to C.M. Yeates, Nov. 14, 1988.

"Relax..." T. Gehrels postcard to L.A. Frank, Nov. 2, 1988.

"If there indeed were as many cometesimals..." T. Gehrels to B. Jakosky, cc L.A Frank, Nov. 14, 1988.

"We express our appreciation..." L.A. Frank, J.B. Sigwarth, J.D. Craven, and C.M. Yeates, 1988. "Telescopic Search for Small Comets in Consecutive Images with the Spacewatch Camera." AGU Poster paper, December.

"The normal criteria for acceptance..." E.M. Shoemaker letter to A.J. Dessler, cc: C.M. Yeates, 1988.

"For your paper..." A.J. Dessler letter to C.M. Yeates, Sept. 9, 1988.

"*Utterly impossible*..." C.M. Yeates letter to A.J. Dessler, Sept. 28, 1988.

"Well, it wouldn't have mattered..." C.M. Yeates interview by P. Huyghe, Nov. 14, 1989.

Scientific American. John Horgan, 1989. "A Snowball's Chance." *Scientific American*, March, p.21.

Run-in with Dessler over a paper. B.R. Clemesha, P.P. Batista and D.M. Simonich, 1988. "Concerning the Origin of Enhanced Sodium Layers." *Geophysical Research Letters*, 15: 1267 (October).

Keeping small comets out of the literature. B.R. Clemesha letter to L.A. Frank, March 5, 1990.

No one believed it. Richard A. Kerr, 1989. "Double Exposures Reveal Mini-Comets?" *Science*, 243: 170 (Jan. 13).

Poster paper on double images. L.A. Frank, J.B. Sigwarth, J.D. Craven, and C.M. Yeates, 1988. *Ibid.*

27 The Turning Point

Warren E. Leary, 1989. "Big Asteroid Passes Near Earth Unseen in a Rare Close Call." *The New York Times*, April 20.

Yeates' paper published. C.M. Yeates, 1989. "Initial Findings from a Telescopic Search for Small Comets Near Earth." *Planetary and Space Science*, 37: 1185 (October).

Consecutive image paper published. L.A. Frank, J.B. Sigwarth, and C.M. Yeates, 1990. "A Search for Small Solar-System Bodies Near the Earth Using a Ground-Based Telescope: Technique and Observations." *Astronomy and Astrophysics*, 228: 522 (February).

Long paper about small comets. A.J. Dessler, 1991. "The Small Comet Hypothesis." *Reviews of Geophysics*, August.

On atmospheric holes in Viking images. L.A. Frank, J.B. Sigwarth, and J.D. Craven, 1989. "Search for Atmospheric Holes with the Viking Cameras." *Geophysical Research Letters*, 16: 1457 (December).

Work of Scott Bolton. S.J. Bolton, 1990. "One Year Variations in the Near Earth Solar Wind Ion Density and Bulk Flow Velocity." *Geophysical Research Letters*, 17: 37 (January).

Vindication (1990-2001)

28 The Rest of the Story

Still dead. John Horgan, 1990. "Death Watch." *Scientific American*, June, pp 22–25.

Fall season's longest subtitle. 1990. "Cross Currents." *Publisher's Weekly*, August 1.

"Savage tale…" Dennis Overbye, 1990. "The Big Splash." *The New York Times Book Review*, November 11.

"Highly readable…" Robert Kunzig, 1990. "Chunks of ice, oceans of life." *New Scientist*, November 17.

"Violently disagreeing…" David W. Hughes, 1991. "The Big Splash." *Sky & Telescope*, May.

Failed to overcome the skepticism. Richard Morris, 1990. "Making a Splash in Astronomical Circles." *San Francisco Chronicle*, October 14.

"Wide ranging..." Anonymous, 1990. "The Big Splash." *Kirkus Reviews*, August 15.

"A provocative work..." Anonymous, 1990. "The Big Splash." *Library Journal*, October 1.

Transient lunar phenomenon. Charles Seife, 1999. "Moon mystery emerges from the X-files." *New Scientist*, October 23.

Novel ways. Peter M. Banks, 1989. "A new means for observation of small comets and other water-laden bodies entering Earth's upper atmosphere." *Geophysical Research Letters*, June.

Enter Stephen Mende. 1986. "Space Shuttle Mission STS-51L: Press Kit." NASA, January.

Great tragedy. Anonymous, 1991. "Obituary: Clayne Yeates; JPL Scientist on Galileo Project." *Los Angeles Times,* April 25.

Bloody nightmare. Michael Marriott, 1991. "Gunman in Iowa Wrote of Plans in Five Letters." *The New York Times*, November 3.

The mother of all small comet critiques. Alex J. Dessler, 1991. "The Small-Comet Hypothesis." *Reviews of Geophysics*, August.

Our paper. Louis Frank and John Sigwarth, 1993. "Atmospheric Holes and Small Comets." *Reviews of Geophysics*, February.

Tex, Dessler's dog. Bruce Tsurutani, 1992. "The Last SPR Dinner Awards." *EOS*, March 17.

29   Polar Rising

Polar mission. "NASA's Polar Mission: Unlocking the Secrets of Earth's Magnetosphere." Goddard Space Flight Center. https://pwg.gsfc.nasa.gov/polar.

Nature's rejection. Letter from Dr. Leslie Sage to Dr. Frank, June 27, 1996.

Science's rejection. Letter from R. Brooks Hanson to Dr. Frank, January 13, 1997.

30   Vindication

The official NASA News press release. William Steigerwald, 1997. "Polar Spacecraft Images Support Theory of Interplanetary Snowballs Spraying Earth's Upper Atmosphere." *NASA News*, Release No. 97-59, May 28.

At the same time. Gary Galluzzo, 1997. "UI's Louis Frank presents proof for revolutionary small comet theory." *General News*, University of Iowa, May 28.

Website on the small comets. 1997. *Small Comets*. The University of Iowa. http://smallcomets.physics.uiowa.edu.

"Riveted..." Kathy Sawyer, 1997. "Cosmic Snowballs Detected Pelting Earth's Atmosphere." *Washington Post*, May 29.

Robert Meier. Faye Flam, 1997. "From space snowballs come theories on rain and seas." *The Philadelphia Inquirer*, May 29.

"Vindicated..." Richard Monastersky, 1997. "Is Earth Pelted by Space Snowballs?" *Science News*, May 31.

Gung-ho. William J. Cook, 1997. "A scientist—and his theory—vindicated." *U.S. News & World Report,* June 9.

Science magazines. Corey S. Powell, 1997. "Return of the Space Snowballs." *Scientific American,* August; Richard Kerr, 1997. "Spots Confirmed, Tiny Comets Spurned." *Science,* May 30; Robert Matthews, 1997. "Not a snowball's chance..." *New Scientist*, July 12.

Chimed in with an editorial. 1997. "Cosmic Snowballs." *The New York Times*, May 30.

Our Polar announcement. 1997. "Polar Spacecraft Images Support theory of Interplanetary Snowballs Spraying Earth's Upper Atmosphere." *NASA News*. Release No. 97-59, May 28.

Later interview with Hoffman. Robert Matthews, 1997. "Not a snowball's chance..." *New Scientist*, July 12.

A small comet may have struck the Mir space station. Bryan Burrough, 1998. *Dragonfly: NASA and the Crisis Aboard Mir*. HarperCollins.

Cyclops. Charles Bullard, 1997. "U of I prof pleased data lift theory." *The Des Moines Register*, August 13.

31 Grand Slam

Heinrich Holland. William J. Broad, 1997. "Spotlight on Comets in Shaping of Earth." *The New York Times*, June 3.

First paper. L. A. Frank and J. B. Sigwarth, 1997. "Detection of atomic oxygen trails of small comets in the vicinity of Earth." *Geophysical Research Letters*, October 1.

Second paper. L. A. Frank and J. B. Sigwarth, 1997. "Transient decreases of Earth's far-ultraviolet dayglow." *Geophysical Research Letters,* October 1.

Third paper. L. A. Frank and J. B. Sigwarth, 1997. "Simultaneous observations of transient decreases of Earth's far-ultraviolet dayglow with two cameras." *Geophysical Research Letters*, October 1.

Fourth paper. L. A. Frank and J. B. Sigwarth, 1997. "Trails of OH emissions from small comets near Earth." *Geophysical Research Letters*, October 1.

Based on these comments. Letters from Charles Kolb to L.A. Frank, December 19, 1997.

Editor's note. Michael Prather, 1997. *Geophysical Research Letters*, October 1.

32 Dessler and His Pawns

They had not all been silenced. "Polar Comets." David Morrison's *NEO News*, June 2, 1997.

The Polar results could be explained by lightning. Email from L.A. Frank to Tom Gehrels, June 14, 1997.

"Interesting times again!" Email from Tom Gehrels to L.A. Frank, June 2, 1997.

But Mike A'Hearn. "Polar Comets." David Morrison's *NEO News*, June 2, 1997.

There was Dessler. Richard Kerr, 1997. "Spots Confirmed, Tiny Comets Spurned." *Science*, May 30; Monastersky, 1997. "Is Earth pelted by space snowballs?" *Science News,* May 31; Corey Powell, 1997. "Return of the Space Snowballs." *Scientific American*, September 15.

"Personally nominate." Alex Dessler email to L. A. Frank, January 12, 1998.

In a "Highlights of This Issue." Michael Prather, 1997. "Looking for Small Comets—one Found." *Geophysical Research Letters*, December 15.

The first critical paper. George Parks et al., 1997. "Does the UVI on Polar detect cosmic snowballs?" *Geophysical Research Letters*, December 15.

Second critical paper. Bashar Rizk and A. J. Dessler, 1997. "Small comets: Naked-eye visibility." *Geophysical Research Letters*, December 15.

Third critical paper. Jennifer A. Grier and Alfred S. McEwen, 1997. "The small-comet hypothesis: An upper limit to the current impact rate on the moon." *Geophysical Research Letters*, December 15.

Fourth critical paper. Timothy D. Swindle and David A. Kring, 1997. "Implications of small comets for the noble gas inventories of Earth and Mars." *Geophysical Research Letters*, December 15.

Fifth critical paper. Mark B.E. Boslough and G. Randall Gladstone, 1997. "An Impact Plume Model for Atmospheric Holes in the FUV Dayglow." *Geophysical*

Research Letters, December 15.

33 Steering Through Rough Waters

All Bob Hoffman quotes in this chapter. 1998. Bob Hoffman interview by P. Huyghe, September.

Edward Tagliaferri. L.A. Frank email to Bob Hoffman, August 2, 1997.

Hoffman tried to set him straight. Bob Hoffman email to Alex Dessler, reply to message, January 8, 1998.

Alex Dessler was giving a talk at Rice University. Alexander J. Dessler, 1998. "Small Comets: A Case History in Pathological Science." Rice University, April 16.

34 Contact Sport

An Associated Press story. 1998. "Space Snowball Fight Rages On." *Associated Press*.

Tom Donahue's remarks. Richard Monastersky, 1998. "Small comet theory melts under scrutiny." *Science News*, June 6.

Intellectual slugfest. Jules Crittenden, 1998. "Critics give comet theory snowball's chance in hell." *Boston Herald*, May 27.

"If it was really true…" Forrest Mozer email to Bob Hoffman, cc: L.A. Frank, January 19, 1998.

"Am I imagining…" Alex Dessler email to L. A. Frank, January 20, 1998.

Mozer clearly had a closed mindset. Lou Frank email to Forrest Mozer, February 15, 1998.

"Independent and blind." *Ibid.*

Mozer stuck his neck out. Forrest Mozer email to Bob Hoffman, cc: L.A. Frank, December 15, 1997.

Mozer's paper. F. S. Mozer, J. P. McFadden, I. Sircar, and J. Vernetti, 1998. "Small comet 'atmospheric holes' are instrument noise." *Geophysical Research Letters*, October 1; F. S. Mozer, J. P. McFadden, I. Sircar, and J. Vernetti, 1999. "Correction to 'Small-comet "atmospheric holes" are instrument noise.'" *Geophysical Research Letters*, January 1.

Mozer also played a supporting role. J. P. McFadden, F. S. Mozer, J. Vernetti, and I. Sircar, 1998. "An instrumental source for the dark pixel clusters in the Polar VIS and UVI experiments." *Geophysical Research Letters*, October 1.

I immediately wrote back to him. Lou Frank email to Forrest Mozer, February

21, 1998.

Hoffman was not happy. Bob Hoffman interview by P. Huyghe. September 1998.

"No errors…" Forrest Mozer email to L. A. Frank, October 16, 1998.

"Oversight." Forrest Mozer email to L. A. Frank, October 20, 1998.

I explained this to Parks. L. A. Frank email to Parks, July 21, 1998.

The small comets also eluded… S. Knowles, R.R. Meier, B.A.S. Gustafson, and F. J. Giovanne, 1999. "A search for small comets with the Navel Space Command radar." *Journal of Geophysical Research*, June 1; L. A. Frank and J. B. Sigwarth, 1999. "Comment on 'A search for small comets with the Navel Space Command radar' by S. Knowles et al." *Journal of Geophysical Research*, October 1; S. Knowles, R.R. Meier, B.A.S. Gustafson, and F. J. Giovanne, 1999. "Reply." *Journal of Geophysical Research*, October 1.

Swapped insults. Jules Crittenden, 1998. "Critics give comet theory snowball's chance in hell." *Boston Herald*, May 27.

Harris must have lost a lot of sleep. A. W. Harris, 2000. "'Atmospheric holes' are not small comets." *Journal of Geophysical Research*, August 1.

As noted in our rebuttal in the same issue. J. B. Sigwarth, L. A. Frank, 2000. "Comment on 'Atmospheric holes are not small comets by A. W. Harris,'" *Journal of Geophysical Research*, August 1.

Using a stratospheric-mesospheric circulation model. B. Hannegan, S. Olsen, M. Prather, X. Zhu, D. Rind, J. Lerner, 1998. "The dry stratosphere: A limit on cometary water influx." *Geophysical Research Letters*, May 15.

Tom Donahue seemed to have changed his mind. Richard Monastersky, 1998. "Small comet theory melts under scrutiny." *Science News*, June 6.

A major paper on the small comets. L. A. Frank and J. B. Sigwarth, 1999. "Atmospheric holes: Instrumental and geophysical effects." *Journal of Geophysical Research*, January 1.

35 Spacewatch Redux

Near-Earth Asteroid Tracking. Steven H. Pravdo, David L. Rabinowitz, Eleanor F. Helin, et al., 1999. "The Near-Earth Asteroid Tracking (NEAT) Program: An Automated System for Telescope Control, Wide-Field Imaging, and Object Detection." *The Astronomical Journal*. V117 N3.

David Rabinowitz reported to Hoffman. David Rabinowitz email to Bob Hoffman, January 12, 1998.

Eugene Shoemaker. Kathy Sawyer, 1997. "Evidence of Cosmic Snowball Starts Scramble for Explanations." *Washington Post,* June 1.

The issue of using the Hubble Space Telescope. 1998. "Astronomers Track Down Asteroids in Hubble Archive." NASA Jet Propulsion Laboratory Press Release, March 10.

On their proposed naked-eye/binocular search for LAFOs. Alan Harris email to Alex Dessler, January 14, 1998, sent to L.A. Frank on January 20, 1998.

Harris replied a couple of days later. Alan Harris email to Bob Hoffman, March 25, 1998.

What Hoffman said to Dessler. Hoffman interview by P. Huyghe.

Bob McMillan's reply to Hoffman. November 1998.

36 Deep Waters

Landmark study. David Deming, 1999. "On the possible influence of extraterrestrial volatiles on Earth's climate and the origin of the oceans." *Palaeo*, 146: 33-51.

"The chief point of my paper." Letter from David Deming to Lee Kump, August 14, 1997.

Deming is not alone. William Fyfe, 1980. "Crust Formation and Destruction." In D. W. Strangway (ed). *The Continental Crust and Its Mineral Deposits—Special Paper 20*. Geological Association of Canada, 20: 77-88.

And wrote in his letter. William Fyfe letter to L.A. Frank, October 1, 1992.

The standard view. Frank Press and Raymond Siever. 1994. *Understanding Earth*. Freeman.

This belief is "blind extrapolation." Armand Delsemme, 1997. *Comets and the Origin and Evolution of Life*, Springer-Verlag.

In a 1993 issue. Raymond Jeanloz, 1993. "The Hidden Shore." *The Sciences*, January/February.

Ever so slowly. E. C. Pielou (1998). *Fresh Water*. University of Chicago Press.

Dale E. Ingmanson and William J. Wallace, 1995. *Oceanography: An Introduction*, 5th edition, Wadsworth.

To illustrate this. K.K. Kelley, 1993. "Water Vapor and Cloud Water Measurements Over Darwin During the STEP 1987 Tropical Mission." *Journal of Geophysical Research*, May 20.

A 1996 paper led by Andreas Engel. Andreas Engel et al., 1996. "The total hydrogen budget in the Arctic winter stratosphere during the European Arctic Stratospheric Ozone Experiment." *Journal of Geophysical Research*, June 20.

Andy Dessler. Dessler, A. E., E. J. Hintsa, E. M. Weinstock, J. G. Anderson, and K. R. Chan, 1995. "Mechanisms controlling water vapor in the lower stratosphere: "A tale of two stratospheres." *Journal of Geophysical Research*, 100 (D11), 23,167-123,172.

There is, of course, hard data. Steven Young, 1997. "Satellite Backs Comet Bombardment Theory." *Reuters*, August 11.

The high readings. Robert Conway, et al., 1996. "Satellite measurements of hydroxyl in the atmosphere." *Geophysical Research Letters*, August 1.

On the one hand. Richard Kerr, 1997. "Rising Damp from Small Comets?" *Science*, August 22.

But he then told Richard Monastersky. Richard Monastersky, 1996. "Reservoir of water hides high above Earth." *Science News*, August 23.

This comment led to a heated reply. Dana Thibault, 1997. "Letters: Hypothesis has no room for belief." *Science News*, October 25.

James Russell of Hampton University. Richard Kerr, 1997. "Rising Damp from Small Comets?" *Science*, August 22.

"It's when we get challenging observations…" *Ibid.*

Penn State microwave measurements. John J. Olivero, 1990. "Water Mysteries on High." *Earth and Mineral Sciences* Vol. 59 No. 1.

Tobias Owen reported. Donald Goldsmith, 1997. "Comet Origin of Oceans All Wet?" *Science*, July 18.

A study by Uwe Fink. Uwe Fink, 1992. "Comet Yanaka (1988r): A New Class of Carbon-poor Comet." *Science* Vol. 257, Issue 5078.

"The comet is really totally different…" 1992. "Surprise Upsets Theories of Comets' Uniformity." *The New York Times*, September 29.

Water…on the Sun. Oleg Polyansky, 1997. "K-Band Spectrum of Water in Sunspots." *The Astrophysical Journal*, October 13.

Not only are there snowballs in hell. John Noble Wilford, 1991. "Photos by Radar Hint at Ice on Poles of Mercury." *The New York Times,* November 7.

Water on Venus. Thomas M. Donahue, R. Richard Hodges Jr., 1993. "Venus methane and water." *Geophysical Research Letters*, April 9.

The presence of water on the Moon. David Perlman, 1998. "NASA Finds Watery World Below the Moon's Surface." *San Francisco Chronicle*, September 10.

Orion is a veritable "water factory." Martin Harwit, David A. Neufeld, Gary J.

Melnick, and Michael J. Kaufman, 1998. "Thermal Water Vapor Emission from Shocked Regions in Orion." *Astrophysical Journal Letters*, April 20.

Saturn, Uranus, and Neptune... 1997. "Europe's Space Telescope ISO Finds Water in Distant Places." European Space Agency Press Release, April 29.

Implications for the origin of water in the solar system. 1998. "Astronomers find water-bloated space cloud." *The Gazette*, Johns Hopkins University, April 20.

Markarian 1. Associated Press, 1994. "Water Found in Distant Galaxy." *The New York Times,* June 3.

37 Last Dance

A prestigious talk I was asked to give. Louis A. Frank, 1999. "Small Comets and Our Origins: The Ecstasy and Agony of the Scientific Debate." University of Iowa's Sixteenth Annual Presidential Lecture, February 21.

"We got zero…" "'Nail in the Coffin' for Bombarding Minicomets." *Sky & Telescope*, May 2000.

"It would seem fair…" Before the Faculty Judicial Panel, The University of Iowa, In Re the Matter of Louis Frank, Respondent, Brief and Final argument of Respondent, Louis Frank, Request to Reopen Record.

The University of Iowa's Press Office had prepared a press release. Gary Galluzzo, 1999. "UI researcher fails to detect small comets." *General News*, University of Iowa, Dec. 16.

"It would seem fair…" Robert N. Downer. Before the Faculty Judicial Panel, The University of Iowa. In Re the Matter of Louis Frank, Respondent. Brief and Final Arguments of Respondent, Louis Frank. "Request to Reopen Record."

I returned some of these calls. Ann McGlynn, 1999. "Critic of comet theory cites studies." *Des Moines Register,* December 21.

Then on December 19, 1999. John Kenyon, 1999. "Frank alleges fraud." *The Gazette* (Cedar Rapids, Iowa), December 19.

38 The Witch Hunt

"I write to set the record straight…" Lou Frank, 1999. "Trust Science." *The Gazette* (Cedar Rapids, Iowa), December 27.

On Christmas day. John Kenyon, 1999. "Allegation of fraud retracted." *The Gazette* (Cedar Rapids, Iowa), December 25.

At the end of December. L.A. Frank email to Jack Fix. December 27, 1999.

Mutel had sent me a copy of the paper. R.L. Mutel and J. D. Fix, 2000. "An optical search for small comets." *Journal of Geophysical Research: Space Physics*, November 1.

I… ended my email with these words. L.A. Frank email to Bob Mutel. January 13, 2000.

On January 26 I received a surprise letter. Letter from Jon Whitmore to Professor Frank. January 26, 2000.

In the complaint. Referenced in: Before the Faculty Judicial Commission, In Re the Matter of Louis Frank, Ethics Policy Violation, Report of the Panel. Summary of Panel's Findings and Recommendations.

"Dr. Frank is a 36-year member of the faculty…" Before the Faculty Judicial Panel, The University of Iowa, In Re the matter of Louis Frank, Respondent, Brief and final Argument of Respondent, Louis Frank, Request to Reopen the Record.

"Confident that the paper is correct…" Robert Mutel and Jack Fix email to L.A. Frank and J. B Sigwarth. April 11, 2000.

The headline read. Jim Jacobson, "Panel to scrutinize professor's ethics: UI physicist could be fired over accusations leveled at colleague." *The Gazette* (Cedar Rapids, Iowa), July 13, 2000.

"Recidivist." Lee Anna Clark and Mark E. Schantz, Before the University of Iowa Faculty Judicial Panel, In the Matter of Louis B. Frank, Academic Officer's Post-Hearing Brief. January 25, 2001.

Exhibit 6, email from John Craven to Jack Fix, Dec. 29, 1999. In the Deposition of John D. Fix, In the Matter of Louis Frank, Respondent. Before the Faculty Judicial Panel, the University of Iowa.

It also came to light during the hearing. Letter from Lyle Muller, city editor of *The Gazette* (Cedar Rapids, Iowa), to Mark Schantz. July 26, 2000.

The panel members released their report. Before the Faculty Judicial Commission, In Re the Matter of Louis Frank, Ethics Policy Violation, Report of the Panel. Summary of Panel's Findings and Recommendations.

"This type of response…" Before the Faculty Judicial Panel, The University of Iowa, In Re the matter of Louis Frank, Respondent. Respondent's Brief to President Coleman Pursuant to UOM Section 29.7j(1).

39 Snowballs in Hell

On the 8th of January 2000. Clair Wood, 2000. "Spanish ice chunks of interest to Iowa astrophysicist." *Bangor Daily News*, February 3.

Jesús Martínez-Frías, Antonio Delgado Huertas, Roberto García, et al., 2000. "Hailstones fall from clear Spanish skies." *Geotimes*, June.

A newspaper column. *Bangor Daily News, Ibid.*

"Once again purports to prove…" Referee #2, Replies to LA. Frank and J.B Sigwarth, "Detection of Small Comets with a Ground-Based Telescope." April 14, 2000.

"Significantly improved." Referee #2, "Referee's Report on the Revised Manuscript No. 20.0054." Replies to LA. Frank and J.B. Sigwarth, "Detection of Small Comets with a Ground-Based Telescope." June 22, 2000.

"You have done a commendable and thorough job…" Referee #2, Replies to LA. Frank and J.B Sigwarth, "Detection of Small Comets with a Ground-Based Telescope." August 4. 2000.

On the 21st of February 2000. L.A. Frank and J. B. Sigwarth, 2001. "The Detection of Small Comets with a Ground-Based Telescope." *Journal of Geophysical Research: Space Physics*, March.

A month later I traveled to Copenhagen. Gary Galluzzo, 2000. "UI researcher to deliver prestigious Niels Bohr Lecture." University of Iowa, *General News*, December 4.

40 Small Comets and the Future

Quip by 16th century rationalist. Carl Sagan and Ann Druyan, 1985. *Comet*. New York, Random House, p. 29.

Appendix 2: How This Book Came to Be

I came across a story. Philip M. Boffey, 1986. "Did Ocean Waters Arrive in Rain of Icy Comets?" *The New York Times*, April 1.

My article. Patrick Huyghe, 1986. "Origin of the Oceans." *Oceans*, August.

It was subsequently adapted. Patrick Huyghe, 1986. "The Oceans Came From Outer Space." *Washington Post*, November 16.

It was published. Louis A. Frank with Patrick Huyghe, 1990. *The Big Splash*. Birch Lane Press.

Obituary of sorts. Louis A. Frank, 2014. "Cosmic Rain: A Maverick's Tale." *EdgeScience*, September 19.

BIBLIOGRAPHY

(A chronological bibliography of papers by Louis Frank on small comets published in the scientific literature. See http://smallcomets.physics.uiowa.edu/pdf)

L.A. Frank, J.B. Sigwarth , and J.D. Craven, 1986. "On the Influx of Small Comets Into the Earth's Upper Atmosphere, I. Observations." *Geophysical Research Letters*, 13: 303 (April).

L.A. Frank, J.B. Sigwarth, and J.D. Craven, 1986. "On the Influx of Small Comets Into the Earth 's Upper Atmosphere, II. Interpretation." *Geophysical Research Letters*, 13: 307 (April).

L.A. Frank, J.B. Sigwarth, and J.D. Craven, 1986. "Reply [to Donahue]." *Geophysical Research Letters*, 13: 559 (June).

L.A. Frank, J.B. Sigwarth, and J.D. Craven, 1986. "Reply [to Rubincam]." *Geophysical Research Letters*, 13: 703 (July).

L.A. Frank, J.B. Sigwarth, and J.D. Craven, 1986. "Reply [to Hanson]." *Geophysical Research Letters*, 13: 985 (September).

L.A. Frank, J.B. Sigwarth, and J.D. Craven, 1986. "Reply [to McKay]." *Geophysical Research Letters*, 13: 979 (September).

L.A. Frank, J.B. Sigwarth, and J.D. Craven, 1986. "Reply [to Chubb]." *Geophysical Research Letters*, 13: 1079 (October).

L.A. Frank, J.B. Sigwarth, and J.D. Craven, 1986. "Reply to Davis and Nakamura et al." *Geophysical Research Letters*, 13: 1186 (November).

L.A. Frank, J.B. Sigwarth, and J.D. Craven, 1986. "Reply to Morris." *Geophysical Research Letters*, 13: 1484 (December).

L.A. Frank, J.B. Sigwarth, and J.D. Craven, 1987. "Reply to Soter." *Geophysical Research Letters*, 14: 164 (February).

L.A. Frank, J.B. Sigwarth, and J.D. Craven, 1987. "Reply to Cragin et al." *Geophysical Research Letters*, 14: 557 (May).

L.A. Frank, J.B. Sigwarth, and J.D. Craven, 1987. "Reply to Wasson and Kyte." *Geophysical Research Letters*, 14: 781 (July).

L.A. Frank and J.D. Craven, 1988. "Imaging Results from Dynamics Explorer 1." *Reviews of Geophysics*, 26: 249 (May).

L.A. Frank, J.B. Sigwarth and J.D. Craven, 1989. "Search for Atmospheric Holes with the Viking Cameras." *Geophysical Research Letters*, 16: 1457 (December).

L.A. Frank, J.B. Sigwarth, and C.M. Yeates, 1990. "A Search for Small Solar-System Bodies Near the Earth Using a Ground-Based Telescope: Technique and Observations." *Astronomy and Astrophysics*, 228: 522 (February).

Louis Frank and John Sigwarth, 1993. "Atmospheric Holes and Small Comets." *Reviews of Geophysics*, February.

L. A. Frank and J. B. Sigwarth, 1997. "Detection of atomic oxygen trails of small comets in the vicinity of Earth." *Geophysical Research Letters*, October 1.

L. A. Frank and J. B. Sigwarth, 1997. "Transient decreases of Earth's far-ultraviolet dayglow." *Geophysical Research Letters*, October 1.

L. A. Frank and J. B. Sigwarth, 1997. "Simultaneous observations of transient decreases of Earth's far-ultraviolet dayglow with two cameras." *Geophysical Research Letters*, October 1.

L. A. Frank and J. B. Sigwarth, 1997. "Trails of OH emissions from small comets near Earth." *Geophysical Research Letters*, October 1.

L. A. Frank and J. B. Sigwarth, 1999. "Comment on 'A search for small comets with the Navel Space Command radar' by S. Knowles et al." *Journal of Geophysical Research*, October 1

J. B. Sigwarth, L. A. Frank, 2000. "Comment on 'Atmospheric holes are not small comets by A. W. Harris,'" *Journal of Geophysical Research*, August 1.

L. A. Frank and J. B. Sigwarth, 1999. "Atmospheric holes: Instrumental and geophysical effects." *Journal of Geophysical Research*, January 1.

L.A. Frank and J. B. Sigwarth, 2001. "The Detection of Small Comets with a Ground-Based Telescope." *Journal of Geophysical Research: Space Physics*, March.

Illustration Acknowledgments

All Dynamics Explorer 1 and Polar images courtesy Louis A. Frank, the University of Iowa, and the NASA Goddard Space Flight Center

All diagrams and charts: L.A. Frank, J.B. Sigwarth/University of Iowa

vii: Tom Jorgensen/University of Iowa

5, *Geophysical Research Letters*, April 1986, American Geophysical Union

6, Louis A. Frank, the University of Iowa, and the NASA Goddard Space Flight Center

11: UAF Geophysical Institute

12: University of Iowa

40: Sjo/iStock

65: The Swedish Institute of Space Physics and Frank, L. A., J. B. Sigwarth, and J. D. Craven, "Search for atmospheric holes with the Viking cameras," *Geophysical Research Letters*, 16, 1457-1460, 1989

98: Kolovos, G., Seiradakis, J.H., Varvoglis, H., and Avgoloupis, S. "Photographic evidence of a short duration: Strong flash from the surface of the moon," *Icarus* 76.3, 1988.

102: NASA/ Malin Space Science Systems

111: Carnegie Institution, processed by University of Arizona and JPL, courtesy of NASA. Image from the Las Campanas Observatory in Chile

138, Tom Gehrels, University of Arizona, Spacewatch Telescope/Stewart Observatory, Jet Propulsion Laboratory

142: Tom Gehrels, University of Arizona, Spacewatch Telescope/Stewart Observatory, Jet Propulsion Laboratory; L.A. Frank, J.B. Sigwarth, and C.M. Yeates. "A Search for Small Solar-System Bodies Near the Earth Using a Ground-Based Telescope: Technique and Observations." *Astronomy and Astrophysics,* February 1990

155: Birch Lane Press/Carol Publishing Group

169, 171, 181: University of Iowa and Face of the Earth™ Copyright © 1996, ARC Science Simulations, arcscience.com

174: NASA Goddard Space Flight Center

236-237: Gregory L. Pickett, Iowa Robotic Observatory; L. A. Frank and J. B. Sigwarth. "Detection of small comets with a ground-based telescope," *Journal of Geophysical Research*, March 2001

APPENDIX I

How to Search for Small Comets
By Louis Frank

Can the naked eye spot a small comet before it disrupts in the atmosphere? Can a pair of binoculars? Can a small telescope? Here is how you can—or why you can't.

First of all, consider the naked, rested eye. The apparent brightness of the small comets at the distance previously detected with the Spacewatch Camera at a range of 137,000 km (85,600 miles) is typically V=19 magnitude. At a target range of 1000 km (625 miles) the apparent brightness is about V=8.3 because you have gained V=10.7. For the naked eye, the threshold brightness has been estimated by Ernst Öpik and Peter Millman in terms of the percentage of objects detected as a function of brightness. Öpik claims 8% and 50% of the objects are detected at V=5 and V=4, respectively. Millman estimates less than 0.5% and 25% at V=5 and V=2, respectively. Clearly, under the best viewing conditions, the threshold of the naked human eye is about V=5. Thus the small comets would be dimmer than the threshold of the naked eye by V=3.3, or a factor of about 20 in light intensity. The naked eye simply cannot detect the small comets at a range of 1000 km (625 miles).

Can the small comets be detected with binoculars? We have previously estimated the detection rate with telescopes owned by amateur astronomers and have shown that the detection rate is about 1 or 2 per night of clear-sky viewing if the telescope is pointed in the proper position just after dusk and just prior to dawn. But let us take a look at the binocular situation for a range to the small comet of 1000 km (625 miles). We will use binoculars with a magnification of 7X and a clear entrance pupil of 50 mm in diameter. For the human eye, the entrance pupil varies but a diameter of 5 mm is typical. Thus the binoculars increase the light-gathering capability of the human eye by a factor of 100, or V=5. As seen through the binoculars then, the brightness of a small comet with V=8.3 is V= (8.3 - 5) = 3.3 and is detectable with the eye.

Consequently, the frequency of detection of the small comets with the binoculars is of interest. The viewing geometry is quite constrained because the objects will not be seen in the darkness of Earth's shadow and certainly not through the sunlit atmosphere. In fact, the atmospheric brightness is not less than V=3.1 and similar to the brightness of the small comet, until about 72 minutes after sunset and during the period 72 minutes before sunrise. We estimate that there are approximately 30-minute windows for detection of these small comets at solar phase angles in the range of 18 degrees to 40 degrees which are contiguous to and on the dark side of the above two windows. Importantly, the binoculars must be pointed at the intersection of Earth's shadow and the 1000-km (625-mile) range. The apparent motions of the small comets will be in the range of about 1 degree/s and will remain in the field-of-view of binoculars with a field-of-view of 10 degrees for about 10 seconds.

Now we must calculate the frequency of the small comet detections with your binoculars. The viewing is through the Earth's shadow so that part of the detection area must be subtracted because, obviously, if the Sun doesn't shine on the comets they are essentially invisible. (See the viewing geometry in the accompanying drawing.) With an apex at your binoculars and the two lines extending to 1000 km (625 miles), a rough estimate of the area in sunlight is approximately 2/3 of the total area of the triangle. If your field-of-view is 10 degrees then the detection area in sunlight is about 6×10^4 km^2. For the present purposes of simple estimation, we use the average interplanetary rate of 1.2×10^{-9} small comets/km^2-s intercepting the above detection area and neglect the effects of Earth's gravitational focusing. Thus the detection rate is about $(1.2 \times 10^{-9}) \times (6 \times 10^4) = 7 \times 10^{-5}$ events/s, or 1 event every 5 hours. Since your viewing time is about 30 minutes each for the pre-dawn and post-dusk samples, one would expect to see one event every 5 days of viewing with these binoculars. The short period for viewing is due to the rotation of the Earth and the rapid geometric degradation of the viewing geometry.

For telescopes with larger apertures the viewing geometry remains the same as shown in the accompanying figure. The viewing geometry is generally parallel to the ecliptic plane. Basically, you can estimate your detection rate by comparing your telescope's characteristics with those for the binoculars as given above. For example, if you have a collecting aperture of 90-mm diameter, then the range of your telescope is approximately proportional to the diameter, or about 1800 km (1125 miles), since the object brightness decreases as r^{-2}. At this distance, for an object with an apparent speed of 10 km/s, the apparent angular speed is about 0.3 degree/second. The event rate is proportional to the fan collecting area, i.e., proportional to the square of the above range times the viewing angle of your telescope. If the field-of-view of your telescope is the same as that of the above binoculars (probably smaller), then your event rate will be $(1800/1000)^2 \times (10/10) = 3.2$ greater than with the binoculars. This is about 1 event every 1.5 hours for the average interplanetary rate. The occurrence frequency is greater than this average for the months of June through the first week of November due to seasonal variations by factors in the range of 2 to 3. The rates are lesser for the other months with January at the minimal rates.

Appendix 1: How to Search for Small Comets

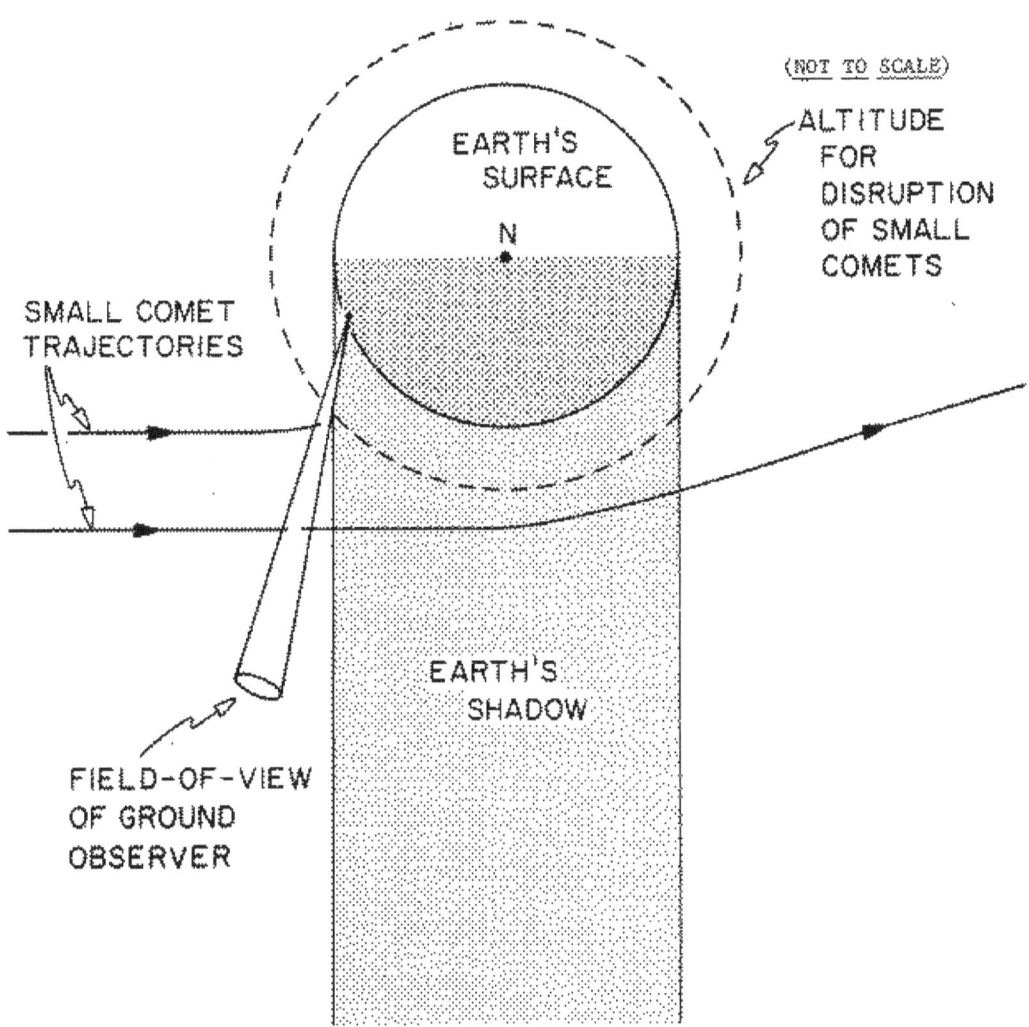

MOST FAVORABLE VIEWING FOR SMALL COMETS BY GROUND OBSERVERS WITH SMALL TELESCOPES

For a summary of the capability of the human eye for the detection of faint meteors, see D.W.R. McKinley, *Meteor Science and Engineering*, McGraw-Hill Book Company, New York, pp. 106-107, 1961.

For the variation of sky brightness throughout twilight, see C.W. Allen, *Astrophysical Quantities*, 3rd Edition, Athlone Press, London, p. 134, 1973.

APPENDIX 2

How This Book Came to Be
By Patrick Huyghe

In the 1980s, I was a freelance science journalist based on the outskirts of New York City. In April 1986, I came across a story in *The New York Times* entitled "Did Ocean Waters Arrive in Rain of Icy Comets?" by Philip Boffey. Being a magazine writer, I pitched the idea to Michael Robbins, who was then the editor of *Oceans* magazine. Intrigued, Robbins greenlit the idea. I subsequently interviewed Louis Frank over the telephone and contacted his critics for their comments as well. It was a time when journalism required "balance." My article entitled "Origin of the Ocean" appeared in the August 1986 issue of *Oceans*; it was subsequently adapted for publication as "The Oceans Came From Outer Space" in *The Washington Post*.

I concluded the article with this paragraph: "The history of science is full of wild ideas, some of which have held up over time, some of which haven't. A few examples should suffice: an ether fills the cosmos; the Earth is the center of the universe; the continents are moving and at one time formed a single landmass; life arose from a few elemental molecules in the Earth's primordial broth; humans are descendants of the first primitive life form. In light of these ideas, Frank's new theory appears rather tame. Whether it's right or wrong, of course, is another matter. But it's some sea story."

Frank must have liked the story. He would later call and ask if I would be willing to help him write a book on the small comets. I agreed.

Some months later I flew out to Iowa City and spent three days in his office at the University of Iowa bombarding him with questions and recording his answers with my tape recorder. I would also get together with Frank at subsequent meetings of the American Geophysical Union where we discussed the latest developments. Over the course of more than a year I then transcribed our interviews, as well as interviews with his critics, and organized the material into chapters, which I would then mail to Frank for his corrections and changes. I wanted the book to be in his own words, his voice. Eventually, the book was finished, and my agent, Harvey Klinger, took over, eventually selling the book to Carol Publishing Group. It was published as *The Big Splash* in 1990.

A few years later Frank contacted me to say that a second spacecraft, Polar, had confirmed the discovery of the small comets by the Dynamics Explorer 1. He wanted my help in coordinating the media outreach with NASA Goddard and the University of Iowa. I agreed, meeting with Frank, NASA scientists, and the people in NASA's audio-visual department at NASA Goddard in Greenbelt, Maryland. With this material, I created a website for the small comets (smallcomets.physics.uiowa.edu). As Frank relates in this book, the announcement of the Polar results at the American Geophysical Union meeting in Baltimore in May of 1997 could not have gone better.

So there was more to be said about the small comet discovery, and Frank asked me to fly out to Iowa again, so that he could fill me in on all the details. He wanted me to write the next chapter of the small comet story. And so the process began again. By 2001 we had produced another 40,000 words to add to the small comet story. Frank wanted to call the book *Cosmic Rain*. Unfortunately, however, the book's original publisher expressed no interest in publishing an updated edition of *The Big Splash*.

Then on April 15, 2003, Frank suffered a stroke that effectively ended his career. In December of 2010, Frank's partner in small comet "crime," John Sigwarth, who by then was a senior scientist in the Heliophysics Division at NASA's Goddard Space Flight Center, died of an aortic aneurysm at home at the age of 49. Four years later, on May 16, 2014, Frank himself passed away. Then, in January of 2021, John Craven, the third author of the original small comet papers, passed away from complications of ALS in Fairbanks, Alaska.

I prepared an obituary of sorts on Frank in the September 2014 issue of *EdgeScience*, a magazine I edit for the Society for Scientific Exploration, which included a short excerpt from the unpublished chapters we had worked on together detailing the Polar findings. There was no response.

The small comets were dead, certainly at least as far as the scientific community was concerned—and, it seemed, for everyone else as well. Or at least I thought so until March of 2021, when I received an email from William Abbott, who had been following the news of the discovery of water throughout the solar system for the past two decades, from the glacial ice in Mercury's polar craters to the crystalline water-ice on Pluto and its tiny satellites. In light of this new corroborating evidence for Frank's small comets, he urged me to get *The Big Splash* republished so that Frank's discovery could no longer be swept under the rug by the scientific establishment. He was right, of course; no one has yet come up with a better explanation for the data Frank cites in support of the existence of the small comets. Bill offered to help in republishing *The Big Splash*, not knowing that additional chapters had been written. I agreed to take up the project.

After obtaining a reversion of rights to *The Big Splash* from the publisher, I contacted Suzanne and Jessica Frank, Louis Frank's daughters, the beneficiaries of his estate. They have graciously allowed us to publish the book you now hold in your hands.

Cosmic Rain is the greatly expanded, second edition of *The Big Splash*. I have made no changes to the original text—published here as Part 1—with the following exceptions. Portions of the original last chapter "Small Comets and the Future" have been incorporated into the new last chapter of the same name for this second edition. And Chapter 28 of the original volume, entitled "Turning Point," now appears in abbreviated form; the material cut from this chapter dealt with several research projects in progress when *The Big Splash* was published. They are now covered and updated in Chapter 28, entitled "The Rest of the Story" of Part 2 and in subsequent chapters. There was no need to rewrite history; the discovery of small comets will do that all on its own.

In summary, I suppose I have become the de facto historian of this controversy, admittedly from Louis Frank's point of view, although I always insisted that Frank summarize the criticisms from his fellow scientists and reply to them. I have done my best to insure the accuracy of this account. But should anyone find factual errors in this telling, please contact me (cosmicrainbook@icloud.com), and I will make sure corrections are made in future editions of this book.

NAME INDEX

Abbott, William, 279
Ackerson, Kent, 54
Acuña, Mario, 163
Adams, Dennis, 45-46
A'Hearn, Mike, 184-185, 193, 194, 214
Ahrens, Thomas, 90, 98
Aiken, Art, 193
Alfven, Hannes, 22, 86
Allen, Lew, 134
Alvarez, Luis, 114-115, 118, 131
Alvarez, Walter, 114
Arnold, James, 100
Arrhenius, Gustaf, 31
Asaro, Frank, 114

Baldwin, Ralph, 93
Banks, Peter, 157
Bauer, Sig, 26-27
Beatty, Timothy, 61
Belton, Mike, 91
Bigg, Keith, 57
Bills, Bruce, 95
Black, Hillel, 155
Blake, Harold, 60
Blondlot, René-Prosper, 85
Bodin, Wes, 48
Boffey, Philip, 278
Bohlin, Ralph, 79, 173, 210
Bohr, Niels, 237
Bolton, Scott, 151
Boslough, Mark, 188-189
Bowell, Ted, 184, 193, 214
Brookshaw, Leigh, 38
Brown, Charles, 173
Burch, James, 150-151
Burrough, Bryan, 173

Campbell, L.L., 203

Casani, John, 89, 133
Carovillano, Robert, 193
Carruthers, George, 173
Caudal, Gerard, 51
Chahine, Moustafa, 134
Christensen, Andrew, 193, 196
Chubb, Talbot, 62, 68-70, 124
Clarke, Arthur C., 59
Cleary, Anne, 159
Clemesha, Barclay, 147
Clifford, Stephen, 95
Combi, Michael, 173
Coleman, Mary Sue, 230
Conway, Robert, 220
Cook, William, 172
Cormier, Kay, 47
Cragin, Bruce, 70-71
Craven, John, 7, 11, 12, 14, 23, 66, 80, 81, 86, 231
Crawford, David, 189
Croskey, Charles, 45

Davis, Paul, 93-94, 124
Delsemme, Armand, 218
Deming, David, 205, 215-217
Deming, Drake, 194
Dessler, Alex, 5, 15-17, 23-24, 34-35, 50, 52, 62, 64, 65, 70, 81, 93, 94-95, 98, 106, 120-125, 128, 131, 132, 136, 145-147, 150, 156, 160-161, 165, 185, 186-187, 188, 196, 198-199, 197, 208, 210-211, 212, 213, 219
Dessler, Andy, 219
Dolan, Jeff, 165
Donahue, Thomas, 16-17, 24-28, 30, 31, 44, 78-84, 86, 88, 89-90, 120, 124, 133, 145, 172, 173, 197, 198, 206, 221
Downer, Robert, 230, 232

Name Index

Duncan, Martin, 110-111

Eberhardt, Peter, 32
Engel, Andreas, 219
Evans, Robin, 210
Everhart, Edgar, 112, 118

Fanale, Fraser, 55, 104-105
Feldman, Paul, 85, 162, 173, 238
Fernández, Julio, 100, 113, 118
Ficklin, Billy, 50
Fink, Uwe, 143, 221
Fischbach, Ephraim, 155
Fix, John, 224, 225-227, 228-232, 234
Fort, Charles, 58-59
Frank, Jessica, 279
Frank, Suzanne, 279
Freeman, John, 101
Frías, Jesús Martínez, 233
Fyfe, William, 217

Galileo, 125, 132, 148
Galluzzo, Gary, 226, 227
Gehrels, Tom, 131-132, 134-135, 136-137, 140-145, 147, 184, 212
Gibson, Frank, 61
Gladstone, Randall, 188-189
Goertz, Christoph, 159
Gold, Thomas, 24, 126, 155
Goldstein, Richard, 193
Gombosi, Tamas, 90
Grier, Jennifer, 187
Grossmann, Klaus, 43

Haines, Richard, 60
Hall, Doyle, 89
Hannegan, Bryan, 205
Hansen, James, 91
Hanson, Bill, 63-64, 70-71, 72-73, 120, 124
Hanson, R. Brooks, 166
Harris, Alan, 184, 197, 203-204, 210-211, 212-213
Harvey, Paul, 156
Harwit, Martin, 222
Heinberg, William, 19

Helin, Eleanor, 131, 193, 209
Hermeier, Barbara, 123
Hills, Jack, 109, 118
Hoffman, Bob, 163, 167, 172-173, 178, 186, 191-198, 200, 209, 211-214
Holland, Heinrich, 37, 176
Holt, Henry, 149
Holt, John, 76
Holt, Steve, 192, 194
Hord, Charles, 92
Horgan, John, 92
Hoyle, Fred, 121-122, 256
Hunten, Donald, 25, 27, 88, 89-90, 132, 134, 147, 184, 222
Hu Zhang-Wei, 59

Ingersoll, Andrew, 91, 104
Ingmanson, Dale, 218

Jeanloz, Raymond, 218
Jefferson, Thomas, 130
Johnson, Robert, 36
Johnson, Torrence, 133, 141, 147

Kelley, K.K., 218
Kenyon, John, 227-228, 231
Kerr, Richard, 85-86, 90, 172, 185, 220
Kessler, Don, 137
Kieffer, Hugh, 104
Kirk-Davidoff, Daniel, 218
Klinger, Harvey, 278
Knauth, Paul, 216
Knowles, Stephen, 194, 197, 204-205
Koehn, Bruce, 184
Kolovos, George, 99
Kring, David, 188
Kuiper, Gerard, 111
Kump, Lee, 216
Kurczewski, Eric, 56
Kyte, Frank, 35, 121-122

Lane, Arthur, 160
Lanzerotti, Louis, 36
Lawson, Alvin, 58
Lewis, John, 27

Lindley, David, 84
Lu, Gang, 159
Luhmann, Janet, 204, 206, 208, 229

Maccabee, Bruce, 57
Maran, Stephen, 167
Masursky, Harold, 90, 103
Mathews, John, 193
Matthews, Robert, 172, 173
Matson, Dennis, 141
Mattingly, Thomas, 99
MacDonald, G. J., 216
McDonald, James, 59
McEwen, Alfred, 187
McFadden, James, 197, 200-201, 206, 208
McGlynn, Ann, 227
McKay, Christopher, 35
McMillan, Bob, 143, 213-214
Meier, Robert, 62, 167, 170, 172, 173, 193, 204-205, 219
Mende, Stephen, 158-159
Mendillo, Michael, 74-77, 133
Meng, Ching, 162, 193, 198
Michel, Helen, 114
Millman, Peter, 275
Monastersky, Richard, 172, 185, 206, 220
Morris, Donald, 35, 37, 106
Morrison, David, 184
Mozer, Forrest, 197, 198-201, 203, 206, 208, 231
Muhleman, Duane, 221
Muller, Richard, 115
Murphree, Sandy, 64-67
Mutel, Robert, 165, 224-232, 234

Nakamura, Yosio, 95-96, 124
Neufeld, David, 223
Neugebauer, Marcia, 160
Ney, Edward, 56
Nicholson, Dwight, 159

Oberst, Jürgen, 95, 98
O'Keefe, John, 98
Olivero, John, 27, 44-46, 133, 216, 220
Oort, Jan, 108
Öpik, Ernst, 53-55, 97
Oró, John, 38
Owen, Tobias, 205, 221

Pack, Dee, 195, 196
Parks, George, 163, 178, 180, 185-186, 199, 201, 206, 208, 217
Paterson, Bill, 227
Paxton, Larry, 194, 197, 202, 206
Pickett, Greg, 234
Pielou, E.C., 218
Planck, Max, 4, 239
Pollack, James, 91
Polyansky, Oleg, 221
Polyzou, Wayne, 229
Powell, Corey, 185
Prather, Michael, 183, 185-186, 187-188, 205
Press, Frank, 96

Quinn, Thomas, 110

Rabinowitz, David, 209
Rampino, Michael, 115
Raup, David, 115
Rawlings, Hunter, 159
Reid, George, 41-42
Rizk, Bashar, 187
Robbins, Michael, 278
Robinson, Robert, 193
Roble, Raymond, 42
Rodolfo-Sioson, Miya, 159
Roush, Ted, 104
Rothenberg, Joseph, 192, 194
Rubey, William, 31, 37
Rubincam, David, 35, 124
Russell, Christopher, 91
Russell, James, 205, 220

Sage, Leslie, 165
Sandel, Bill, 80, 90
Sawyer, Kathy, 167, 210
Schantz, Mark, 230-231, 232
Scotti, Jim, 134, 136, 138, 143, 197, 211-213

Scudder, Jack, 163
Sepkoski, Jack, 115
Shan, Linhua, 159
Shemansky, Donald, 89, 90, 91
Shoemaker, Eugene, 131, 136, 145, 146, 147, 209-210
Siever, Raymond, 217
Sigwarth, John, vii, 7, 12-14, 16, 23, 45, 66, 70, 71-73, 75-76, 80, 99, 121, 137, 141, 142, 150, 159, 160, 165, 172, 202, 204, 224, 226, 227, 230, 234, 235, 279
Sisler, Wade, 167
Skorton, David, 228
Smith, Bradford, 110
Smith, Robert Alan, 159
Solomon, Susan, 141-142
Soter, Steven, 124, 126, 127-128, 130, 131, 133
Stapelfeldt, Karl, 210
Steigerwald, William, 167
Stewart, Ian, 81, 92, 172, 214
Stothers, Richard, 115
Swindle, Timothy, 188

Taft, Laurence, 126-128, 131, 132
Tagliaferri, Edward, 195-196
Terrile, Richard, 110
Tetenbaum, David, 76
Thibault, Dana, 220
Thomas, Norman, 149
Thomas, Paul, 38
Torr, Doug, 163
Torr, Marcia, 163
Tremaine, Scott, 110
Tsibliyev, Vasily, 173
Tsurutani, Bruce, 161

Van Allen, James, vii, 12, 20, 24, 83, 89, 91, 121, 150, 165
Velikovsky, Immanuel, 85
Vogan, E.L., 203
Vondrak, Richard, 193

Waite, Hunter, 176, 177, 178, 182-183
Wallace, William, 218

Wasson, John, 35, 121-122, 124
Wegener, Alfred, 22
Weisburd, Stephie, 15
West, John, 56
Wetherill, George, 81
Whelan, Martin, 193, 194
Whipple, Fred, 47
Whitmire, Daniel, 115-116, 118
Wickramasinghe, Chandra, 121
Wilde, Pat, 32
William of Ockham, 81
Winglee, Robert, 185, 188, 200, 201, 208
Winograd, Isaac, 216-217
Wiśniewski, Wieslaw, 143
Withbroe, George, 172-173, 175, 193, 194
Wolbach, Wendy, 37, 117

Yeates, Clayne, 133-147, 150, 159, 160, 209, 211-213, 224-225, 229, 235, 238
Zel'dovich, Yakov, 96
Zisk, Stan, 105

www.ingramcontent.com/pod-product-compliance
Lightning Source LLC
Chambersburg PA
CBHW040927240426
43667CB00024B/2974